PLANT PROPAGATION
IN PICTURES

PLANT PROPAGATION IN PICTURES

ADRIENNE AND PETER OLDALE

DAVID & CHARLES
NEWTON ABBOT LONDON
NORTH POMFRET (VT) VANCOUVER

ACKNOWLEDGEMENTS

On pp 84–96 you will find a guide to propagating most of the plants found in our gardens and homes. This guide is based on one originally published in *Propagating Out Of Doors* published by Wm Collins & Sons Ltd, and we would like to thank them for their permission to use it once again.

The plates on the pages listed below are reproduced by kind permission of Pat Brindley: 20, 21, 22, 23, 33, 37 (*d*), 38 (*a*), 39 (*g*), 40 (top right), 42, 45, 46, 47, 51 (*g*), 53 (*f,g*), 57 (*e,f*), 66, 67, 71 (*f,g*).

ISBN 0 7153 6875 3

Library of Congress Catalog Card Number 75-4062

© Adrienne and Peter Oldale 1975

Set in 11 on 12pt Imprint
by Wordsworth Typesetting, London
and printed in Great Britain
by Redwood Burn Limited Trowbridge & Esher
for David & Charles (Holdings) Limited
South Devon House Newton Abbot Devon

Published in the United States of America
by David & Charles Inc
North Pomfret Vermont 05053 USA

Published in Canada
by Douglas David & Charles Limited
132 Philip Avenue North Vancouver BC

CONTENTS

INTRODUCTION

This book shows how to increase your house or garden plants. Most commonly-grown kinds can be easily propagated provided the right method is chosen. Elaborate equipment is not essential – a bright windowsill, a few pots and polythene bags, oddments of wire, and above all, a supply of good quality potting compost – are all that is usually needed. Plants to be propagated should be well-grown and in good health, so such problems as watering, fertilising, potting, correct compost mixtures, and growing temperatures, must be considered first of all.

Seed sowing is the method most often chosen for many favourite flowering plants. Given care with soil preparation, sound seed – and a little luck with the weather – you can succeed with hardy sorts quite easily. By planting seeds in boxes, indoors, in a frame or greenhouse, even quite tender kinds can be grown.

However, even simpler than seed, and more likely to give 100 per cent success, is to propagate by division. A large bulb or root is split into several sections which rapidly develop into full sized plants after replanting. A few plants, such as ivy, even produce roots from their stems above the soil. Cutting these up into short lengths and potting them will also produce new plants. Layering is a method in which we encourage stems above the soil to form roots by packing moist soil and moss round them. Cuttings are perhaps the most widely used propagation method for shrubs and many houseplants. A piece of stem, usually with one or two leaves left attached, is cut off completely and stood in compost till roots form.

All these methods are shown in step-by-step pictures and the final part of the book shows a group of useful aids you can make yourself – a potting bench, sieve, and propagatiny case.

You will notice that we make little mention of greenhouses. This is not because they are not desirable – they make almost all propagation work much easier – but because most newcomers to gardening do not own one, and we want to show that this need not prevent them setting to work. If you do have one, the methods we show can all be successfully used, and the greater control of temperature, light and moisture will be all to the good.

There are many people who are thought by their friends to possess green fingers. The cuttings they take, the seeds they sow, all grow and flourish, seemingly without effort. In fact, this success usually comes simply from careful work, attention to detail, and the realisation that plants are alive. They have need for food, air, light and water, just as we do. If one or more of these is lacking, sickness and poor growth are inevitable. By applying our methods you will be ensuring that the plants you propagate will have the best chance of survival. Not all of them will live – not even the professionals expect this – but the chances of success will be greater, and a hobby that can be most frustrating in failure will become most satisfying in success.

BASIC SKILLS IN PLANT CARE

CHOOSING YOUR TOOL KIT

You are not likely to have to buy many special tools to start propagation. A spade, fork, rake, and trowel are the main ones you need. However, there are some small items which will make life easier for you, especially as you progress.

Secateurs (fig *a*) – A small, sharp pair of secateurs is often of great help in taking cuttings from hard wooded trees and in the making of layers. Unless you also have a large amount of pruning to be done there is little to be gained by purchasing expensive kinds. Get a reasonably priced pair. Keep them well dried and cleaned after use and the cutting surfaces smoothed over with grease. Rust, and the resulting damage to the cutting edge, is the worst thing that can happen to such tools.

Propagating knife (fig *b*) – There are many patterns of small, folding knives which gardeners favour for propagating work. The differences between them are very much of fashion and personal taste. Almost any small knife will serve provided it can take a keen edge. A specialist gardener's grafting knife will have an excellent shape, good steel blades, and probably a slim, tapered ivory blade which is used for certain grafting operations. Even a first-class tool of this kind is not likely to cost very much and since it is essential to have a knife that keeps a good edge you should get as good quality as you can afford.

Razor tool. Many gardeners now use this type of sharp cutting tool in the home and workshop. We have found them very satisfactory for taking cuttings and so on. The edges provided by the replaceable blades are as sharp as the very best of propagating knives but the tools themselves are likely to be much cheaper, though somewhat clumsy for smaller work.

MATERIALS FOR PROPAGATING

You do not need elaborate materials for propagation work. Plain *polythene sheet* is useful stuff and is indeed very cheap. It can be bought in various thicknesses and in large sizes from builders' merchants. Use the thinner sort for giving warm protection to young plants or shielding them from wind and rain. A large thicker sheet is useful for keeping working tops clean, especially where some of the

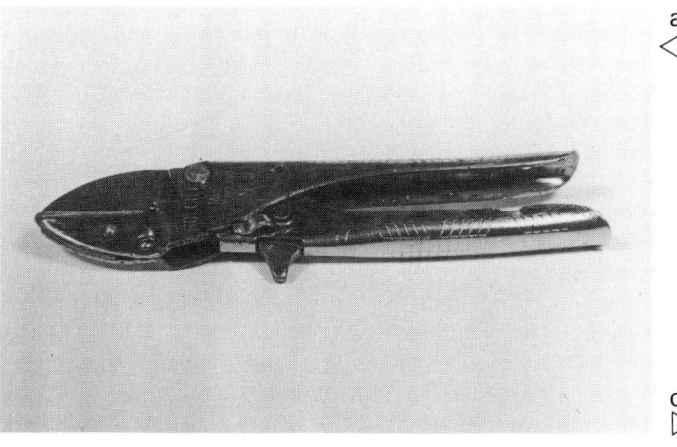

a

c

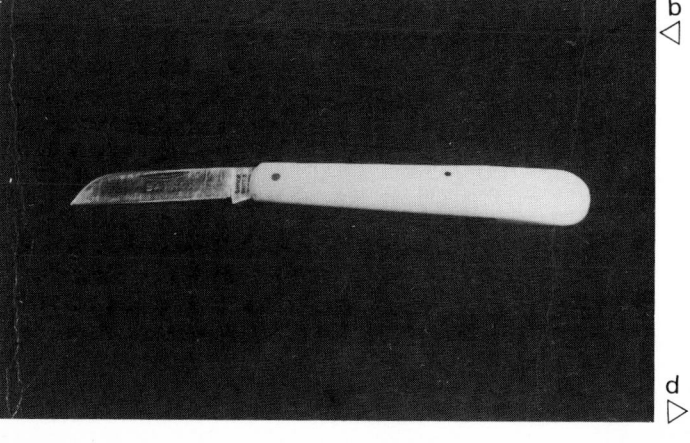

b

d

potting work is being done indoors. Plain polythene bags make ideal propagating covers for cuttings and seeds (page 19).

String and thin wire of various thicknesses is always useful in the garden. Short lengths of plastic covered wire are good for quick and firm tying of cuttings to stakes, or holding layers to their pegs. *Raffia* (fig c) is a traditional material for some jobs such as the tying of grafts, though patent plastic clips are also much used.

Plant pots (fig d) nowadays are almost always made of plastic. They are so light and easy to handle and clean that they have become almost universal in modern gardens. The old clay pot can still be found in millions and although heavy and fragile is equally serviceable, provided it is well washed after every use, preferably using a proprietary fungicide soap. The most useful sizes are those about 3½in (10cm) in diameter, but you will also require some smaller pots – 2in (5cm) diameter – and one or two broader ones for bigger plants.

An entirely different sort of plant pot are the *fibre pots*. These are made of sheet fibre – often compressed peat – and nowadays sometimes impregnated with plant food chemicals. They are intended to save plants the shock of transplanting. They can be simply put out in the open ground, or into larger pots, without taking the plants out. The pot fibre becomes penetrated by the plant roots and is eventually absorbed by them and by the soil nearby. Plants grown in these pots often develop more rapidly than any other type.

For seeds, especially of the more tender kind, you may need *boxes*. Besides the traditional wooden box, which is serviceable and sometimes obtainable or made very cheaply, you can now buy excellent plant *trays* in plastic. These are really better than anything else, being cheap, light, easy to clean, and unlikely to carry disease. You can also buy translucent covers for them to form miniature "propagating cases" within which even very tender seedlings can be grown successfully.

A few simple chemicals are useful too such as *rooting hormones* which stimulate root development in cuttings and layers, and of course ready-prepared composts (see page 10). There are many *insect-* and *disease-repellent sprays* too, though with clean and careful work these should be rarely needed. *Grafting wax* is another special material for more advanced work, used to seal off newly made grafts from the air. With all these materials, it is important to follow the manufacturer's instructions on methods very carefully, especially with regard to the strength of solutions. Too much or too strong chemical can often do immense damage.

Larger equipment such as *propagating cases* and *garden frames* are useful but not essential at first. On page 74 we show how you can make your own. This has the advantage of saving money, and that you can make them to the exact size needed to fit into an odd corner of your home, greenhouse, or garden.

Finally, do not forget the *watering can*. It is better to have two of these, in different colours, keeping one colour for water and the other for liquid fertilizer. A fine *pump spray* for water is very desirable whenever plants are grown under cover.

With these basic tools, you can go ahead and propagate almost every plant you are likely to find in your garden and home.

BEFORE YOU BEGIN

Before starting on any method, let us look at the basic needs of plants. We can then see in what way we must help a young plant to grow successfully.

The principal needs of plants are light, air, water, and plant food chemicals. They also need to be kept at a certain temperature. This varies from plant to plant, depending upon which part of the world it originated. Clearly, plants from the tropics will require warmer and usually more humid conditions than those from high mountains. Some types need brilliant light, others comparative shade. No one of these factors is more important than the other, for without any one of them nearly all plants will die. What is essential is to balance them, to provide the plants with the optimum amount of each.

All green plants need *light* to act upon their leaves to produce plant food. Without it, the leaves cease work, the plant can no longer convert the soil chemicals into plant food and it will gradually die. However, plants differ widely in the amount, intensity, and daily duration of light they need. A plant whose wild habitat is at the foot of trees in dense tropical jungles will be used to very shady conditions and need much less light than a desert plant from a place where sunshine is practically continuous throughout every day of the year. The vast majority of British outdoor plants have become largely adapted to temperate conditions and this reduces their more extreme needs. Even so, if the mature plant is a shade lover, the young plant also will require shading; if the mature plant is a sun lover (such as the popular *pelargonium*) then the young cutting or seedling will also tolerate drier, brighter conditions.

When you are propagating outside the natural daylight is uncontrollable but you can still adjust the plant's position from a shady to a bright spot or vice versa. You can choose a cooler, northern or eastern aspect, rather than a hot, directly southern facing place. Indoors, whether propagating on a window sill or in a greenhouse, you must watch the question of light even more carefully. It is always very much darker inside our homes than outside. Although the human eye adjusts rapidly to this darkness and the contrast often does not seem as great as in fact it is, any keen photographer knows the light in even a well-lit room may be less than one-fiftieth of the light outside.

Plants propagated in ordinary rooms must be put in the brightest position possible, though not usually in the direct midday sun. This scorching heat for an hour or two each day is likely to cause problems in a number of directions. It may dry up the plant by direct radiation, causing it to lose water and rapidly making the leaves flag. It may raise the soil temperature to such a degree that it dries out completely. A west window on the other hand is often very good. It will be fairly bright during the day and only face the sun towards evening when the scorching rays are much reduced in power. Rooms upstairs are often lighter than ones downstairs, being free from the shade of surrounding buildings or trees. Also, the temperature upstairs is likely to be less variable. In general therefore, for all except a known desert plant, choose a position that will give the maximum possible light but where the young plant will not be scorched by the heat of midday.

Water is the second great need of all plants. It is the pressure of water in the sap which keeps the leaves and stems plump, firm, healthy and extended to the sun. Without water, leaves flag, stems become shrivelled and the plant collapses and dies. Strangely enough, too much water in pot plants can also cause very similar effects. Waterlogged conditions prevent essential air getting to the roots and the plant suffers. Roots do badly also where the water supply is low and the air supply high. Most plants prefer a soil which has very roughly equal water and air spaces in it. Seedbeds must always be well-drained, but with a texture that retains some moisture. With pot plants, well-prepared compost (page 10) will always have one constituent, often *peat*, which can absorb water and keep it available to the plant. It is also used in the garden as although it has practically no plant food value, its spongy yet airy texture holds both water and air in balance near the plant roots.

Over watering of young plants is in fact the beginner's most common fault. Newly propagated plants lose water only slowly from the leaves. At first the loss is quite small, but as growth starts, rather more moisture is breathed out. This must be replaced. Outside, seedlings can be watered in the evening (direct sun on water spattered leaves can burn them). To maintain the water supply in pot grown plants, the pots can be regularly watered and sprayed but it is better to try to prevent the leaves losing water at all. This may be done by enclosing them in a polythene bag, which will keep the air inside very humid indeed, so reducing the loss. In fact, it may never be necessary to water plants or seedlings protected like this.

Incidentally, many gardeners prefer to use soft rain water. This is certainly very pure but there is little practical difference for amateurs if you use ordinary tap water. An exception is if your water is alkaline; it should not then be used for lime-hating plants like rhododendrons.

The water *temperature* is often much more important. It is always desirable to use water which is at least as warm as the soil round the plant roots. Otherwise these may react to a sudden deluge of ice cold liquid by slowing down their activity and so root development. There is no need to be continually watering. In fact, this is likely to do more harm than good. Judgement of when or when not to water is a definite gardening skill, but if in doubt give too little rather than too much. Happily, most of our garden plants are very tolerant of errors of this kind and the methods of propaga-

tion which we show will often succeed even though the water supply may be rather irregular.

Air is another essential need for all plants. Without it they cannot obtain the nitrogen and oxygen which are part of their living cycle. A constant flow of fresh air is a great asset to a young, developing plant. Outside of course, there is no problem, only indoor pot-grown plants may suffer. However, newly struck cuttings and seedlings naturally have a much lower air requirement than adults. They can often manage well for a considerable time on the small amount of air within a polythene bag or under a fairly tight glass cover. In this case the lack of fresh air is more than compensated for by the retention of humidity around the developing leaves. As they grow, however, it is most important to give the plants gradual access to the fresh air. Start by opening the cover for a few minutes each day, increasing the daily period until the plants are at last "hardened off" to natural conditions.

This hardening off process is an essential one and also one which gives a lot of trouble to amateurs. Many home gardeners strike cuttings successfully and grow seedlings well up to a certain point. Then, as they begin to transplant outside, failures take place. By gradually introducing the plant to the open air, little by little over a period of a week or a fortnight, many of these losses can be avoided. The use of fibre pots is also a help in reducing such transplanting shocks.

Chemical foods are needed by all plants in small quantities. Do not be misled by talk of plants growing in plain water without soil. All such systems of growth have arrangements made for providing chemicals in suitable proportions. Often these chemicals are themselves dissolved in the water; sometimes they are sprayed over the leaves. Take care too not to be troubled too much by the various (often fierce!) arguments about *artificial* or *natural* fertilizers for the soil. It is true that there are considerable differences in the effects on soil of, for example, farmyard manure as against pure raw chemicals like commercial farm fertilizers. However, these differences from the point of view of the plant are often irrelevant unless the chemicals are applied too heavily. This is because all plants without exception feed on raw chemicals, not upon the natural manures. Within the soil itself small organisms such as bacteria break down natural manures into chemicals which can then be absorbed by plant roots. These chemicals are exactly the same as those provided by the so-called artificial fertilizers. The difference is simply that much of a dressing of chemical fertilizer is immediately available, whereas with other manures there is always a little delay, whilst the soil organisms do their work.

In either case the quantity of chemical needed is very small indeed. The plant obtains by far the most of its weight and growth from the atmosphere and the water in the soil. This is particularly true for young cuttings. These can often survive until fully rooted in a compost which contains no plant food whatsoever! Even raw sand will do, or plain water. The cutting lives on the food stored in its own leaves and stems while it develops a new root system. These roots will then push out, searching for chemical foods, until it is transplanted into soil or compost. They will then take over the food provision of the whole new plant. In spite of this toughness though, cuttings started in a balanced compost, with chemicals, usually do better.

Seedlings, being so tiny, naturally need at first very little plant food in the way of chemicals. Most seeds carry a small food supply on which the seedlings live whilst they push out their first leaves and roots. However, the soil within which they then find themselves should ideally contain a little plant food in carefully balanced proportions.

Seedlings are amazingly sensitive to differences in compost. Professional gardeners use differently mixed composts for fresh seed, for mature seedlings and for cuttings. This is because of their differing food needs. The compost for first seeding always contains much less chemical than that for the more mature plants. Much research has gone into the development of these composts, which are now commercially available on a large scale for amateur use. It is far better at first to purchase ready-mixed, professionally made compost for your pots and seed boxes. Even the minutest trace of impurity can cause great damage to young seedlings. Similarly the absence of a tiny trace of a particular chemical may result in stunted growth or complete failure. Later, as experience is gained, you may choose to mix your own composts.

Warmth for young plants can be a difficult question. Naturally, few plants will be happy for long in sub-freezing temperatures, and though they may survive, growth is nearly non-existent below a temperature of 45–50°F. Above these minimum temperatures however, because of their various origins, plants differ widely in the heat they need to maintain vigorous growth.

Naturally cuttings or seeds grown outside are dependent for warmth on the weather, so the season of sowing or taking the cuttings is very important. The soil must have had time to warm up sufficiently and it is important that the remaining part of the growing season be long enough to allow the plant to develop a mature leaf and stem structure, enabling it to withstand the harder weather following. Those plants listed as being hardy for outdoor propagation fall into this group. Beware though of treating suggested seeding or propagating dates as hard and fast rules. For example, if one gardener sows his seed in May with success, it does not necessarily follow that you, in a different place with a different aspect will achieve the same results by following his rule. If you live in a place with late, cold springs a delay in seeding will be necessary to allow the soil to warm up. If on the other hand you live in a more favoured area you may find that you get better plants by sowing earlier, even taking the risk of a few losses from frost in order to take advantage of the long growing season ahead to achieve bigger and better plants in the end. Similarly, successive years differ in their temperatures, the May of one year being substantially colder or warmer than of years before or after. A gardener must keep his eye on the overall temperature, the weather pattern, the amount of cloud and so on. Even the different soils affect the temperature problem. Clay and other heavy, damp soils take much longer to warm up in spring than light, sandy types.

Propagating indoors is of course another matter. Here the temperature can be controlled directly. Even so, if the plant is intended to be put out later into the garden, it must reach suitable maturity and be hardened off correctly in time for it to develop fully in the garden before the onset of bad weather later in the year.

In propagating indoors or outdoors therefore, temperature is associated with timing and this can be best gained by experience. Use this book as a guide to start you off on the right lines, but be prepared to alter things to suit the special needs of your own area, soil type and climate.

COMPOST MAKING

When young plants are being propagated their successful growth will depend a great deal on artificial methods, especially when the work is done indoors in pots or boxes. Before moving on to a detailed look at propagating by seeds, cuttings, layers, grafting and so on, a few simple techniques can be studied.

Compost is the name given to two rather different gardening materials. One is rotted down garden rubbish made outside in a compost heap. This becomes a black, rich soil, very useful for improving garden seed beds. The other is compost used in pot plant culture and greenhouse work. This is a mixture of soil, sand, peat, leaf mould, and chemicals in proportions varying with the type of plant. This is the sort usually meant when gardeners discuss 'making up compost'.

Because garden plants have originated in different parts of the world, their compost needs are very different. Those from forests like a deep rich soil produced by centuries of rotted fallen leaves. Others come from deserts, swamps, jungles, shallow soiled plains or mountains and need similar soils. A few, such as azaleas, will not grow in soil containing lime. Specialists have developed compost mixtures to suit most plant types but only a few of these mixtures are important. With only minor variations one of these will prove satisfactory for most plants.

Many such composts are based on loam. This is simply fertile garden soil of medium weight. To this, various proportions of sand (to give good drainage), peat (to improve texture) and chemicals are added. Loam used in making compost is usually partially sterilised. A proportion of the bacteria living in it are killed, leaving alive the most beneficial. This may be done by chemical treatment (using proprietary products purchased from nurserymen) which is perhaps simplest for amateurs. Professionals often use heat for sterilisation. This involves passing hot steam through small quantities of soil, which often also has the incidental effect of increasing the supply of nitrogen fertilizer in the soil.

Often the place of loam is taken by *vermiculite*. This is a granular, light-weight material which has no value as plant food but gives the compost an ideal, crumbly texture, well drained and airy. Plant food is provided by adding carefully measured proportions of raw chemicals. Composts of this kind grow perfect plants but the mixing must be very accurately done. Too much chemical can kill a plant as certainly (and much more quickly) as no chemicals at all. Most such "no soil" composts are bought ready-mixed.

Sand for composts must not be too fine, or it may cause caking. Coarse, washed sand similar to that used for concrete, and containing pebbles up to about $\frac{1}{8}$in (3mm) in diameter, may be chosen. Most sands are chemically inert, having no plant food value. Their object in composts is to ensure good drainage and prevent consolidation.

Peat is the remains of dead plants, and like sand has practically no value as a fertiliser. It acts mainly as a sponge in composts, retaining water and providing an ideal, moist atmosphere throughout the pot. By adjusting the proportion of peat, composts holding more or less water can be produced. It is often bought in large dry bales. These must always be broken up finely before mixing, and the crumbled peat soaked and drained.

Chemicals are bought in small quantities, of horticultural grades. It is important that the amounts needed be carefully measured and evenly mixed into the compost. Otherwise, pockets of extremely concentrated chemical may result and plants put there may well die.

Store all composts under cover, or heavy rain may destroy their texture and wash away valuable plant foods. If you wish to stock large quantities, keep the chemicals separately from the other ingredients (which can be mixed in bulk), mixing them up finally in small quantities as required.

A few composts specify a proportion of farmyard or horse manure. This must on no account be fresh. Only well-rotted material, at least six or seven weeks old, should be chosen.

When mixing composts, all the bulky main materials are measured by volume, not by weight. The chemicals, which are very much smaller in quantity may be given as weight per volume of compost.

Do not attempt to make improvements in compost mixtures. In particular, increasing the proportion of chemicals can have unpredictable and usually damaging effects. In fact for best results it pays the newcomer to buy composts ready-mixed. These are prepared with properly sterilised soil or vermiculite, sand, peat, and chemicals, in the exact proportions needed. Some gardeners though enjoy such mixing work so for their guidance we have included a few compost recipes suiting plants of different types.

Mixing Composts

For home mixing, few composts are as successful as the world-famous John Innes potting and seeding composts. Each is based on loam, a fertile, medium-weight sifted soil, peat, and coarse, clean sand. These are mixed thoroughly in the proportions given below, and stored. Before use, a small quantity of fertiliser is added, either plain ground

chalk and superphosphate of lime, or a mixture known as John Innes Base.

Incidentally, a bushel is commonly taken as a measure in garden work. It is the amount of material that would fill a box 22in (55cm) long, 10in (25cm) wide and 10in (25cm) deep.

COMPOST FOR SEEDS
    2 parts Loam
    1 part Peat
    1 part Sand
    To every bushel, add
        $1\frac{1}{2}$oz superphosphate of lime
        $\frac{3}{4}$oz ground chalk, limestone or whitening

COMPOST FOR POTTING AND CUTTINGS
    7 parts Loam
    3 parts Peat
    2 parts Sand
  To every bushel, add
        $\frac{3}{4}$oz ground chalk, limestone or whitening
        4oz John Innes base

TO MAKE JOHN INNES BASE FERTILISER
Mix 2 parts hoof and horn meal
     2 parts superphosphate of lime
     1 part sulphate of potash
Store in a dry container for use as required

It might be asked why one does not simply use garden soil, which after all grows most plants well outside. The reason is that in the warm, moist and rather airless conditions of a small pot, any bacteria, fungi and pests may rapidly increase. Even fertile soil may bring with it problems of this kind, and may collapse into an unhealthy mass. Proper mixtures of sand, peat and sterilised loam or vermiculite avoid these problems. Good compost is the foundation of success with potted or boxed young plants.

POTTING AND BOXING

A plant in a pot has little soil to feed on, compared to one growing in open ground but even a small pot full of good compost can contain enough chemicals to feed a large plant. The traditional clay pot is porous; air and moisture can pass through the sides therefore providing more airy conditions, but the soil is liable to dry out fairly rapidly and need more frequent watering. Plastic pots conserve water but exclude air, so need a lighter, more open compost. You can grow good plants in either, though professionals are turning more and more to plastic. The fibre pot allows the plant roots to grow until they start to pass through the pot walls when the whole thing is then inserted into a larger pot, or put out in the open ground. The young roots are thus never disturbed or the growth checked caused by normal re-potting. Fibre pots are available in many sizes, including tiny square ones joined in strips. These take up little space on the greenhouse bench or windowsill.

House plants are sometimes grown in more decorative containers, bowls, miniature troughs, etc. These will be just as serviceable, from the plants point of view, provided drainage holes are made in their bases. This is vital. Very few plants will tolerate stagnant water about their roots, excluding air. Remember, even the plant parts underground must all breathe – or die! As waterlogging kills pot-plants rapidly, good drainage is essential. The traditional drainage method is to put a layer of broken pot shards or small clean gravel in the base of the pot, under the compost. This allows the water to drain away freely and yet prevents the compost from clogging the drain holes. The open-textured compost which does not cake together may be used without any base drainage material (though holes in the pot itself are of course still needed). "No soil" composts drain even more easily. The vermiculite these contain instead of sterilised soil is a light, flaky, granular material, sterile and without value as plant food but giving an excellent open texture. Of course, any compost must be firm enough to provide a good anchorage for the roots without being so solid as to restrain their growth.

The soil in pots of established plants may become solidly packed, and caked on top produced by surface watering. Break up this top layer with a pointed tool so that water can penetrate evenly. A better solution still is to move the plant into a slightly larger pot.

The diameter of pot used obviously depends on the size of plant, but some species will tolerate a much smaller pot than appears possible. For example, a tall *Ficus elastica* (rubber plant) can be grown in a pot which seems ridiculously small. Always choose the smallest reasonable size of pot. If a plant cannot fill its pot with roots the surrounding mass of wet soil may become stagnant and cold. For the same reason, never re-pot a plant into a much larger pot. Choose one only 1in (2.5cm) or so greater in diameter.

Always pot up plants at the same depth they were previously. Never bury the stem of a young plant or it may rot off at soil level. Leave ¼in (6mm) of space above the compost up to the rim of the pot. This can then be filled up at each watering.

The planting itself is relatively simple. Place some compost over the drainage material, and stand the plant roots on this. Adjust the amount to bring the plant crown to the correct level. Then add more compost around the root ball, pressing it gently around and between the fine roots. Fill gradually, keeping the plant central in the pot.

Potting is half a craft and half an art. The plant must be firmly held, yet be free to thrust its roots out; well watered, yet not drowned; with air near its roots, yet not dried out. Fortunately, most of our plants will survive even if the potting is done with less than professional skill, yet it is in this art that most of our green-fingered friends excel.

Seedboxes, whether of wood or plastic, are merely large, wide and shallow pots. They need drainage material (or a very open compost), adequate moisture and a firm yet open texture in their filling. Consequently the work of preparing them is much the same as for pots.

Potting in Practice

(*a*) Drain holes in plastic pots are sometimes blocked by thin plastic "flashings". Clear these before use. Old pots, especially those of clay, must be thoroughly washed before re-use.

(*b*) Cover the base of the pot with clean broken stones or chips of old clay pots to a depth of roughly $\frac{3}{8}$in (10mm).

(*c*) Scoop the pot full of moist (but not soaking wet) compost.

(*d*) Overfill the pot, draw a strip of wood flat across its top and strike the filled pot several blows on the potting bench. This should settle the compost $\frac{1}{4}$in (6mm) or so to give space for easy watering later.

c
▷

a
◁

b
◁

d
▷

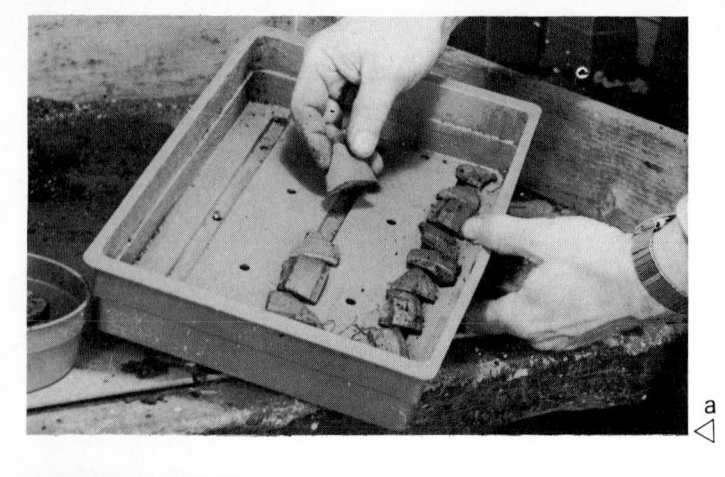

Seedbox Filling

Seedboxes are filled in much the same way as pots.

(*a*) Cover any drain holes with pieces of broken pot or small flat stones. This extra drainage is needed if the compost is on the heavy side.

(*b*) Sieve the compost to remove the coarser particles . . . .

(*c*) . . . . and spread these $\frac{1}{2}$in (12mm) deep over the base of the box.

(*d*) Overfill the tray with the finer sievings.

(*e*) Draw a strip of wood across to level the compost . . . .

(*f*) . . . . and use a pressing board to squeeze the compost firmly down $\frac{1}{4}$in (6mm). Seeds usually need the slightly closer texture that this pressing produces. It must not, of course, be so heavy as to consolidate the soil completely!

a

b

c

e

d

f

WATERING

All plants need water but many potted and boxed plants are killed by being given too much. Injury by under watering is much more unusual. Unfortunately, there is no simple rule as to when and how watering should be done. The amount needed depends in the first place on the type of plant. One whose wild home is a swamp or river bank will need much more water than a desert plant. Plants from high rainfall parts of the tropics will need more than those from northern mountains.

The temperature is important, too. Plants in hot rooms breathe out much more water from their leaves than similar plants in a cool place. Warmth also stimulates growth, which increases water needs still further.

The season also alters water intake. Spring is a season of rapid growth and high water need. Summer often sees the flowering period and warm days causing a steady water loss from leaves. On the other hand, plants often cease growing in winter and therefore use very little water indeed. (It is almost true to say that plants which are not actually growing require to water at all. Simply maintain a moist atmosphere around their roots.)

Given all these variables it is difficult to lay down hard and fast rules about watering adult plants. However, it is easier with young seedlings and cuttings, for their water needs are much less. For propagation work start by using compost that is evenly moist. Water the dry-mixed compost carefully before putting it into the pots. It should just hold together when squeezed in the hand but instantly break apart when tapped with a finger. On no account must it remain a wet, soggy bundle. Probably no further watering will be needed for some time after potting. When it is necessary, plunge the filled pot or tray into water up to its rim. Do not flood it. Let the water find its own way up through the drainage holes to darken the upper surface. After this set the pot or tray aside to drain. Watering from below like this is always preferable to top watering. It prevents consolidation of the soil surface and makes certain that even the lower parts of the tray are moistened.

Young plants and seeds use so little water that they can go without watering for weeks, provided the air round them is humid. However, in warm and dry rooms (especially those with central heating) the soil may dry out rapidly, not through the plant using the water, but by ordinary evaporation. This moisture loss must be made up so that the plant roots are kept surrounded by moist air. All such watering must be lightly done. It is better by far to reduce the need by plunging the complete pot or tray into deep, moist peat so that evaporation is reduced or eliminated. Alternatively, the whole plant, pot, leaves and all, can be enclosed in a sealed polythene bag. This stops almost all water loss.

When top watering larger, growing plants, completely fill the ¼in (6mm) space left at the top of the pot. Place the spout of the watering can directly into the top of the pot, keeping your thumb over its tip. Then release just enough water to allow the pot to fill up. Never water established plants in driblets. Always give a good soaking and then leave the plants to dry out almost completely before watering again. A little water every day is usually bad practice.

Soil around growing plants may become packed hard. This makes it difficult to water thoroughly. Either the water pours straight through the pot between the dry soil and pot side or it simply floats over the surface without soaking in. In the first case, lift the root ball completely from the pot and gently crumble away its hardened sides, freeing a few roots. Then re-insert the plant in the pot, using a stick to press fresh compost round its sides. Then water by submerging it entirely in a deep bucket till bubbles cease to rise. In the second case, break up the surface crust with a pointed stick and remove about ¼in (6mm) of soil. Replace this with fresh compost. Then water thoroughly, preferably by immersion.

The rules for watering are then:

1 Make sure that your compost is evenly moist before potting.
2 Water all pots and trays from below, at first.
3 If possible, keep young plants in a humid atmosphere.
4 Always water larger plants generously and then allow the plant practically to dry out before watering again.
5 Do not give water to a regular timetable. Plants differ in their needs throughout the seasons. One which requires a weekly or even daily dose during summer may require nothing at all for three months in the winter.
6 Get to know your plants, their background and original homes, and water accordingly.

Watering Methods

(*a*) Plunge newly filled plant pots rim deep into water. Do not allow this to overflow the surface. Leave them there until water has risen to darken the compost.

(*b*) Drain them thoroughly by standing them on mesh trays or inverted sieves.

(*c*) Similarly, plunge filled seed trays (before sowing) into shallow water till the surface darkens. Allow to drain freely afterwards.

(*d*) To help to avoid over watering, use moisture-testing devices. These are comparatively cheap and use electronic methods to indicate whether the soil is dry, moist or wet. Some types even indicate if the plant needs fertilizer.

Keeping young plants growing during holidays can be a problem. However, even when they must be left unattended it is still possible to provide them with adequate water. If they are then stood in a warm, bright but not sunny position, in as even a temperature as possible, most kinds will survive for several weeks. Water is however essential, and we show on the facing page two methods, well tried by experience, for providing this:

a ◁

b ◁

c ◁

d ▷

First Method

(*a*) Place two bricks on a table or other suitable surface and stand a large, open water container such as a bucket on them. Fill this nearly to the brim with water.

(*b*) Make "wicks" of rolled flannel strips and soak them in the water. Trail the wet wicks over the edge of the bucket, keeping one end in the water. The flannel will slowly draw water up from the bucket by capillary attraction.

(*c*) Secure the other ends of the strips into the soil with wire. Raise the bucket on bricks so that the water level is above the pots, enabling the plants to use up all the supply of water.

Second Method

(*d*) Place several lengths of wood across a shallow tray or dish of water.

(*e*) Wicks of wet flannel are then thrust through the drain holes of each pot and the other ends put into the water so that each plant can draw enough water up for its needs.

b

c
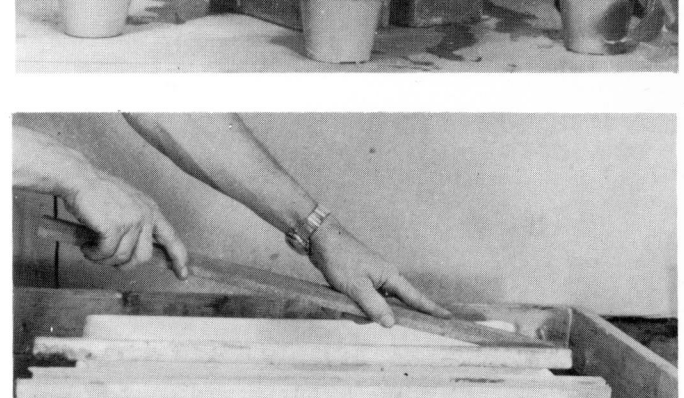

a

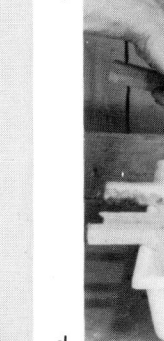

d

e

a ◁

b ◁

HUMIDITY

Many young plants prefer to have humid air, heavy with water vapour, around their leaves. However, rooms in houses tend to have rather dry air. This is especially so with central heating. One way of helping plants in such rooms (or indeed in greenhouses) is to "double pot" them. This is a simple operation which can be very effective. The aim is to provide a "cloud" reservoir of moisture vapour immediately around the plant, and at the same time reduce water-loss from the pot:

(*a*) Select a pot roughly 2in (5cm) greater in diameter than the one containing the plant. Scatter clean broken stone or gravel as drainage material over the base, and add a layer of well-moistened, granulated peat. Make this deep enough to make the soil level the same in both inner and outer pots. Lower the plant container into the outer pot and press it down firmly into the peat. Then, fill up the gap between with more wet peat, well pressed down with the fingers.

(*b*) The peat gives off a humid vapour which rises around the leaves. Keep the peat wet at all times.

Standing pots on pebbles in a deep saucer, and adding water till this is just below the tops of the stones, is another method. Fine spraying with pure water is helpful to nearly all plants, except those with furry leaves.

c ◁

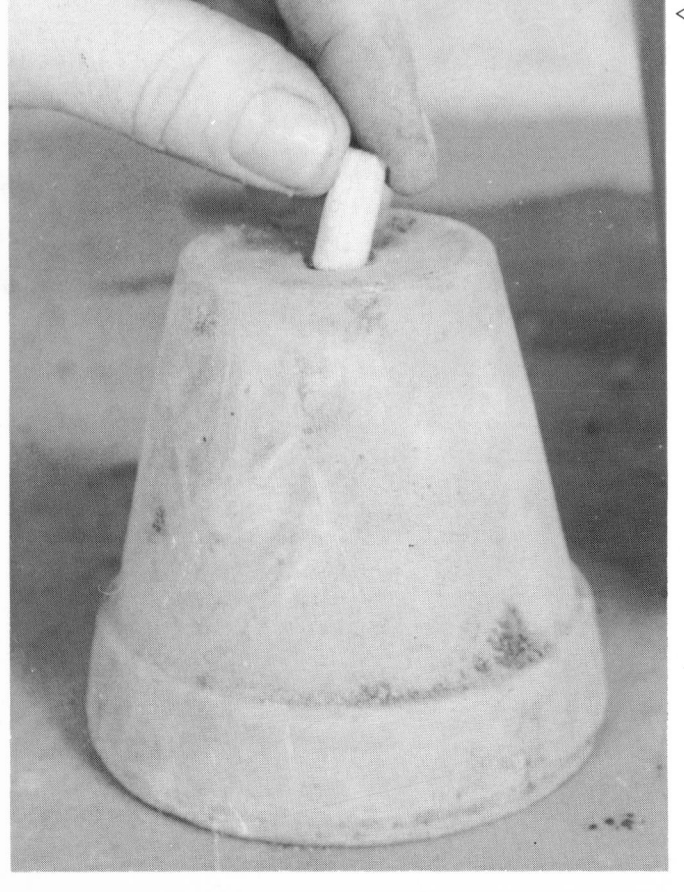

d ▷

A few bog plants, and many ferns do best if propagated in even wetter conditions. For this you can easily improvise a suitable high-humidity propagator from a couple of plant pots and a polythene bag which will maintain very moist conditions for weeks, without attention:

(c) Plug up the drain hole in a small *clay* pot with a piece of wood.

(d) Stand the clay pot within a larger plastic one, using potting compost to bring their rims up level.

(e) Fill the gap between the two pots with compost, pressed gently down with a length of wood, and insert your seeds or cuttings here.

(f) Fill the inner, clay pot with water. This will rapidly soak into the porous sides, so refill it after an hour or two, when it has become saturated. Then stand both pots in a large, unperforated polythene bag and draw this up around them.

(g) Pinch the top of the bag together and blow gently into it to inflate the upper part of it away from the leaves of the cuttings.

(h) Seal the bag with a rubber band. Inside, the water in the clay pot is gradually absorbed by the compost and the atmosphere will be maintained at a very high humidity for several weeks.

g ▷

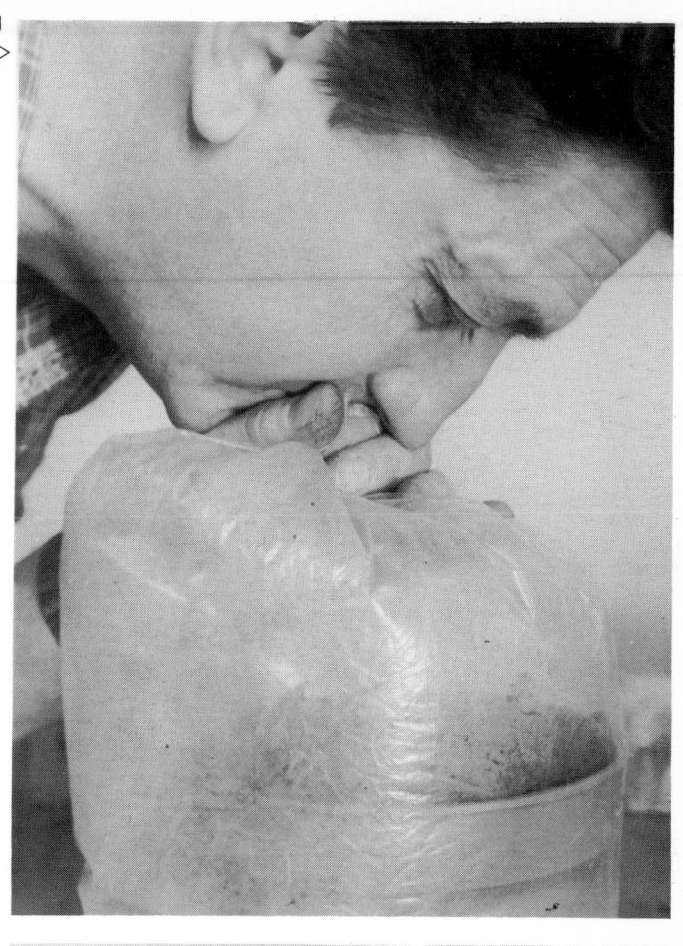

e ◁

f ◁

h ▷

CHOOSING THE RIGHT MOMENT

All plants respond to the passage of time and follow a distinct seasonal pattern. This starts with development of new growth, which then matures and later flowers. Fruit follows and then comes a rest period. Their propagation, too, is usually best done only at certain seasons. Of course, the various kinds differ in this, but it is usual to select a plant's natural vigorous growing period for propagation. Response to the challenge of propagation is then at its highest and success will be more likely.

To some extent a plant can be made to change its seasonal clock by altering the timing of the light it receives. Electric lights may extend the apparent day length, shading reduce it. However, this takes a great deal of skill and time and is used mainly in professional nursery work.

Out of doors hardy seeding is often done in spring. The ground is rapidly being warmed up by the sun, the heavy rains of winter are over, and the soil is draining, leaving it airy and light. We can look forward to longer, warm and bright days and moist evenings – ideal weather conditions for young and delicate seedlings.

A few seeds germinate best if sown immediately after they have ripened, in late summer or early autumn. These types pass the winter as small seedlings. Many hardy kinds that have become adjusted to the British climate will tolerate this treatment. However, there are many others which could not endure the long cold winter and therefore these must be started from seed under protection in winter, and put out finally in late spring.

When working indoors with half-hardy kinds we can of course control the conditions, but even here season does play its part. We must be sure that by the time the young plant is ready for transfer outside the weather will be suitable. There would be little use in growing plants which came to the planting out stage in the middle of winter!

Tender plants that never leave the protection of house or frame may do better still if sown at particular times of year, though this is indeed much less vital, since all conditions of light, warmth, etc. can be fully controlled.

The simplest of non-seed methods – division of the roots – can be done most successfully just after flowering. The plant then has a week or two of growing season in which to recover, followed by its natural rest. Alternatively, division just as growth starts may be done, especially with plants that flower very late in the year. Dividing a plant actually in bud or flower is not normally satisfactory, but if done the flowers must be removed, or new roots may be very slow in forming.

With vegetative propagation, (by cuttings, layering, grafting etc.), the test is usually the presence of suitable cutting material or shoots for layering. Speaking very generally, shoots without flower buds make the best cuttings and layers. These are usually available after flowering. However, spring shoots, even with buds, often grow away rapidly due to the exceptional vigour of the plant at this season.

As a broad general rule, if you have a healthy plant, growing strongly and providing suitable material for propagation, it will usually succeed, no matter what the calendar may say. Many of our commoner houseplants can in fact be propagated all year round.

(*a*) Soft cuttings are taken from fuchsia all year round.
(*b*) Yellow flag iris are propagated by division in summer.

a ◁

b ▷

(*c*) Daffodils are increased by division.

FEEDING

Feeding with chemicals is not usually needed for young plants. A well-mixed compost should contain all the food that the cutting or seedling needs until it is transferred to its final pot. If you feel however that the young plant needs a little extra boost, use a proprietary liquid food and never exceed the dosage recommended by the manufacturer. It is always better to give too little and too weak than too much and too strong!

PESTS AND DISEASES

Young plants are not subject to many of the principal adult plant diseases. Indoors you may suffer losses of seedlings from fungus attack. This is best prevented by thorough cleaning and watering of the prepared pots and trays beforehand with a commercial fungicide.

Attacks by spider and mite can be treated by a Derris spray. Insecticide in aerosol spray-guns is probably the simplest and easiest to use inside the home but as with fertilisers the manufacturers instructions must be carefully followed. In no case must the dosages recommended be exceeded.

Actual attack on outdoor seedlings may be made by slugs, caterpillars, mice – even rabbits. Deterrent chemicals are widely available and reasonably effective. Surrounding a small seedbed with fine-mesh netting, sunk 8in (20cm) into the soil, is not expensive and discourages rabbits, moles, cats – or children!

PROPAGATION METHODS

PROPAGATION FROM SEEDS

The most widespread method plants use for propagation is seeding. Nearly all plants do produce seed and in the vast majority of cases this seed is fertile. The fact that the offspring from these seeds are very slightly variable is actually of value. Nurserymen have used this fact to develop the thousands of varieties we now enjoy in cultivation.

When a plant produces seeds there are nearly always two parents (though some plants can also fertilise themselves) who have slightly different characteristics. The combination of the two produces a seedling which itself is not quite the same as either of its parents. It has characteristics of both, but in different proportions.

From time to time, even more important changes called mutations may take place. They may actually change the appearance of a plant in a startling and new way. Our gardens owe a great deal to mutations and because of them we have been able to develop wide ranges of colours, sizes of flowers and so on. The reasons for them are complicated, and some kinds of plant seem to mutate more easily than others. Take the example of the rose. Many millions of pounds are spent every year in growing hybrid roses from seed in the hope of producing an unusual colour, scent or habit. Often only a single seedling is selected out of many thousands grown. Yet in the background history of all the huge range of colours, habit, size, leaf shape, of climbers, ramblers, bushes, tea roses, floribundas and so on, there are only a few actual rose species. The differences have been

(*a*)   Just one variety of the rose – "Peace".

produced by breeding. Most commercially sold seed does produce similar seedlings, though some kinds are known to be more variable than others. You cannot then be sure that they will breed "true" – whether the seedlings will be like their parents or not. However, this only adds to the interest. There is always the possibility of raising some new and splendid variety! This has happened, and in Britain a wonderful range of new lupins was developed by an amateur called Russell.

Sowing seeds as a method of propagation is not, indeed, the easiest way. Division of plants (page 34) is much more certain of success. Seed though remains universally popular. By no other method can you obtain so many fine plants at so low a cost.

Broadly speaking, seeds divide into three different types. The first and simplest of these are hardy seeds which can be sown out of doors and grown on without protection in the garden. Secondly, there are half-hardy seeds which may be sown under some protection in early spring and then transplanted out into the garden to grow on, after the frosts have gone. Thirdly, there are seeds of plants which must be sown in heat and continue to need protection throughout the whole of their lives. These include many house and greenhouse plants.

A seed is in fact a complete plant, with its own food supply, respiration system and growth potential. Whether or not it will grow depends on several important factors. As with all plants these include moisture, warmth, air, and, later on, light. Chemical food at first is less essential. Nearly all seeds contain within themselves a small amount of plant food so they do not immediately need more to start to grow. You can even germinate many seeds on a totally sterile medium such as cotton wool or in raw sand. However, once the seedling has started to grow it will of course have to find more food quickly before its store is exhausted. Water and warmth are needed at once. In moist conditions, seeds will absorb water through their outer skins and start to grow if the temperature is high enough. Without warmth, chemical reactions within even a moist seed are very much slowed down and may stop altogether. Once these two factors of warmth and moisture are combined, the seed will split open its outer coat and start to thrust a tiny root downwards and an equally tiny stem upwards. This is a crucial stage for the plant. Before its basic food supply is exhausted its roots must reach a source of chemicals and its rising stem must reach sufficient light to start the leaf processes without which it cannot convert these chemicals into tissue. Access to air then becomes vital, both for roots and stem.

For successful seeding therefore we must provide first, a moist and humid atmosphere with adequate warmth. The

c. soil must contain accessible chemicals as the seeds develop, but must not contain too much as strong chemicals can actually inhibit growth. The important factor of access to air is sometimes overlooked. If a seed is totally immersed in water, excluding all air, its development will be very slow if it takes place at all. Well drained soil allowing air to penetrate it and help the chemical work of the roots is essential for success. Last but not least – the seed needs light. As the first leaf unfolds, we must give light, or the stem will extend further and further up searching for it, and becoming weak and "drawn" in the process.

The methods we may choose to provide these vital conditions can vary. We provide warmth by the sun or artificial heat. We provide water by moistening (but not water-logging) the soil or pots of compost. We provide air indoors through ventilation, though always making sure that this is not so cold as to inhibit growth. We provide food by making sure the soil or compost has the necessary chemicals in the right proportions. We provide light by the sun or by placing germinating seedlings under artificial lamps.

Fully hardy seeds are sown outside, directly into the soil. It is essential to cultivate the ground well before seeding and also to remove weeds as far as possible. The less competition that seeds find as they develop, the better their chances of survival. Deep, rich soil is best produced by the method known as double digging.

(b) To propagate lupins, sow seeds outside or take soft cuttings.

(c) Sweet William is grown from seed and is biennial.

(d) Columbine seed is sown outside and plants propagated by division.

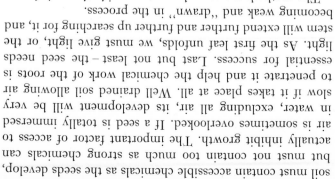

Soil Preparation by Double Digging

This is quite simple in principle but is not a quick operation. Do not attempt too much at a time. It is better to take
several weekends over the digging of a large seedbed, rather
than do a poor job in one day due to tiredness.

The basic idea is to improve the condition of the soil in
depth. The upper 6 or 8in (15 or 20cm) of soil is usually
fairly fertile and is known as topsoil. Beneath it will be
layers of "sub-soil" which may be much poorer and contain fewer plant foods. By digging vegetation and fertilisers
into this bottom layer we can improve drainage and
aeration to a considerable degree.

(*a*) Double digging is especially valuable on overgrown
land, with heavy weed or grass cover. This can be
skimmed off and deeply buried where it will not re-grow
again. Start by skimming away a strip of the grass along
the length of your plot and setting the skimmings aside.

(*b*) Similarly remove a trench 8–10in (20–25cm) deep
and 12in (30cm) wide of the best topsoil. This also must
be wheeled away to the other end of the plot for use
towards the end of the job.

(*c*) Clean out and level the base of the trench with the
spade. Simple double digging is then done by forking
over the base of the trench which loosens up the subsoil
and so improves the drainage. This alone will result
gradually in the production of a deeper layer of fertile
soil.

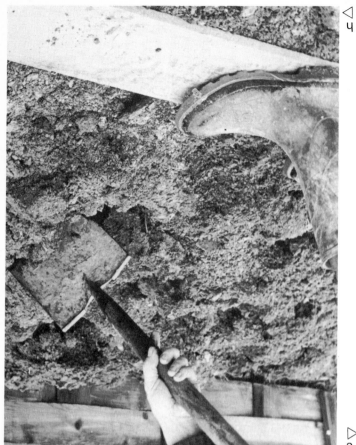

(d) Better still though, is to incorporate manures, particularly natural heavy organic manures such as stable and farmyard manure with straw. Turn this as deeply as possible into the trench bottom.

(e) Next skim away the grass on the next strip, adjacent to the trench. Do not take these skimmings away, but turn them upsidedown into the bottom of the trench and then chop up the turves and weeds thoroughly with a spade. More manures can be added now.

(f) Next dig the top soil 12in (30cm) back from the trench side . . . .

(g) . . . . and turn it over into the first trench. This soil will thus fill up the first trench and leave exposed a second trench. This is then dug over and fertilised in the same way. Repeat this process right across the plot using the grass and soil from the first trench to fill up the last one, leaving an even, well-cultivated surface.

(h) When working over cultivated land, it helps to tread on broad planks laid over the soil surface. This reduces soil consolidation, a valuable point in wet or heavy land.

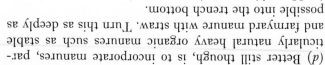

Seeds in Boxes

Even when a plant is hardy and could be grown from seed sown outdoors, we may still choose to grow it in boxes under some sort of protection. This enables us to provide fertile, well-mixed compost and to keep close control over the development of the seeds from the beginning. It also means that we can sow much earlier in the year, to give mature, flowering plants in early summer.

You can make the best use of boxes for seeds if you have a garden frame or greenhouse. Within this protection they can be cared for more easily. But you can do nearly as well by improvising a frame from a few lengths of wood or a number of bricks made into a rectangular box about 8in (20cm) deep. Even if this is only covered with transparent polythene it makes a primitive cold frame which will give quite a lot of protection, though not of course against severe frost. A glass covering, more difficult to make, is even better, and gives rather greater protection.

(a) When sowing, it helps to mix the fine seeds with some sand. It is then easier to see when the light-coloured sand is evenly distributed.

(b) Cover the seeds with a fine layer of very sandy soil, applied through a small-mesh sieve. Many seeds need this light covering, not so much to help germination but because without it their strong roots may lift the whole seed bodily above the soil surface, instead of the root pushing down into the compost beneath.

(With large seeds you can easily space them apart evenly instead of scattering broadcast. Place a few on a piece of glass as long as the length of the tray. Then rest the edge of the glass along the first seed line. Push the seeds off the edge of the glass at the proper spacing.) Accurate and immediate labelling of boxes is essential.

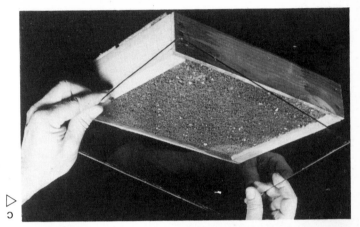

(c) Cover the sown box with a sheet of glass. Alternatively, slip the box into a large polythene bag, with a thin strip of wood laid across the box centre to prevent the polythene dropping and touching the seedlings.

Many seeds germinate best in complete darkness. Cover the box with newspaper. However, the moment there is any sign of life, which may be only after a few days, this paper must be removed or the young seedlings will push up thin stems, seeking light, and become weak and drawn.

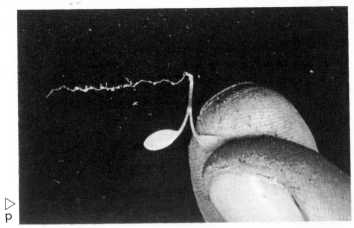

Lift the glass every day and wipe it dry of the moisture which will have condensed underneath. This moisture comes from the compost. Heavy drops from it can injure the seedlings. Tilting the tray a little will encourage the drops to flow to the box edge instead of dropping on them. After germination has started the glass can be gradually lifted away and after a few days, removed permanently. At this period give the seedlings as much light as possible. Place the box high, near the frame or greenhouse glass, or in the brightest place on the window sill that does not get the scorching midday sun.

PRICKING OUT

The next job (with scattered seed) is to prick out the seedlings, lifting them and replanting them further apart in other boxes. This job should be done as early as possible, when the seedlings appear almost too small to handle. Never leave it until they look thoroughly established or you will have greater losses. Many seedlings cast up a single pair of leaves first and it is at this stage that transplanting can best be done. Incidentally, never hold seedlings by their stalks, only by their leaves.

The biggest risk though, at this stage, is a disease called "damping off". This is an attack by a certain type of fungus. Prevention is much better than cure for this, and is fortunately easy. Simply soak the prepared box in a chemical called Cheshunt compound before or just after transplanting. This, combined with allowing the seedlings plenty of air, will reduce the risk of disaster.

Of course not all the seedlings will come up at the same time, so transplant those that are ready, and leave the others to develop over the succeeding few days.

(d) The two-leaved seedlings are very tiny, though the root may be long in proportion. Always handle them by the leaves and with the utmost care.

(e) When transplanting, make a hole with a pointed stick in the compost, deep enough to accept the entire root. Lower the seedling into the hole so that its leaves are just above the compost surface. Then gently press fine compost about the roots and stem till the seedling is held firmly (f).

When completed the boxes should have seedlings in neat straight lines, spaced out so that each has enough soil and light to grow. The usual spacing is about 1½in – 2in (5cm). Once they have developed in these boxes they can be replanted into pots or out of doors.

SPECIAL POTS THAT EASE TRANSPLANTING
Transplanting always checks a young plant's growth. One way of overcoming this has been the development of the peat fibre pot. Plants grown within pots like this are never transplanted. As they outgrow the pot the roots will pass through the fibre. Then the whole thing, pot, roots and all, can be planted bodily in a larger pot, without the plant feeling any check at all.

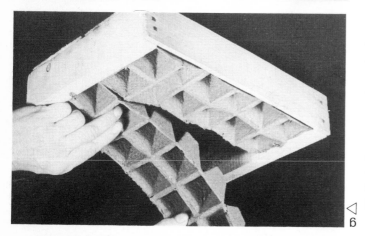

(g) A useful type of fibre container for seedlings consists of strips of tiny square pots which can be arranged to fit inside a normal seed tray. Fill them in exactly the same way as a box, but the compost used must have a much greater than normal proportion of peat, say up to about 30 per cent, to give it greater water retaining capacity. Simply smooth the compost flush with the surface of the pots.

(h) Press a shallow hollow in the surface of the compost ready to receive the seed and slide single or small groups of seeds into place. Then cover them and water by immersion as shown earlier.

a ◁

Seeds from Fruits

(*a*) Plants which produce soft fruits often have fertile seeds embedded in them. One of these is the well-known "Christmas orange" which produces a tomato-like fruit.

(*b*) Split this, and tiny seeds will be found embedded in the flesh. Squeeze them out on a piece of glass. Allow them to dry and store in a dry jar. Sow them by sprinkling a few on top of a filled pot, lightly covering them with compost.

(*c*) Cover the pot with glass and newspaper. After germination, remove the paper and later the glass and prick the seedlings out wider apart to grow on.

Ferns and Cacti

The easiest way to propagate ferns is certainly the division method shown on page 34. Many ferns develop a closely woven mat of roots from which the finely divided fronds arise. By cutting these roots into separate pieces a number of new young plants can be obtained. However it is possible, and very interesting, to grow ferns from seeds, or *spores*.

The fern as we know it in the garden does not produce seed of a normal kind. Its life history is much more remarkable. Instead most mature ferns produce a fine, brownish powder consisting of thousands of powdery spores. Each grain of this dust, if it falls into a moist, dark place, can develop into a *prothallus* – a tiny, flat and usually rounded plant which looks rather like a moss or liverwort. It clings flat to the soil and is practically invisible at first, appearing more like a fine, green film over the soil.

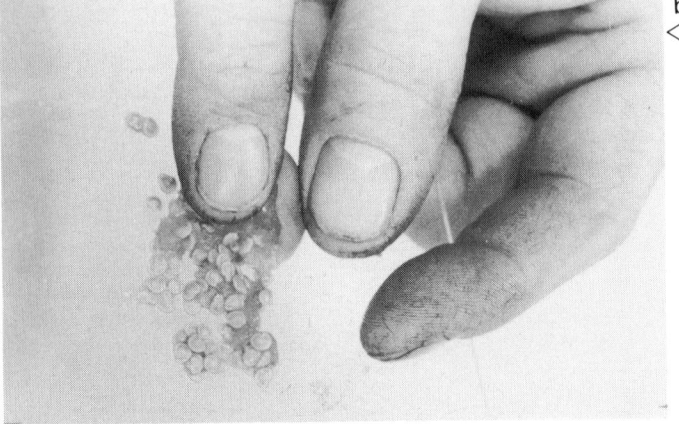

b ◁

Although so small, the prothallus is a complete plant. It grows and develops male and female organs which unite to produce fertile seed. After this the prothallus throws up the familiar fern fronds which then rapidly increase to full size. When mature, these fronds will produce more spores to continue the cycle.

To propagate ferns from spores we must provide for all these developments. Start by collecting the spores, usually in late summer. They are visible as fine brown powder clinging to the edges of the fronds or in lines underneath them (fig *d*). Hold a sheet of white paper under the frond and tap it lightly and the powder will be deposited on the paper. Alternatively you can cut off the frond and put it into a dry paper bag where, after a few days, a brisk shake will release a large number of spores.

c ◁

Spores are far too fine to bury in the normal way. Instead, take a small pot of peaty sterilised compost watered very thoroughly indeed. Plenty of moisture is essential, much more than for normal seeds. Ensure that the whole of the plant pot is soaking, and allow it to drain for a few minutes. Then sprinkle the spores very finely over the surface. It is practically impossible to sow too finely. Do not cover them at all but place a small piece of glass over the top of the pot and then stand the pot itself in a shallow dish of water. Alternatively, pull an unperforated polythene

d ◁

bag over the pot before standing it in water. The object is to keep the spores in a very humid atmosphere on top of the compost. Most fern spores prefer to be well shaded during this part of their lives, so place the pot out of doors or in a dark, cool shed, or cover it with thick paper. Warmth is not necessary.

Over the next few weeks do not disturb the pot at all, even though water will condense under the glass or inside the polythene bag. A fine, greenish film will appear over the soil showing that some of the spores have germinated. Once these start to develop they will gradually expand to form tiny round, flat, leaf-like objects which are the prothalli spreading over the soil surface. Remove the paper and move the pot to a lighter position keeping the glass or polythene bag in place. During all this time the pot must on no account be allowed to become dry, even if this means re-immersing the whole thing in water up to the compost surface.

e ▷

The prothallus stage lasts several months during which time the pot needs no attention. Eventually small fronds which are exactly like miniature versions of the mature ferns will be thrown up. Once a number of these have appeared they can be picked out just like seedlings and grown on in pots. Keep all the pots moist and covered with glass until the transplants are well established (fig e). Some growers transplant by gently lifting out ¼in (6mm) squares of prothallus-covered compost and spacing them out over the surface of prepared pots. The fronds then develop and of course are left in position.

Though most ferns that we grow are quite hardy and need no heat either at sporing or in later growth, it is often more convenient to do this work under cover. A cold cellar, shed, or the back of a garage is quite adequate for growing ferns and from a few spores collected in summer and autumn you can grow many hundreds of young plants within about eighteen months.

Cacti can also be raised from seed, given time and patience, in a very sandy compost (fig f). It is difficult though to grow most cacti on to maturity, so it is best to leave this until you have gained experience with other, easier types of plant.

f ▷

A GUIDE TO HARDY AND HALF-HARDY ANNUALS AND BIENNIALS THAT CAN BE GROWN FROM SEED.

AH – annual, hardy – can be sown outside
AHH – annual, half-hardy – needs protection on seeding
BI – biennial – sends up leaves the first year, flowers
the following year

	LATIN	ENGLISH		LATIN	ENGLISH
AH	*Adonis annua*	pheasant's eye	AH	*Helianthus annuus*	sunflower (annual)
AHH	*Ageratum houstonianum*	ageratum	AH	*Helichrysum*	everlasting flower
BI	*Althaea rosea*	hollyhock	AH	*Iberis amara*	candytuft
AHH	*Amaranthus gangeticus*	love-lies-bleeding	AHH	*Impatiens balsmina*	balsam
AHH	*Antirrhinum*	snapdragon	AH	*Ipomoea augustifolia*	morning glory
AHH	*Arctotis grandis*		AH	*Jasione montana*	sheep's-bit scabious
AH	*Aster*	aster	AHH	*Kochia scoparious*	mock cypress
BI	*Bellis*	daisy	AHH	*Lathyrus odoratus*	sweet pea
AHH	*Calceolaria mexicana*	slipper flower	AH	*Limnanthes douglasii*	
AH	*Calendula officinalis*	marigold (pot)	BI	*Limonium*	sea lavender: statice
AHH	*Callistephus chinensis*	china aster	AH	*Linaria reticulata*	toad flax
BI	*Campanula medium*	Canterbury bells	AH	*Linum grandiflora*	flax
AHH	*Celosia cristata*	cockscomb	AHH	*Lobelia erinus*	lobelia
BI	*Celsia cretica*	cretan mullein	BI	*Lunaria biennis*	honesty
AH	*Centaurea cyanus*	cornflower	AH	*Lupinus nanus*	
AH	*Centaurea moschata*	sweet sultan	AH	*Lupinus hartwegii*	
AH	*Centaurium venustum*	kentaurion	AH	*Lupinus hirsutus*	
BI	*Cheiranthus*	siberian wallflower	AH	*Lychnis coronaria*	campion
AH	*Chrysanthemum coronarium*		AH	*Matricaria maritima*	double May-weed
AH	*Chrysanthemum segetum*	corn marigold	BI	*Matthioli incana*	brompton stocks
AH	*Clarkia elegans*	clarkia	AHH	*Matthioli incana*	ten-week stocks
AH	*Coreopsis drummondii*	tickseed	BI	*Myosotis*	forget-me-not
AH	*Coreopsis tinctoria*	tickseed	AHH	*Nemesia strumosa*	nemesia
AH	*Cosmos bipinnatus*	mexican aster	BI	*Papaver*	Iceland poppy
AH	*Delphinium ajacis*	larkspur	AHH	*Petunia hybrids*	petunia
BI	*Dianthus barbatus*	sweet william	AHH	*Phlox drummondii*	phlox (annual)
BI	*Digitalis*	foxglove	BI	*Polyanthus*	
AH	*Dimorphotheca*	star of the Veldt	AHH	*Salpiglossis sinuata*	
AH	*Eschscholzia caespitosa*		AHH	*Salvia splendens*	salvia (red)
AH	*Eschscholzia californica*	californian poppy	AH	*Tagetes erecta*	african marigold
AH	*Felicia bergeriana*	kingfisher daisy	AH	*Tegetes patula*	french marigold
AH	*Felicia fragilis*		AH	*Tegetes tenuifolia*	
AH	*Gilia tricolor*		AH	*Ursinia*	
AH	*Godetia amoena*		BI	*Verbascum olympicum*	evening primrose
AH	*Godetia grandiflora*	godetia	AH	*Verbena hybrids*	
AH	*Gypsophila elegans*	gypsophila (annual)			

(*a*) The early flowering incurved variety of chrysanthe-mum, "Ensign", is an annual hardy plant that can be grown from seed.

Primrose (*b*), London pride (*c*), bergenia (*d*), and *Anemone japonica* (*e*), are all propagated by division see page 34.

PROPAGATION FROM ROOTS

The simplest of all plant propagation is division. Most of us are only too familiar with garden weeds which, no matter how carefully dug out, leave behind morsels of root which promptly send up leaves again! A root that is cut often has the power to re-grow. Plants can often therefore be increased simply by chopping them in pieces!

This division is specially suited to plants that have groups of vertical stems or leaves springing from soil level. Michaelmas daisies, phlox, *hemerocallis* and so on, can often be separated into tiny groups of stems with a few roots attached. Primulas and many other small plants grow naturally in clumps of leaf rosettes, each with roots, that can be easily separated. Where plants grow into large clumps it is usually the outer portions that develop into the best fresh plants. It is good practice to discard the centre parts. There is little risk of failure as the young divisions already have complete roots, and if properly potted will only suffer a short check.

However, even plants which cannot easily be divided may still be propagated from roots by cutting these up into short lengths and planting them. From the upper tip of these root cuttings new buds and stems will develop.

Yet other plants produce roots from their stems. These aerial roots may be mainly used for gripping and climbing as with the familiar ivy, but, if the stems are cut and planted in soil, they will change into proper feeding roots.

The beginner should certainly try propagation by division. The job is quick, the risk small, and the results very satisfying.

Root Division – Bulbs

(*a*) Bulbs are very easy to propagate. Large, mature ones can be easily pulled apart and planted up individually. They may however take some years to reach flowering size.

(*b*) Many bulbous lilies can be propagated by pulling off base scales and planting them in sandy compost in a warm place. Insert them blunt end downwards. Hyacinths may produce very small "bulblets" the size of rice grains around the main bulb base, which will also grow on when separated.

Root Division of Perennials

(*c*) All the primula family can be divided after the flowers have faded. The leaves grow together in clumps having their own attached roots. Crumble away the soil to expose the leaf bases and gradually lever apart the individual rosettes.

(*d*) From a single good clump you may get as many as six or eight healthy young plants which will rapidly grow on.

a ◁

c ▷

b

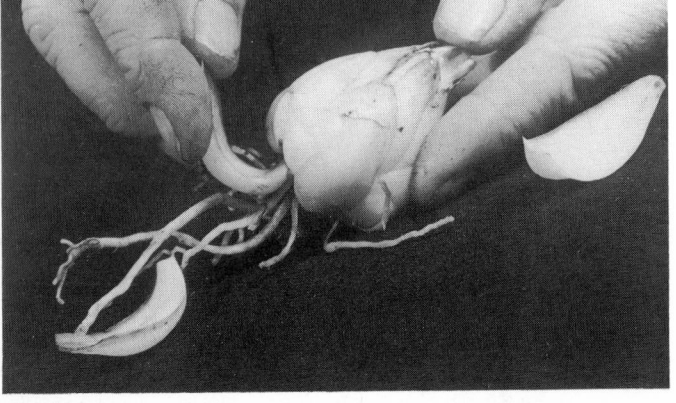

d ▷

(*a*) Large plants may benefit by having their leaves shortened before replanting, to reduce water loss until roots re-form.

(*b*) The heavy iris "roots", really a creeping, thickened stem, can be slit into small sections, each with a leaf, and grown on. The outermost, youngest parts make the best plants.

Root Division of House Plants

(*c*) The chlorophytum or spider plant is one of the most popular of all house plants. Like many house plants (aspidistra, ferns, *peperomia*, *sanseveria*, saxifrage, etc.) it is easy to propagate by division, since its stems spring in bunches from soil level.

(*d, e, f*) Gently separate the various sections into single sprays and repot these in 3½in (9cm) pots. They will suffer hardly any check and quickly grow on into larger plants. Even sprays without roots will often succeed if the stem base is planted firmly. Such pieces are called offsets and are a kind of halfway house between root division, in which each part contains stems, leaves and roots, and cuttings which have no roots at all.

c ▷

d >

a ◁

e ▷

b ◁

f ▷

35

Aerial Roots

HEDERA

(*a*) Most varieties of ivy are extremely easy to propagate. The climbing stems freely develop aerial roots to cling to their host tree.

(*b*) Cut them into short sections or even single leaves, and plant them in ordinary compost.

(*c*) A single tray will give you two dozen or more new plants. For quickest effect, plant several such cuttings, after growth has begun, around the edges of a 3½in (9cm) pot and train them all up a central supporting cane.

MONSTERA

The *monstera* (fig *d*) is a plant that also develops long aerial roots, though less prolifically than ivy. However, these are rarely used for propagation. Instead, the creeping stem is divided into sections (rather like the iris) and planted in a warm, humid atmosphere. The technique is similar to the rooting of *ficus* leaves (page 58). A propagating case or frame is really desirable.

Although this is one of the largest of house plants, propagation by seed is widely used commercially, as seedlings tend to grow on quicker.

c
▷

d
▷

a
◁

b
◁

Root Cuttings

Gardeners have for centuries propagated plants from root cuttings. These are simply short lengths of root, planted in good soil and encouraged to form leaves. Stem buds develop at their upper ends and these grow to form complete new plants. Root buds develop at the lower ends.

We usually choose roots of about pencil thickness, though naturally this does depend on the species. You need a root which has reached maturity without having become woody and hard. The length of root can then be further cut up into pieces 1in–3in (2.5–8cm) long depending again

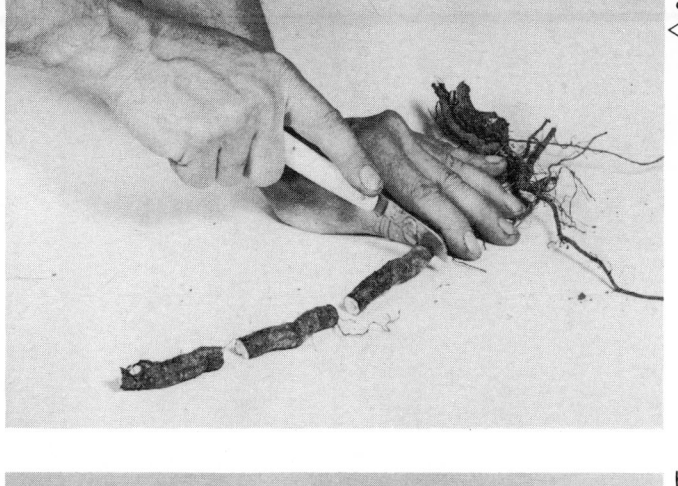

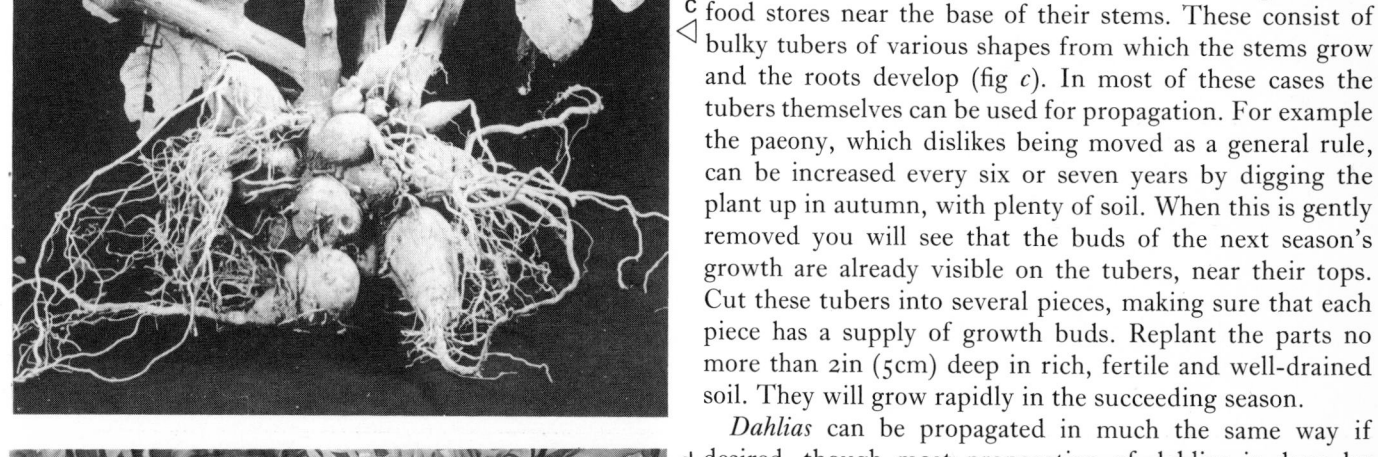

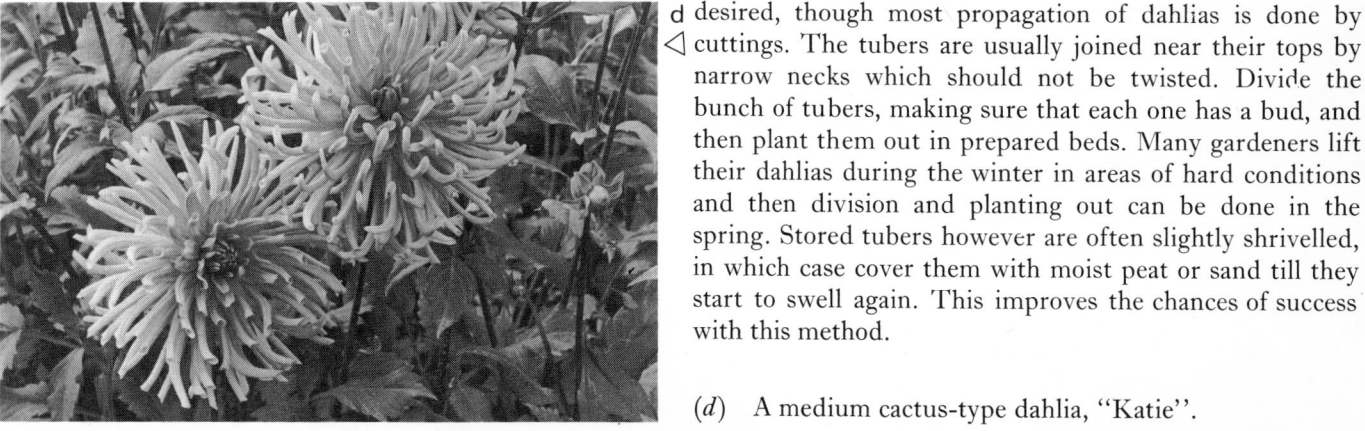

a on the size and type of plant (fig a). Some plants bleed a little from the cut ends, especially during the growing seasons, but this does not seem to affect their rooting capacity. They are then planted in ordinary well-drained soil, 2 or 3in (5 or 8cm) deep. A spadeful or two of coarse sand mixed with the soil at the planting positions will help a great deal. The stem bud will always develop from that part of the root which was nearest to the crown of the mother plant. The new roots will on the other hand always develop from that part of the cutting which was nearest to the root in the mother plant. This will be so even if the cutting is planted upside down. However if a stem bud does develop at the bottom of the cutting it obviously has a harder job pushing its way to the surface after making a U-turn! It is better to put in root cuttings either horizontal b or the right way up. To make this easier, make the lower cut of the root cutting at an angle (fig b). Then when planting, this pointed end can be thrust downwards.

Although you can propagate by root cuttings in the open air, they are easily lost in these circumstances, either by over-vigorous hoeing or perhaps simply by forgetting where they were put in the first place. It is often a better plan to fill deep boxes with good compost and place these out of doors in a sheltered position where you can easily keep an eye on the cuttings and make sure that they are not disturbed until growth has started.

Propagation from Tubers

Dahlias, paeonies and many other plants develop swollen c food stores near the base of their stems. These consist of bulky tubers of various shapes from which the stems grow and the roots develop (fig c). In most of these cases the tubers themselves can be used for propagation. For example the paeony, which dislikes being moved as a general rule, can be increased every six or seven years by digging the plant up in autumn, with plenty of soil. When this is gently removed you will see that the buds of the next season's growth are already visible on the tubers, near their tops. Cut these tubers into several pieces, making sure that each piece has a supply of growth buds. Replant the parts no more than 2in (5cm) deep in rich, fertile and well-drained soil. They will grow rapidly in the succeeding season.

Dahlias can be propagated in much the same way if d desired, though most propagation of dahlias is done by cuttings. The tubers are usually joined near their tops by narrow necks which should not be twisted. Divide the bunch of tubers, making sure that each one has a bud, and then plant them out in prepared beds. Many gardeners lift their dahlias during the winter in areas of hard conditions and then division and planting out can be done in the spring. Stored tubers however are often slightly shrivelled, in which case cover them with moist peat or sand till they start to swell again. This improves the chances of success with this method.

(d)   A medium cactus-type dahlia, "Katie".

PROPAGATION FROM STEMS

Propagation from stems is certainly the most popular and widely used of all vegetative (non-seed) methods for increasing plants. Just as new stems may spring from cut roots, so fresh roots will often develop from cut stems. In both cases the fresh growth arises from the *cambium layer*. This is a layer of actively dividing cells immediately below the surface of roots and stems. New growth from this layer takes place most rapidly if it is kept moist, dark, and is exposed by the outer skin being damaged or removed. These conditions are obviously met when we cut up roots, but we can also provide them with stems, by several different methods.

For example, many plants grow low branches that may become covered with soil. Rooting then takes place there, in the dark moisture, while the stem is still attached to its mother plant. This is known as *layering* and is certainly one of the easiest of stem propagation methods. The stem can continue to draw food from its parent until a completely fresh root system is formed. Then, the parental stem can be cut and the young plant grown on independently. Even with stems above ground level, similar results are achieved by air-layering. Bundles of wet moss are bound round a wounded stem and sealed within plastic sheet. After some time roots form in the dark moss and the stem below can be cut to free the young plant for potting up.

Growing plants from cuttings, stems that are completely separated from the mother plant, is a little more difficult. The new plant has to live for a time solely upon the food stored in its own stem and leaves. It also loses water from its leaves, causing them to flag. This moisture must be replaced by water drawn up the cut stem, or absorbed from surrounding moist air. In practice, the cuttings are taken and placed at once in good moist soil, compost, sand or plain water. The whole stem, pot and all, may then be covered by a glass frame or cloche or surrounded by a plastic bag which help to prevent water loss and keep a humid atmosphere round the leaves. If heat can be provided, rooting can be very rapid indeed – only a few days in some cases. After-

a
◁

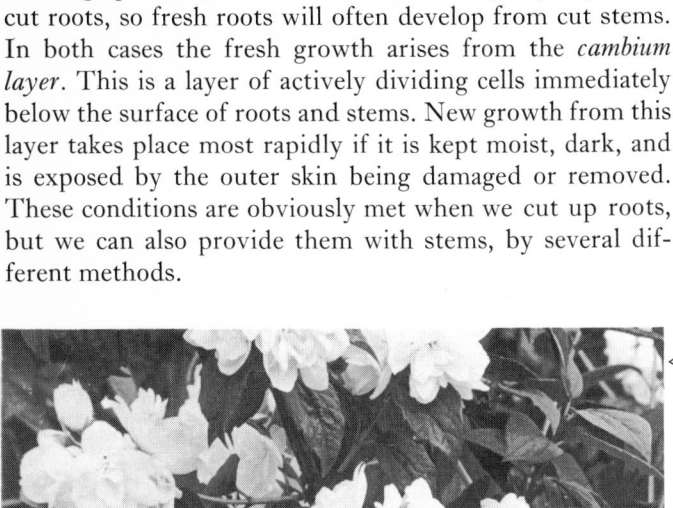

(a) Cuttings of all types can be taken from Philadelphus.

b
◁

c
▷

wards, the covering is gradually removed for the cutting to become accustomed to the open air.

The main requirements are healthy mother plants, fertile soil or good compost, evenly controlled temperature and humidity – and patience! Plants vary in speed of growth and cuttings may take anything from a few days to several months to form roots. Once started though, the young plants are usually fairly easy to grow on.

e ▷

Stem Layering

(*b*) *Philodendron scandens* (the sweetheart plant) is a popular climbing houseplant and one of the easiest of all to propagate. Like ivy, it often develops aerial roots from leaf joints, and layering is almost always successful.

(*c*) Bend the growing tip of a shoot into a pot of compost, burying it at a leaf joint.

f ▷

(*d*) Squeeze the compost gently around the stem and hold it down firmly with a notched edge peg or label. As rooting takes place the tip will develop and the stem leading from the mother plant can be cut away.

(*e*) It is also possible to root two or more parts of the same stem by looping this down into several pots at intervals.

(*f*) The maximum number of young plants are produced if entire stems are laid over compost in a seed tray. Peg down every leaf joint. Each one will root and then can be snipped free with scissors and grown on.

(*g*) *Camellia williamsii* is propagated by stem layering.

d ◁

g ▷

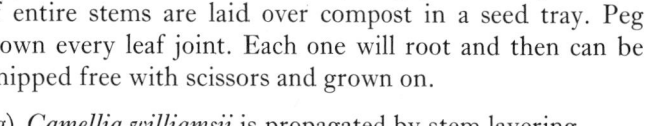

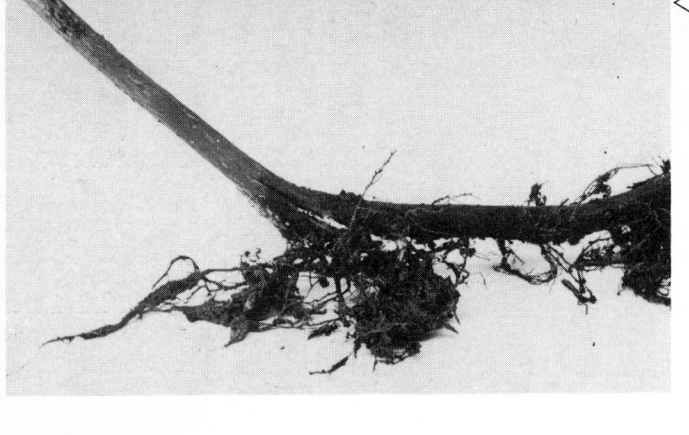

a ◁

RHODODENDRONS BY LAYERING

Garden shrubs such as the rhododendron can be layered directly into the surrounding soil if they have branches that grow near the ground. Plants with long trailing stems, such as honeysuckle and clematis, can also be treated in this way. If you prefer to layer the plants into pots sink them into the ground at some convenient point:

(*a*) After some months, roots will have formed, especially if the stem is wounded at the point it was buried.
(*b*) The stem to the mother plant can then be cut through
(*c*) . . . . . leaving a complete and independent plant to be grown on.

Pot Layering – Rhododendrons

For plants with thicker and less flexible stems, not easy to hold down, another method often called pot-layering can be tried. This method, which went out of use some years ago, has been revived following the introduction of the plastic pot, which can be easily cut with scissors or saw. Rhododendrons or similar shrubs can then be layered directly into them. The new roots can be allowed to fill the pot, which can then be transplanted easily after cutting the mother stem.

b ◁

c ▷

(d) First saw two slits ½in (12mm) apart, part way down the sides of a 3in (8cm) pot. A fine tooth saw or scissors will do this.

(e) Bend the strips outwards and clip them away to make an open slit at each side of the pot.

(f) Soak sphagnum moss in water and place a small quantity in the bottom of the pot. Then three-quarters fill the pot with a good quality potting compost.

(g) Choose a mature but not too woody stem and make an upward slit into the bark about 1in (2.5cm) long.

(h) Insert a sliver of wood into this slit to hold it open.

(i) Lay the stem through the slots in the pot with the wound exactly in the centre of the compost.

(j) Cover the stem with more compost and lay sphagnum moss on top.

(k) Tie the moss and stem securely into place so that they cannot shift. Some gardeners also wrap a polythene bag round the pot, to conserve the moisture.

(l) Rest the pot on the ground until rooting takes place. The pot can then be cut clear of its mother plant.

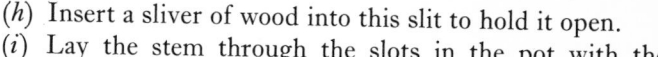
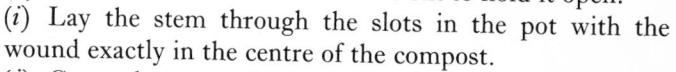

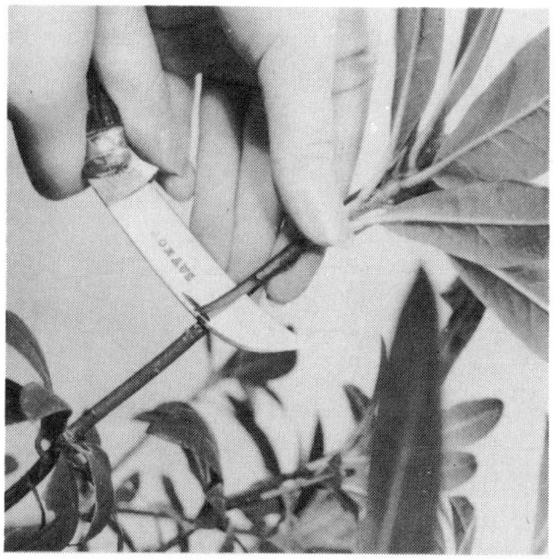

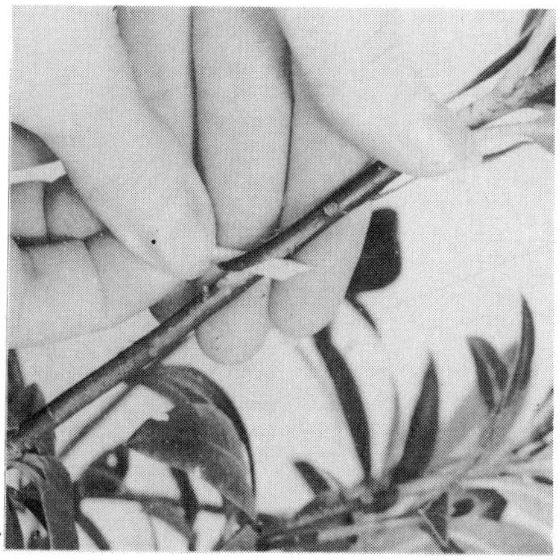

There are many plants, such as the popular *ficus* or rubber plant, whose stems cannot be easily bent down to soil level, yet which could not easily withstand the weight of pot-layering. Air layering can then be used.

The principle of air layering is to wound the bark slightly and surround the stem with moist moss or soil held in place by plastic sheet. Roots then form within the moss and the stem below can be severed.

(Facing page)
(*a*) In most cases roots develop best just below a leaf. Start by slitting gently through the skin of the stem . . . . .
(*b*) . . . . and peeling away ¼in (6mm) wide section of the bark, half way round the stem.
(*c*) Remove the leaves near this slit, to provide 3 or 4in (8 or 10cm) of clear stem.
(*d*) Squeeze a handful of wet sphagnum moss firmly around the wounded stem.

(*a*)   The willow is easily propagated by layers and cuttings.
(*b*)   Air layering is used to propagate *Magnolia sieboldii*.

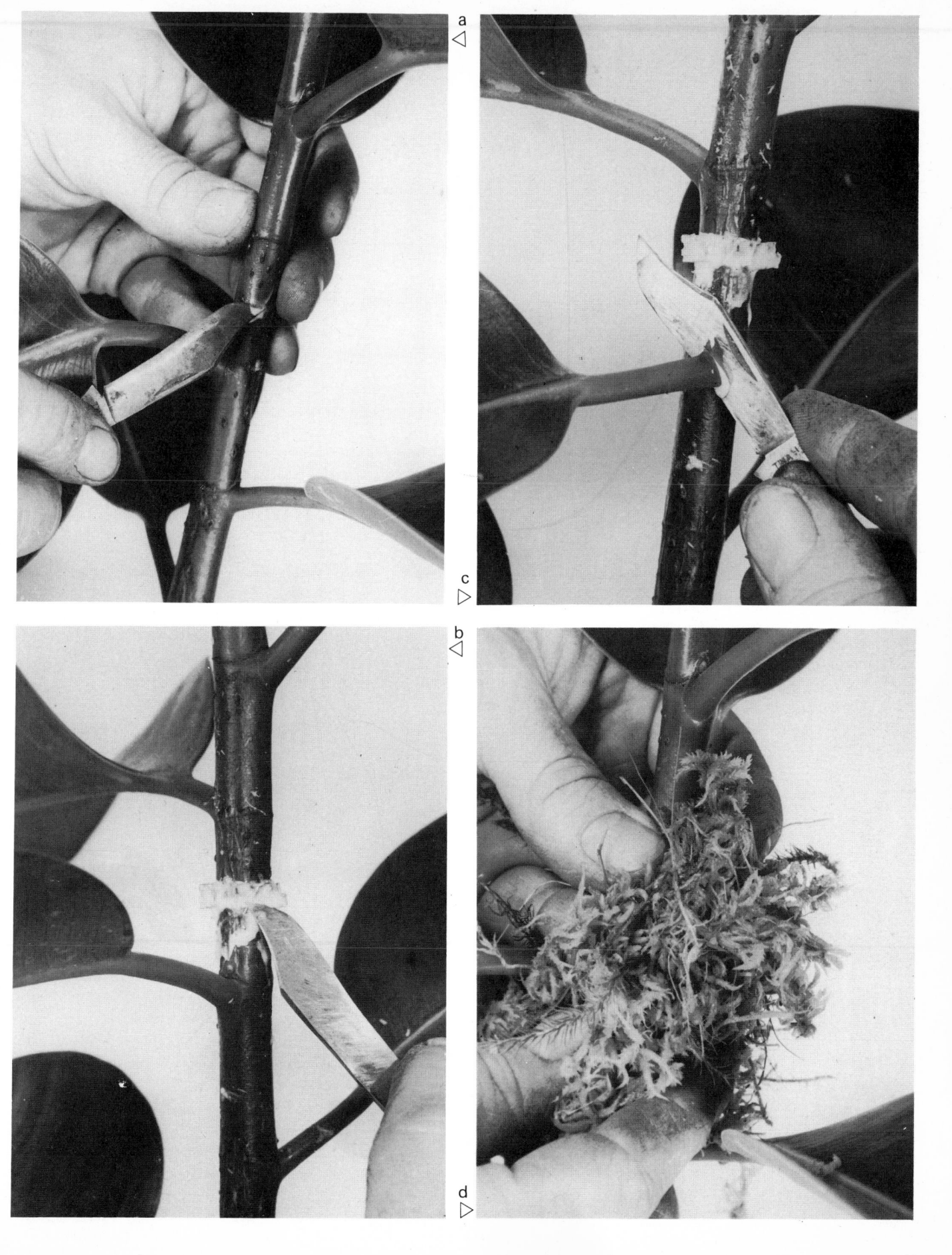

i △

(*e*) Tie this lightly into position . . . .

(*f*) . . . . then wrap a polythene bag around it to form a secure and airtight container.

(*g*) Seal the top and bottom with string . . . .

(*h*) . . . . and leave the "layer" in position until roots have formed. (They may be visible through the plastic.) The stem immediately below can then be severed and the plant's upper part potted in a $3\frac{1}{2}$in (9cm) pot. Keep the new plant close, in a warm, humid atmosphere, for a few days.

(*i*)   Cornus – dogwood – is propagated by layers.

a ◁

Propagating from Suckers

Many shrubs form thickets of branches growing up from ground level, often spread over a large area. When you are dealing with shrubs like this, you have a ready-made, simple method of propagation. All you have to do is dig down and find the connection of an outer stem to the centre of the mother plant. Cut this connection and dig up the stem complete with a few attached roots. This *sucker*, as it is called, is like a ready-rooted layer. It has been in part dependent on the mother plant for its food supply. Now it has been separated it will rapidly develop more roots of its own and become a truly independent plant.

With difficult types you can start by making this cut several months before actually moving the young plant. The sucker is then not disturbed till it has time to form fresh roots, ready to withstand the shock of full transplanting. In other cases though such as with many berberis varieties and the popular sumach, which are rapid in growth and throw up many suckers, you can dig down, cut the sucker free, lift it and replant, all in one operation. Only experience will tell you which of your plants will tolerate the rapid treatment, but broadly speaking those which are rapid growing can be moved quickly, whereas those which are slow growing prefer to be cut, left and then removed later.

Naturally, having the mother plant connection cut in this way is still a great shock to the young plant. The leaves may flag and occasionally the plant will die. However, suckering types of bush usually produce so many that you can tolerate losses of this kind. You can reduce the risk with leafy suckers by pruning off about half of their leaves after making the division. This reduces water loss and the strain on the roots in the early days.

One warning now. Some plants in our gardens throw out suckers which are of a much less desirable kind. Some kinds of laburnum or lilac, for example. You cannot propagate roses or lilacs from the suckers that bush up from below ground level. The reason is that they grow from the roots of the plant; and this is often not the same as the flowering part of the plant. The rose suckers have quite distinctly different leaves and are useless for flowering purposes, although buds can be grafted on (page 66).

In professional work, to save cost and to speed up

(*a*)   Propagate *Phlox subulata benita* from suckers in spring or autumn.

propagation, nurserymen frequently take a common variety of a plant and then graft to its top a stem of the much more valuable kind. The two parts then grow together, the upper part of the plant being of the better sort but the lower part and roots remaining the common type. The results of these curious combinations are that the grafted stems may grow with extra vigour, may have a longer flowering period or produce better blooms or fruit. The suckers though, that spring up from the roots of these grafted plants, are all from the common stock, and so much less valuable. If you propagate them you will only produce a plant of the common sort. Apart from these grafted plants, the many un-grafted sorts that do throw out suckers from their roots give a simple and easy way of propagation.

The use of suckers for propagation is of course a logical development from the division methods shown earlier, and a method you can also use for many herbaceous border plants. Many of these, such as the *Phlox subulata benita* (fig *a*) throw out small stems with a few roots at the sides of their bases. These can be cut free and grown on.

Stem Cuttings

Cuttings are probably the most widely used of all propagation methods except seed. Indeed, there are few plants which cannot be increased in this way. However, not all are easy to propagate. Some require close, warm conditions such as are usually only found in heated greenhouses. Others may take several months to throw out a single root. Still, many of our hardy perennial plants and shrubs and most houseplants can be increased by cuttings. Roses in particular can be grown in this way without difficulty.

Compared to a layer, the cutting has a much more difficult time. A layer remains connected to its mother plant and can draw upon this for food and water until its own roots and leaves are fully established. The cutting, on the other hand, lives entirely on the food still contained within its leaves and stems. It is essential that it should quickly form enough roots to draw in water and food, before this original store is exhausted. For this reason rapid rooting must be encouraged, and the loss of water must be slowed down.

Most of us realise, from watching the opening of cut flower buds, that even a plant cut free from its roots can still grow. Flowers will open and leaves expand, but it is uncommon for roots to form. This is partly because flowers are rarely kept long enough for them to do so. Mainly though it is because a flowering stem tends to concentrate all its efforts on the development of the flower and its following seed. This prevents the stem from supplying food for the growth of roots at its base. Indeed, when they are flowering and fruiting most plants are so concentrated on this effort that it is best not to propagate them at all. It is different when a stem from which a flower has been (*b*) One way of propagating *Clematis montana* is from suckers.

removed or on which a flower bud has not formed is placed in plain water. Roots are often then thrown out. Roses in particular can frequently be started into root, simply by taking stems and standing them in water in a light (but not sunny) place. The growing roots are clearly visible and the stems can later be potted up.

Roots grown in water are however rather more fragile than those grown in ordinary soil, and losses often occur at potting up, when the plant must change from a water culture to soil or compost growth. With such easy rooting cuttings as roses it is far better to start out directly in soil or sandy compost. One interesting point about rooting cuttings in water is that you will see that the first roots nearly always are thrown out at a joint in the stem. The reason is that at these joints (where leaves grow, or have grown but later fallen) the plant forms chemicals which act as hormones in the encouragement of root formation. The maximum effect of this appears to be about $\frac{1}{8}$in (3mm) below each joint, so the customary advice when taking cuttings is to make sure that their base is cut off $\frac{1}{8}$in (3mm) below a leaf joint or "node". This is a good general rule but not absolutely universal. Some plants such as the clematis (fig *b*) seem to prefer a cutting which is taken half-way between the buds.

The development of roots on a cutting has two definite stages. First comes the formation of a "callus" of cork across the cut end of the stem. Sometimes this expands to form a thickened knob at the base of the cutting. Second, pushing through this callus, roots will be seen developing which eventually grow sufficiently to form an entirely new root system. All these new roots form from the green, sappy cambium layer so it is a common practice to expose this as much as possible by peeling away a sliver of the outer bark or by pulling the cutting from the parent stem, bringing with it a small portion of the larger stem's bark. From these much larger areas of cambium there is more chance of rapid and vigorous rooting.

The size of cuttings varies widely, from only 1–2in (2.5–5cm) for many heathers and small rock plants to 12in (30cm) or more for substantial shrubs. As a rule, small cuttings are easier to root than large ones, always provided that the leaves and stem contain enough remaining food material for the plant to live on till rooting starts. This presents problems, because although leaves must be left on to provide food for growth they will also continue to breathe out water. Since the cutting has no roots this may cause the plant to flag and die. To overcome this you can cut off some or all of the leaves, so reducing the water loss. This is a reasonable method to use outside, where covering the plant with airtight seals is impractical. Such cuttings out of doors and taken in summer may have leaves remaining. Later in the year, deciduous shrubs are propagated by taking leafless cuttings. Because they have no leaves, water loss is practically nil. They are simply pressed down into

prepared trenches in the ground and left. A callus will form during the winter and in spring, new roots can be expected to form.

When cuttings are made in pots however, you can let more leaves remain complete on the cutting and then enclose the whole plant and its pot within a polythene bag, stretched over a wire frame. The air within the bag rapidly becomes saturated with moisture breathed out by the plant but this finds its way back into the compost and from there into the plant again. This forms a continuous cycle of water sucked up the stem, breathed out from the leaves into the air, condensing on the bag sides and running back down into the soil to be sucked up once more. Amazingly enough, plants can live for months, even years, in such sealed containers using over and over again the same air and water. This method has much to commend it. Cuttings under sealed covers require practically no attention. Simply stand them in a bright place but not exposed to hot sun. (In direct sunshine, the temperature within the bag may rise rapidly to such a degree as to injure the plant beyond rescue.)

Gardeners divide cuttings into three main types: *soft cuttings*, *half-ripe cuttings* and *ripe cuttings*.

Soft cuttings are made with the fresh young growth of the current year, have green, fleshy stems and usually need some form of warmth and protection. *Half-ripe* cuttings come from stems which are partly mature, with bark developing on the outside. They are usually taken much later on in the season, and may be grown outside, though often in the protection of a cold frame. *Ripe cuttings* are from fully mature stems having complete bark. These are usually the leafless cuttings described earlier, and are grown outdoors, often directly in open soil. Speaking generally, soft cuttings are taken in spring, half-ripe cuttings in late spring or summer and ripe cuttings in autumn or winter.

Since the outdoor cutting requires less preparation and attention than others, we will start by looking at this method, taking the popular rose as an example. Many modern bush roses are in fact commercially grafted on wild rose stocks, so that only the stems and flowers are of the cultivated variety, the roots and stubby main stem just at ground level being the wild or common sort. You can nonetheless still take cuttings from such roses, using their upper stems. Most roses can then be grown on to form their own roots quite satisfactorily.

The success rate for rose cuttings is quite high. You can even root them in plain water. They do however take two or three years to come to full maturity, to the same size that a grafted rose attains after about only eighteen months. Climbing and rambler roses can also be easily grown from cuttings. Many of these are not in fact grafted at all. Also, many of our favourite kinds of shrubs can be increased in the same simple way.

RIPE HARDY CUTTINGS

(*a*) Choose a shoot which has firm, ripe bark. (With roses this stem may have borne a flower at its tip.) Cut away any flower and the two leaves immediately below it. Then pull the stem off, right at the base where it joins the main stem. Bring with it a piece of the bark of the main stem, which is normally called a "heel". The object of doing this is to have a large area of cambium exposed and pressed against the soil.

(*b*) Shorten any long "tail" to about 2in (5cm).

(*c*) Rose and many shrub cuttings are usually between 9in (23cm) and 12in (30cm) long. Cut them off just above a leaf and remove all but the uppermost two leaves. When taking off the leaves do not damage the tiny bud that you will find at the base of each. It is from these buds that growth will

take place. The completed cutting, ready for insertion, will then look like this.

(You can use rooting hormone powder for roses, though their success rate even without it is quite high.)

Next choose a sheltered spot out of doors with deep, rich soil. Scoop out a straight-backed trench about 8in (20cm) deep. Well-drained conditions are important. Roses will not root successfully in very wet soil. To make sure of this, spread a 2in (5cm) thick layer of plain, coarse sand along the trench bottom and press the base of each cutting down into it.

(*d*) Rest the stem of the cuttings against the rear of the trench with the two remaining leaves just above soil level. Press the heels of the cuttings well into the sand. Then tread the soil back into the trench, burying the cuttings for half their length in the soil. They will not be disturbed by wind, but after frost in winter (which tends to expand and lift the soil) go along the row and tread this back firmly. Incidentally, the leaves will fall off as soon as the winter commences but this is quite natural and will not retard the rooting. During winter a callus forms over the cut end of the cutting and rooting starts early in the following spring.

Do not attempt to transplant cuttings, no matter how well rooted, until the subsequent autumn. They can then be planted out either into a nursery bed for a further year or directly into their flowering positions.

Large numbers of shrubs, trees and hedge plants can be

a ◁

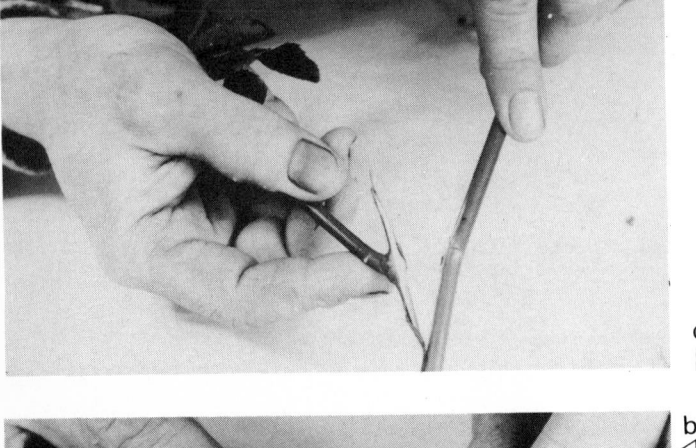

c ▷

b ◁

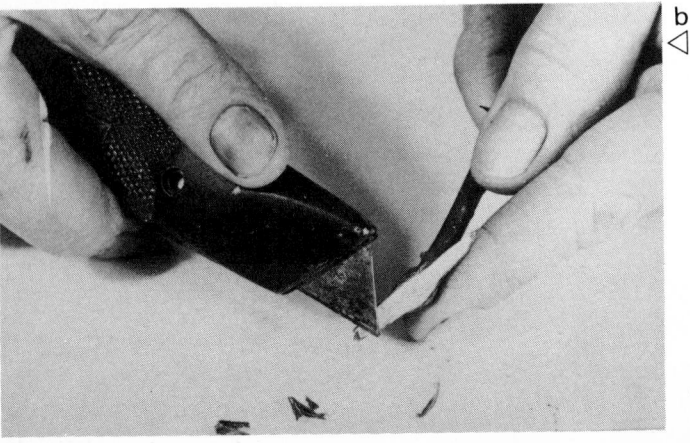

d ▷

propagated with cuttings of this kind. Because their leaves fall in winter they are often known as leafless cuttings, even though there may be leaves on them when they are first taken.

SHRUBS FROM HALF-RIPE CUTTINGS

The next sort of cutting tends to retain its leaves for some time. These are the early summer half-ripe cuttings. Placed in small, well-drained pots, they will root rapidly. The list of plants which can be propagated in this way is very wide and includes many of the most popular flowering shrubs, trees, and climbers which we see in our garden, not to mention heathers and many rock garden shrubs and alpine plants.

Naturally in view of the wide range of size of these plants, the size of the cuttings themselves will vary. There is no hard and fast rule as to how long a cutting should be. You can take as a guide a length equal to four leaf joints. You can of course still use a heel, described in the previous section, but the disadvantage of this is that you can then only get one cutting from each stem, no matter how long this is. Since the stem of an actively growing shrub may be several feet long this would be very wasteful. Instead cut each branch up into several cuttings. Simply slice it into sections, cutting immediately below a leaf joint in every case.

You can often combine summer pruning with the taking of cuttings. This is because most flowering shrubs are pruned immediately after flowering. Those which flower in very early spring are pruned in late spring and these prunings can be used at once as cuttings. Those that flower later in the year will have been pruned in the previous winter and so will provide fewer cuttings. Here you will be better advised to wait till late summer before taking them.

The pots of cuttings can be stood outside, on a sunny window sill or in an improvised frame. In every case good ventilation is essential at all times and the soil in the pot must be kept gently moist. Most cuttings do not require frequent or liberal watering. The few leaves remaining will not breathe out much water. They do however appreciate even and consistent moisture about their roots. Preserving moisture by packing round the pots with peat will help, and of course, larger pots take longer to dry out than small ones. With these, you can fit several cuttings round their edges, instead of just one in the middle.

(e) For example these long, slender stems of Jasmine can be cut into several different parts. (Many gardeners prefer shoots which have not flowered but this is not a universal rule.) Choose healthy stems, but cut off and throw away the very soft growing tips.

(f) Then cut the rest of the stem into pieces four leaf joints long, each cut being made $\frac{1}{8}$in (3mm) below one of these joints. Trim away the lower two leaf pairs but do not damage the small buds at each leaf base. The cuttings will look now rather like this, the surplus stem at the top having been cut away just above the upper leaf pair. If

49

hormone rooting powder is used, dip the moistened base of the cutting into the powder.

(*g on previous page*). Fill a pot with well-drained compost containing a good deal of sharp sand. Slide the cuttings down the edge of the pot in a ring just so far apart that the leaves do not touch each other. It is a fact which is not yet entirely explained that most cuttings root best if they are placed at the side of the pot rather than put in its centre. Next, the pot with its cuttings can be put on a bright (but not sunny) window-sill, or in a sheltered position outside.

A cold frame is ideal for starting off many of our popular shrubs and you can improvise one by placing a few bricks or pieces of wood in a rectangle and covering these with a sheet of glass or even polythene stretched on a wooden frame. However, adequate ventilation must be given in all these cases.

(*a*) Although most plants prefer to have their cuttings taken with several leaves and below a leaf joint, the clematis is something of an exception. With this attractive plant you take cuttings halfway between a leaf joint and only one pair of leaves is necessary. This naturally increases the number of cuttings you can get from a single plant. In order to expose more of the cambium layer, pare away a thin sliver of the bark about 2in (5cm) along one side of the cutting base.

(*b*) To increase the number of cuttings to the maximum you can even slice right down the centre of the stem with a razor blade, so producing two half cuttings with a leaf on either side, as shown later for pelargoniums. These leaf cuttings are then planted around pots in the usual way, with their leaves pointing outwards and the buds carefully arranged exactly at the surface of the compost.

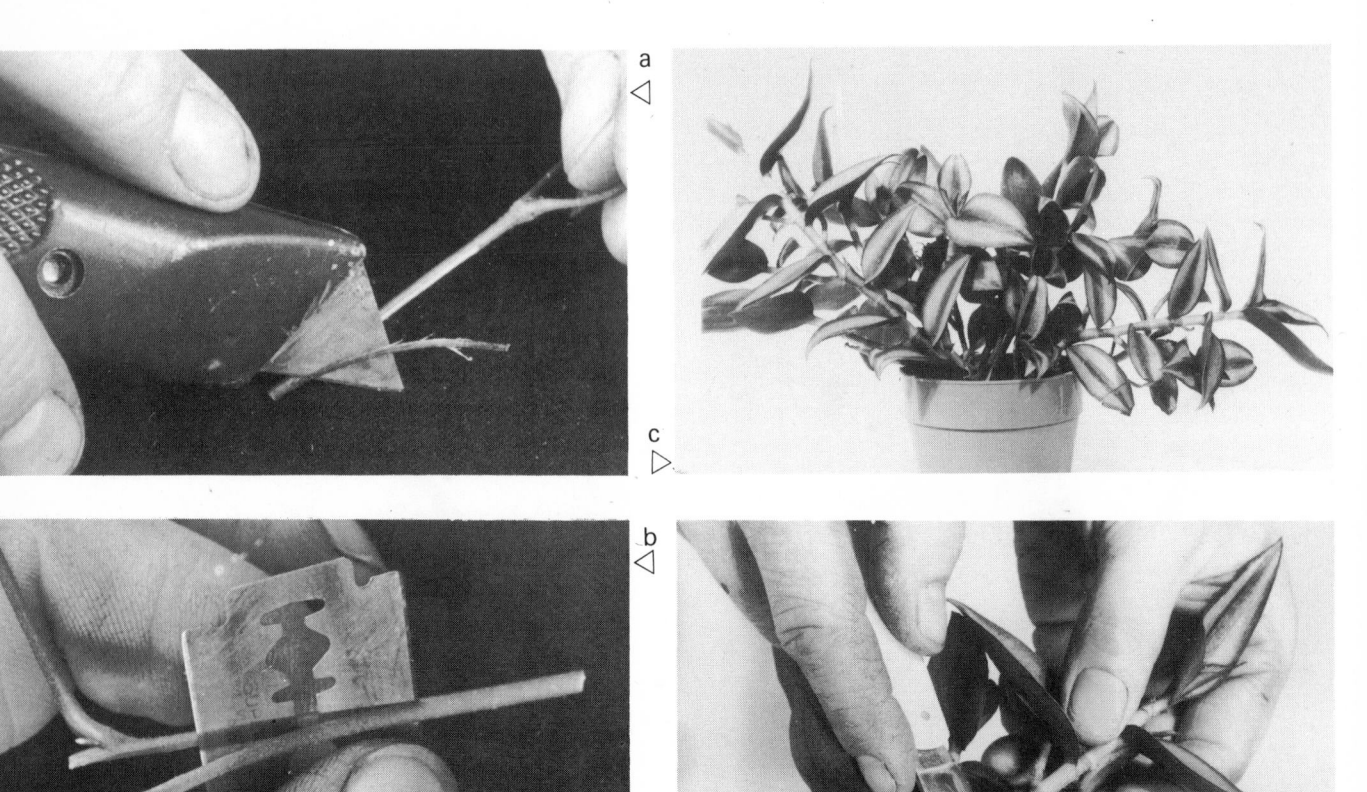

a ◁

c ▷

b ◁

d ▷

SOFT CUTTINGS FOR HOME AND GREENHOUSE

The hardwood cuttings described earlier may take several years to come to maturity. The half-ripe cuttings are somewhat quicker but the quickest of all are certainly soft wood cuttings. These may be ready for planting out after only two or three weeks. Houseplants in particular are often propagated like this, whilst many popular outdoor perennials such as dahlias, asters, etc. are increased in the same way.

The main difficulty with soft cuttings is to prevent them losing all their water, flagging and dying before roots have time to form. To prevent this they are kept in moist, humid and warm conditions, so speeding up growth and reducing water loss. However, these very conditions themselves encourage the growth of mould fungi of various kinds which may attack the leaves and cause them to rot. Because of this risk it is best to use commercially prepared sterilised compost in the pots. For this purpose no-soil compounds based upon vermiculite are excellent.

The season for taking soft cuttings is usually the spring, but this is again far from a hard and fast rule. In general, you can propagate at any time that cutting material is available, and when you can maintain the temperature and light requirements of the plant. Many houseplants can in fact be propagated all year round.

SIMPLE SOFT CUTTINGS

(*c*) *Zebrina* (*Tradescantia*) is a most popular houseplant, and extremely easy to propagate from soft cuttings.
(*d*) With a sharp knife cut away vigorously growing stems just below a leaf joint. Make the cuttings about 2in (5cm) long.
(*e*) Pull away the lower leaves which would otherwise be beneath the soil leaving intact the uppermost two leaves and growing point.
(*f*) Hormone rooting compounds are not needed for this very easy rooting plant. Simply insert the cuttings round a 2½in (7cm) pot. Many plants root best if set at the edge of the pot rather than in its centre. No other protection need be given. Stand the pots on a bright (but not sunny) window-sill and within a few weeks the cuttings should have attained almost the size of the mother plant.
(*g*) Both soft and hard cuttings can be taken from honeysuckle.

e

f

g

Pelargoniums. The well-known geraniums or pelargoniums (*f*) need slightly more care than *Zebrina* but are still quite easy to grow from cuttings. They are typical of many houseplants, requiring a little protection but not much extra heat. Although all require warmth, this is no more than that usually attained in a house room. Try to avoid wide variation in heat throughout the day. It is better to have a lower overall temperature that is even, than a high temperature during part of the day, with low temperatures at other times.

Watering must be restrained. The pelargonium group prefer rather dry conditions in any case. A few growers even allow freshly made cuttings to flag before planting. However, this seems merely to indicate the remarkable vigour of the species rather than an improvement in technique.

(*a*) Make cuttings from firm, vigorously growing new shoots that are not thick and fleshy. Remove any developing flower buds and all lower leaves.

(*b*) Use a very sharp knife or razor to take off the cutting cleanly about ⅛in (3mm) below a leaf joint. (Blunt tools injure the tissues and hinder rooting.) Make each cutting approximately 3in (8cm) long.

c

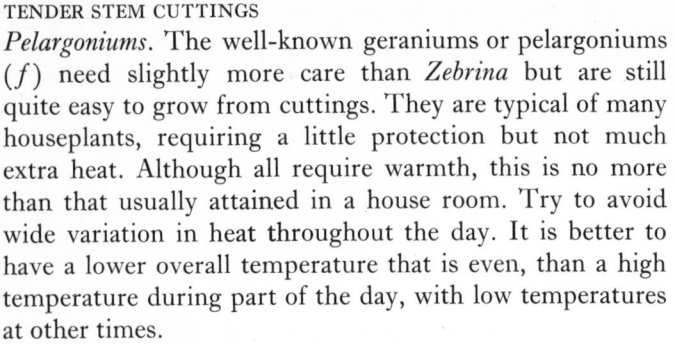
a

b

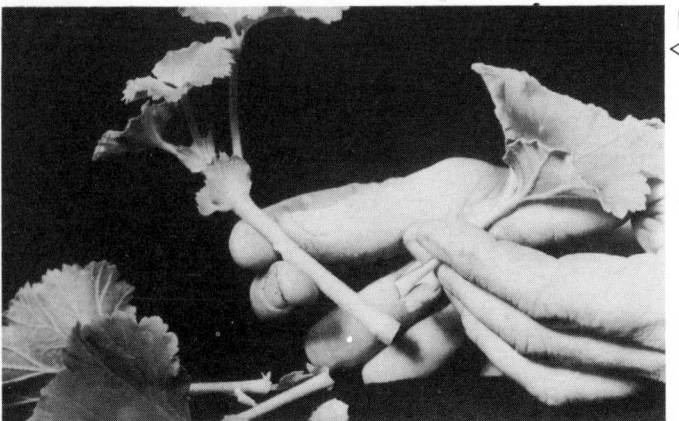

d

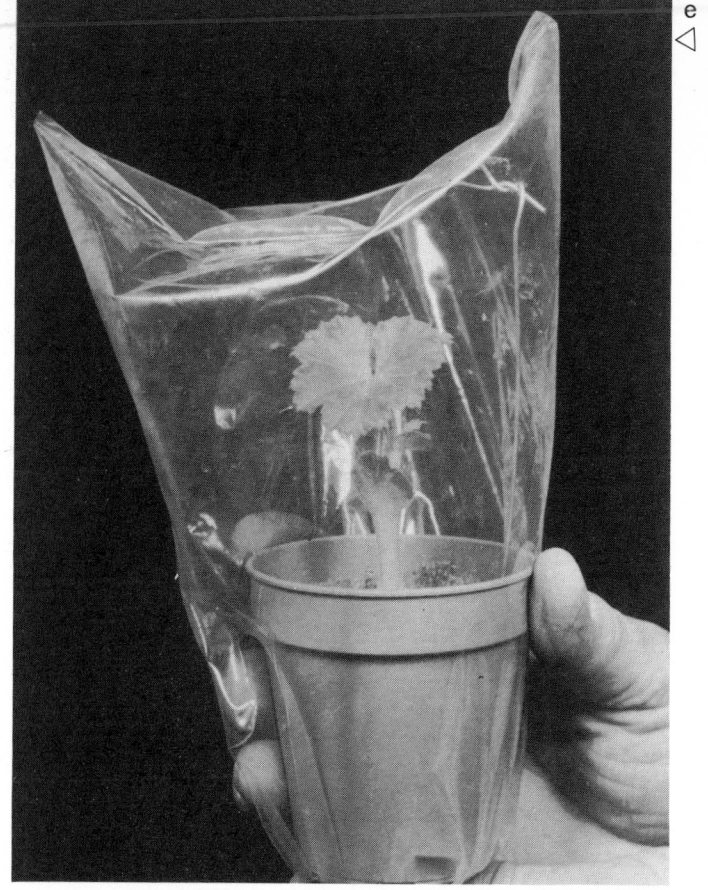

e ◁

g ▷

(c) If hormone rooting compounds are to be tried, apply these to the lower end of the cutting. Then simply place this into the side or centre of a 3½in (9cm) pot. Press the cutting firm, so that it stands upright. Make certain that its lower end is pressed against soil, not "hung" in an air space.

(d) Give protection by bending a length of stiff wire to a loop shape. Thrust its lower end down the pot side and draw an ordinary, unperforated polythene bag over the loop.

(e) Twist the polythene tightly underneath the pot. Stand this on a bright windowsill. Provided the compost was properly moist, further watering will not be needed till rooting starts. This is indicated by new growth from the leaf buds. Then gradually release and remove the bag, avoiding sudden changes of conditions.

(f) Pelargoniums

(g) Soft cuttings are taken to propagate delphiniums.

f ◁

SOFT STEM CUTTINGS NEEDING HEAT

Ficus. Many of our most popular house and greenhouse plants come from hot climates and cannot be propagated easily without some form of heat. If you can make or buy yourself a propagating case this will make your task much easier. "Bottom heat" is best. This is simply heat that is arranged to rise upwards through the soil. It is often provided by electric soil warming cables embedded in the peat and sand covering the base of a propagating case. Alternatively, the pots and trays can be placed above some other source of heat such as electric tubes, radiators or paraffin stoves. The heat must be even and, if possible, continuous day and night for the period of rooting. Humid conditions must be maintained around the leaves for best results, easiest with a propagating case but also possible with the polythene bag method.

(*a*) The best cuttings of *ficus* are made from the growing point. Cut this off about 6–9in (15–23cm) below the tip and ¼in (6mm) below a leaf.

(*b*) Remove all the lower leaves, leaving only the top two and the growing tip.

(*c*) Dip the cutting base in rooting hormone and plant it in the centre of a 3½in (9cm) pot. Do not use a larger pot. *Ficus* prefer rather tight root conditions. Slide a thin cane down into the pot beside the cutting.

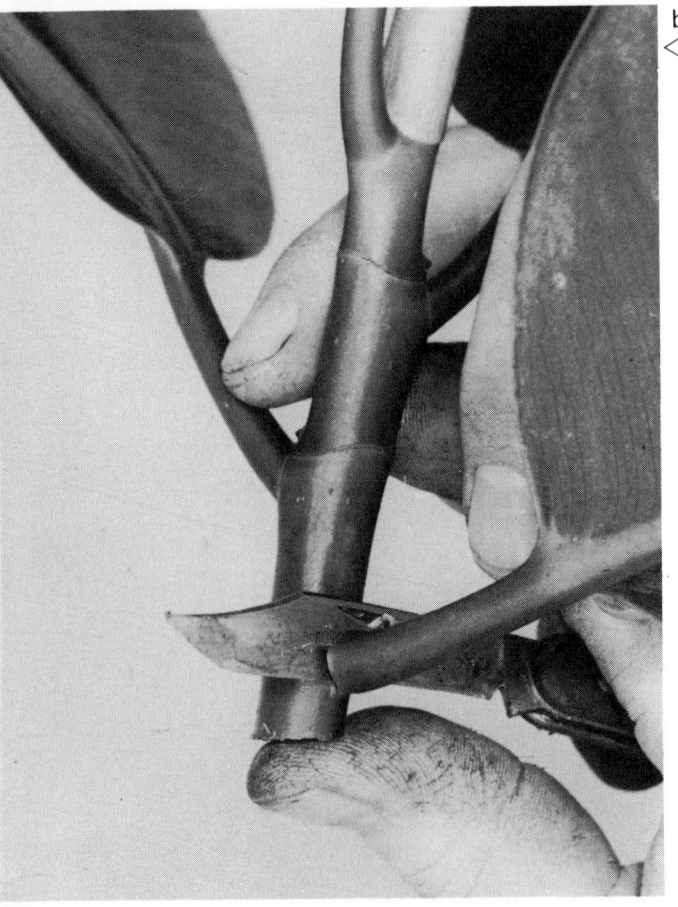

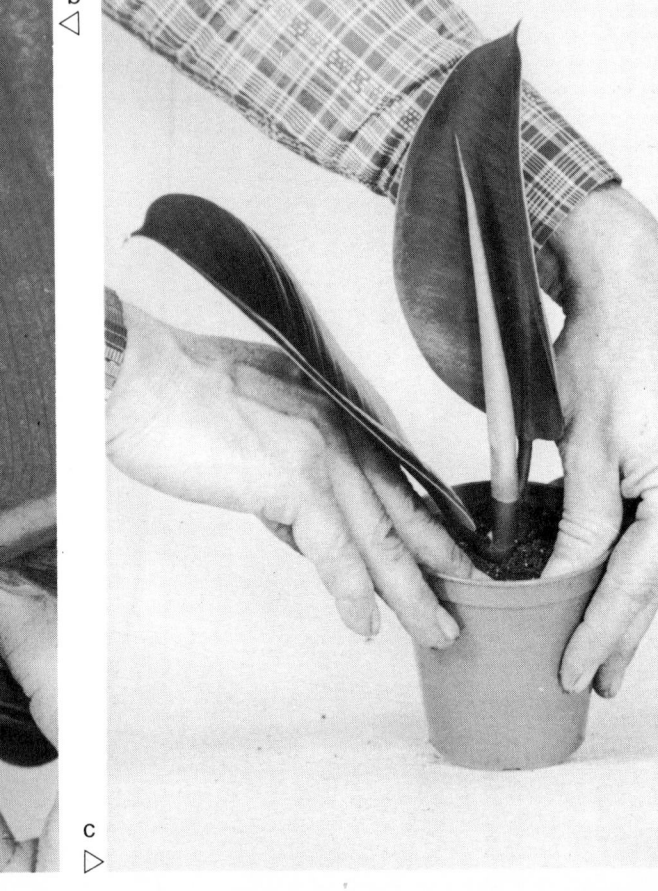

b

c

(*d*) Roll the leaves and hold them to the canes with rubber bands. Otherwise they will flag and fall over to lay flat which may injure them and provoke rotting. Rolling also reduces water loss from the leaves.

(*e*) Erect a stiff wire loop over the top of the complete cutting.

(*f*) Draw over it an unperforated polythene bag and twist this tight underneath. Keep the cutting evenly warm until growth has started, indicated by the gradual development of the growing point. The sheath that encloses the new leaf will fall away and must be removed at once. If it does not come free naturally, peel it gently away once it is loose. Leaving it causes rot. Then release the elastic band, and gradually remove the polythene bag.

e
▷

d
◁

f
▷

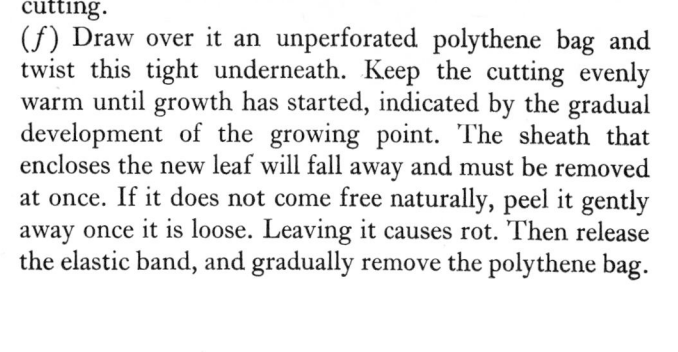

PROPAGATION FROM LEAVES

Just as stems can be used for propagation so can leaves, especially those of houseplants. However, they present more problems. Many leaves do not have a sufficiently active cambium layer to produce roots easily. Also, the single leaf must support growth unaided until roots develop.

Leaf propagation is used when rapid increase of a small stock is needed, or for plants not providing suitable stem-cutting material. The methods used vary. The simplest is where a very short piece of the main stem is cut, complete with a single leaf. This is essentially a very short stem cutting and is grown on in a similar manner to those described in the previous section.

A variant of this, used with plants that grow their leaves in pairs opposite each other, is to make split-stem cuttings. The stem between the leaf pairs is slit longitudinally, giving cuttings, each with one leaf and a portion of stem.

Leaves complete with their own stalk, but without any of the main stem, can also be used. This method suits plants whose leaf stalks spring from a low crown, as in most *peperomias* and the *saintpaulias*. There is therefore no stem with which to make normal cuttings.

Leaves alone, without stem or stalk, can also be made to produce roots. The *begonia* family in particular, but also others, such as *streptocarpus*, are often increased in this way. Many large leaves of begonia, if simply laid flat on moist soil, will produce roots from their veins which develop rapidly into complete young plants.

b
▷

c
▷

a
◁

d
▷

Probably the greatest risk when using leaf propagation is that of the leaves rotting before growth starts. As a general rule (begonias apart), the leaf blades must be kept clear of the soil, and moisture not allowed to rest on them. Watering, if called for at all, must be from below, by immersion. Leaves flagging and showing signs of rot must be removed at once or the trouble will spread to others in the tray or pot.

Plants from leaves are smaller, initially, but growth is swift in favourable conditions and from a single mother plant a large stock, perhaps totalling hundreds, can be developed in a season or two.

Single Leaf Cuttings

Cissus. This simple method gives the largest number of cuttings from a single mother plant. It is merely a very short stem cutting, bearing however only a single leaf. *Cissus antartica* or kangaroo vine (fig *a*) is a hardy and popular houseplant, relatively easy to propagate in this way.

(*b, c*) Cut the stem into single leaves, each having $\frac{3}{8}$in (9mm) of stem remaining.

(*d*) Press each stem section into moist compost, making it firm around the stem but making sure that the leaves themselves do not touch it.

Development may be slow and the young plants small, but they should soon grow satisfactorily. Impatiens, Pilea, Beloperone and Hibiscus can also be grown in this way.

(*e*) To propagate *Hibiscus syriacus* "Blue bird" take leafy cuttings.

(*f*) Leafy cuttings are taken from *Hydrangea petiolaris* in spring and summer.

f
▽

Split Stem Cuttings

Zonal Pelargonium (*geraniums*)

(*a*) When plants have leaves opposite each other on their stems, you can obtain the maximum number of young plants by making split-stem cuttings. Cut the stem ¼in (6mm) below and above a leaf.

(*b*) Then split the stem down the centre to give two half-stem cuttings.

(*c*) Insert these thin cuttings at the sides of a 2½in (6cm) pot and grow on under polythene as shown earlier.

Besides giving more .new plants, the splitting exposes more of the cambium layer for development of roots. This may improve the chances of success. Plants vary, however, in their stem's response to such severe cutting so experiment before completely cutting up a valued plant.

Tender Plants from Single Leaves

Ficus. Single-leaf propagation is especially useful for increasing larger and more expensive plants such as *ficus.* Deep boxes of well-drained compost are best, and heat is essential.

(*d*) Cut the stem into 1in (25mm) sections, each carrying a healthy leaf. The cut surfaces bleed a white juice but this can be ignored as it will soon stop.

(e, f) Plant the cuttings 1–1½in (25–40mm) deep in a compost box kept at a fairly high temperature, 70°F minimum, by bottom heat and give each leaf cutting its own supporting cane.

(g) Apply an elastic band to hold the leaves rolled round the cane. This prevents them sagging and thus rotting.

(h) Keep the box warm for about six weeks until rooting has taken place, shown by development at the tips. Then unfasten the bands and grow the plants on but do not transplant too soon. Wait until prolific roots are found when the compost is disturbed.

(NB For small-leaved *ficus (Ficus pumila)* make four-leaved cuttings in pots kept at a high temperature.)

f ▷

d ◁
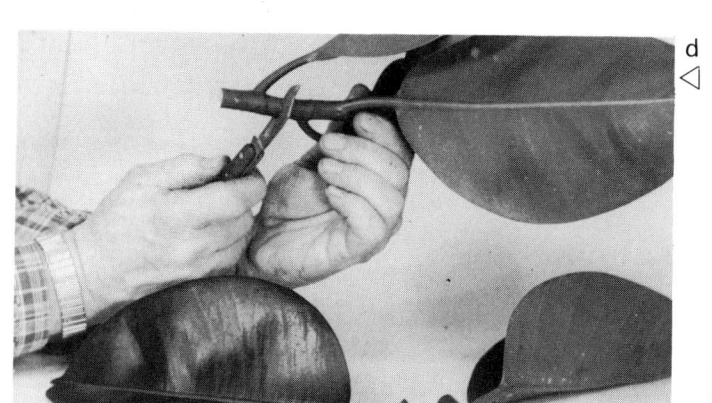

e ◁
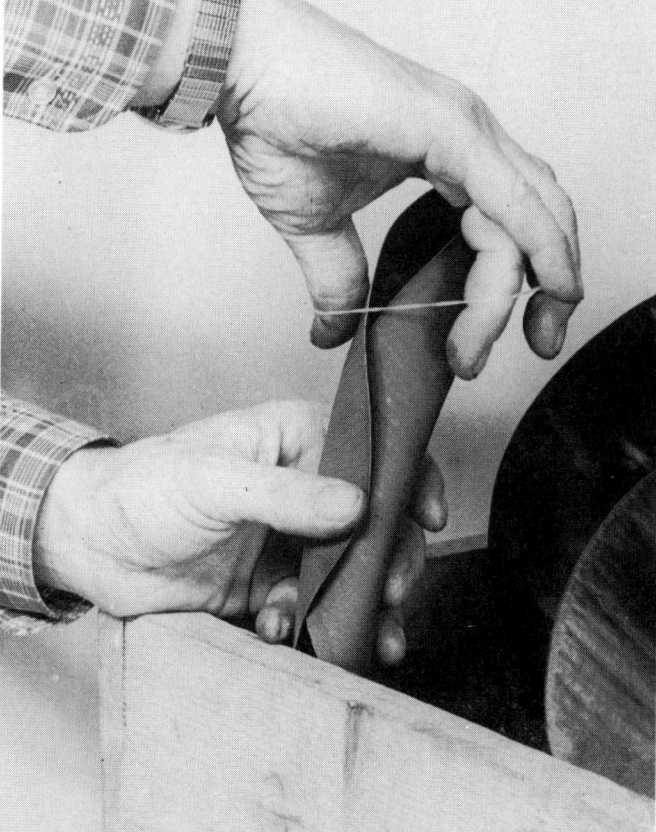

g ▷

h ▷

Peperomia. Many houseplants throw up leaves from soil level, on short stalks, or grow them in rosettes so that there is no stem to make cuttings from in the normal way. Instead, single leaves, complete with their stalks, can be planted. Roots usually form from the stalk, as in the african violet (*saintpaulia*), but many also form near the leaf blade itself. The mother plant will soon replace the removed leaves. Pruned plants, in fact, often grow on more prolifically.

(*a*) Cut away the leaves and stalk using a very sharp knife.

(*b*) Insert the stalks in a tray of compost, keeping the leaf blades clear of the soil.

(*c*) To keep water loss down, give extra protection by erecting a polythene tent. Bend two stiff wires in hoops and insert one at each end of the tray.

(*d*) Draw a polythene bag entirely over the tray.

(*e,f*) Seal the bag end with an elastic band.

The sealed tray can be left unattended in a bright, warm place until development takes place. It will not usually require watering, but if this is essential, remove the bag and immerse the whole tray, keeping all water from the leaves themselves. Allow it to drain for an hour before replacing the plastic.

c ▷

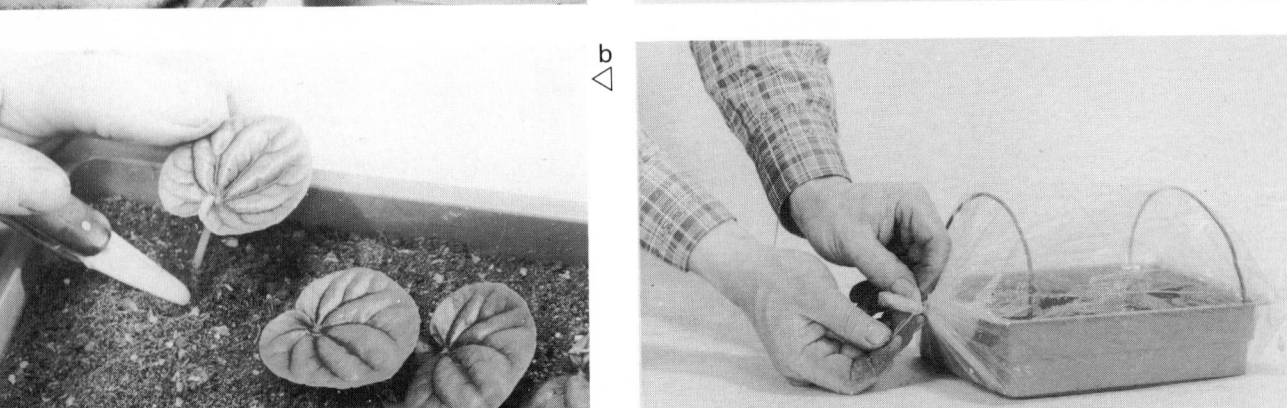

d ▷

a ◁

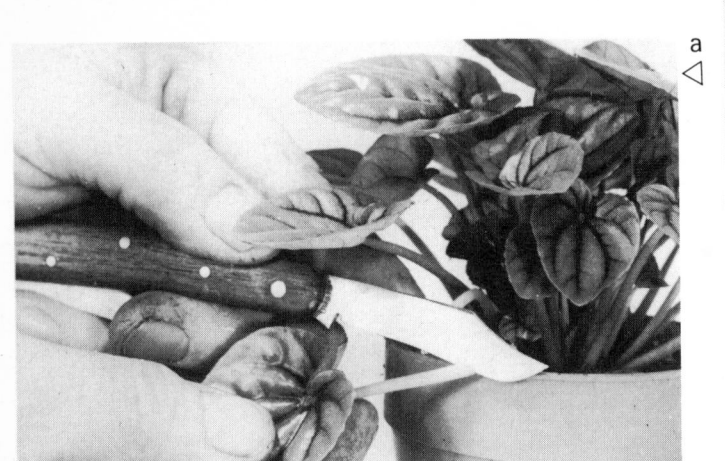

b ◁

e ▷

f ◁

African Violet. Few flowering plants are such favourites in the home as the african violet (fig *a*). It is fairly easy to root their leaf cuttings under improvised protection.

(*b*) Select leaves which are large, healthy, and clean in appearance. Cut them off near their bases taking care not to bruise the stems.

(*c*) Fill a small pot with well-drained compost and extra sand added, or even with plain coarse sand. Slide the stem of the cutting into the centre of the pot.

(Page 62)

(*d*) It helps in propagating these plants to slide a large jam jar or similar over the top of the pot, leaving it there until the stem roots. The glass retains the moisture in the air around the leaf and helps to prevent drying out.

(*e*) Better still plunge the pot and its covering jar into a tray of wetted peat. Then, no further watering is likely to be needed until growth starts. This is shown usually by the appearance of small leaves from near the base of the stalk.

If you propagate these cuttings in plain sand, they must be moved to compost once rooting has taken place as sand does not contain any plant food.

The rooting of african violets presents few problems, but growing on may be difficult, so take considerably more cuttings than you need. Carefully maintain them in a moist, warm atmosphere until growth is well established.

b ▷

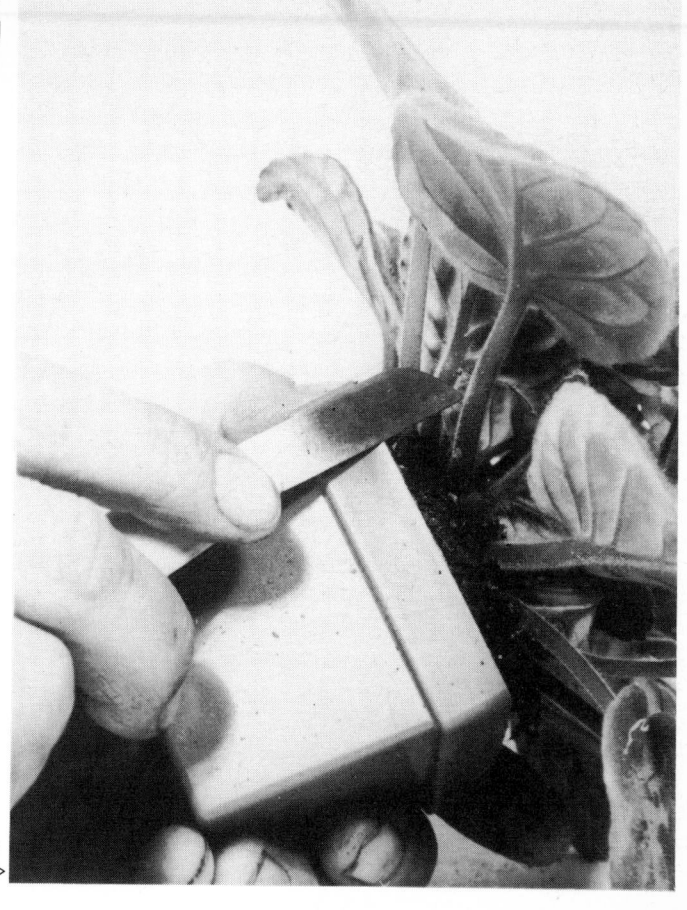

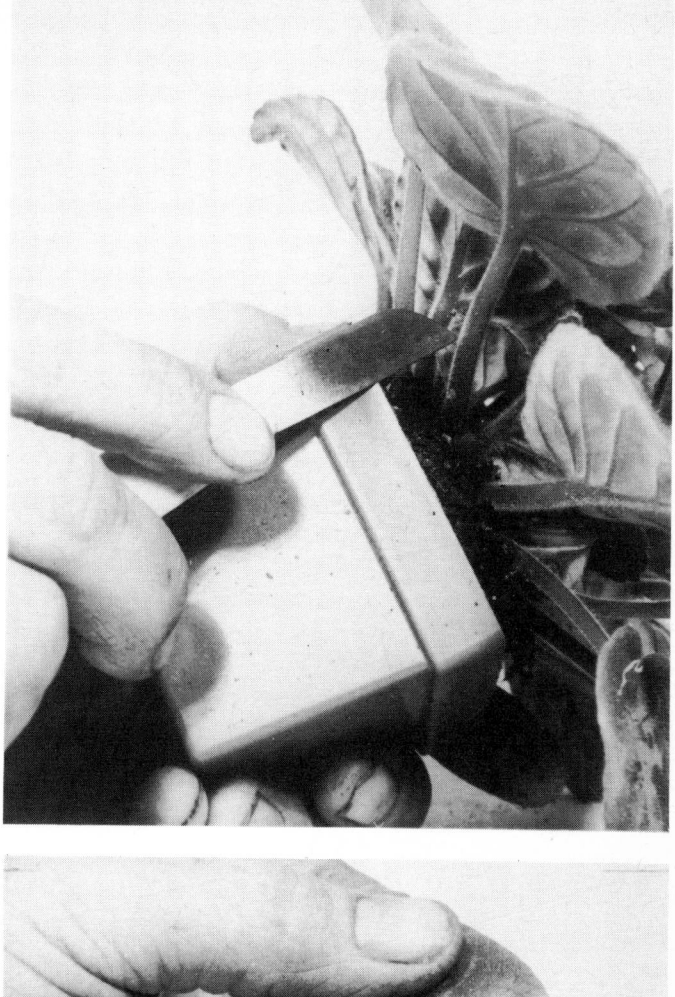

a ◁

c ▷

a ▷

Leaf Layering

Begonia. In this method whole leaves are laid flat upon the compost surface. New roots and young plants form from the leaf veins. This is comparable with the stem-layering shown earlier. It is only suited to a limited number of plants, but these include the immensely popular leafy begonias.

(*a*) Always choose a leaf that is mature, undamaged and growing well. Cut it off with a short length of its stalk.

(*b*) About 1in (2.5cm) from the leaf stalk nick all the radiating main veins with the tip of a sharp knife.

b ▷

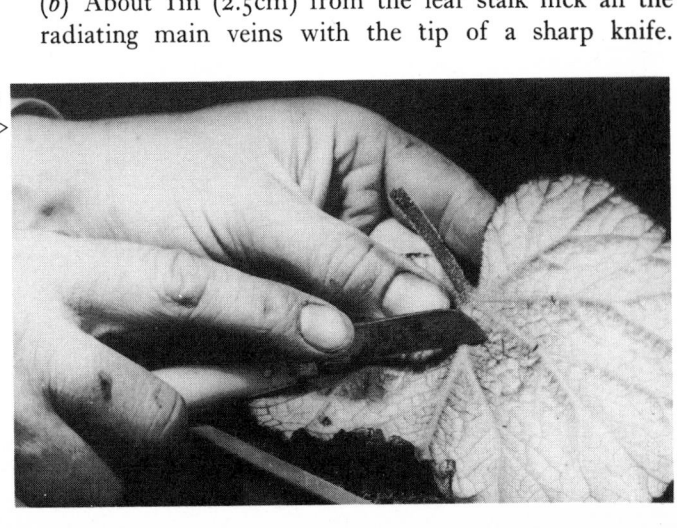

c ◁

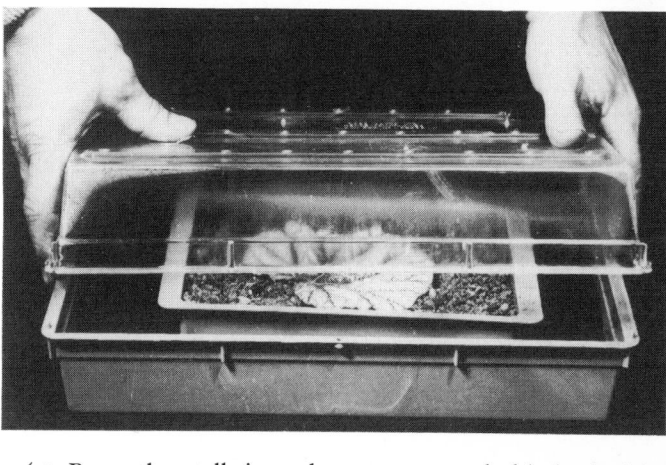

e ▷

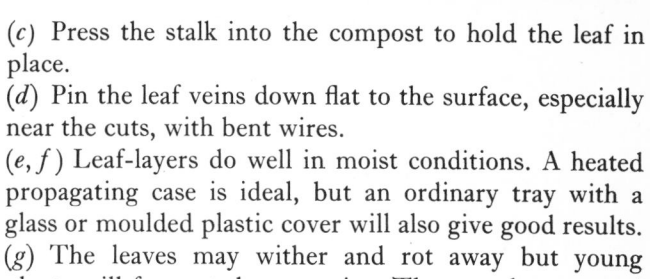

f ▷

d ◁

(c) Press the stalk into the compost to hold the leaf in place.

(d) Pin the leaf veins down flat to the surface, especially near the cuts, with bent wires.

(e, f) Leaf-layers do well in moist conditions. A heated propagating case is ideal, but an ordinary tray with a glass or moulded plastic cover will also give good results.

(g) The leaves may wither and rot away but young plants will form at the cut veins. They can be removed when large enough and grown on in small pots.

g ▷

Begonia leaves (fig *a*) offer themselves to a remarkable variation of leaf-layering. Small sections of cut up leaf laid on the surface will root down into the soil.

(*b*) Start with a full grown, healthy leaf. Slice it into strips roughly ½in (12mm) wide.

(*c*) And then into squares. Ideally, each piece should have a small length of vein.

(*d*) Lay these pieces flat over the surface of a tray of moist compost. (The *Begonia masoniana* prefers to have the cut vein inserted into the compost, not just lying on the surface.)

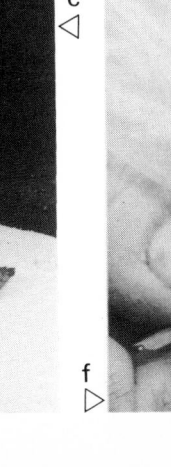

(*e, f*) Draw a polythene bag over the tray, making it airtight by twisting the end.

(*g*) Place the tray in a bright but not sunny window and leave it undisturbed. Do not move the tray unnecessarily whilst rooting is taking place or the leaf sections may be disturbed.

(*h*) The young plants are almost unbelievably tiny at first and their roots are easily broken.

(*i*) When the small plants have developed, carefully lift them free with the tip of a pointed stick . . .

(*j*) And plant up in very small pots.

i ▷

g ◁

h ◁

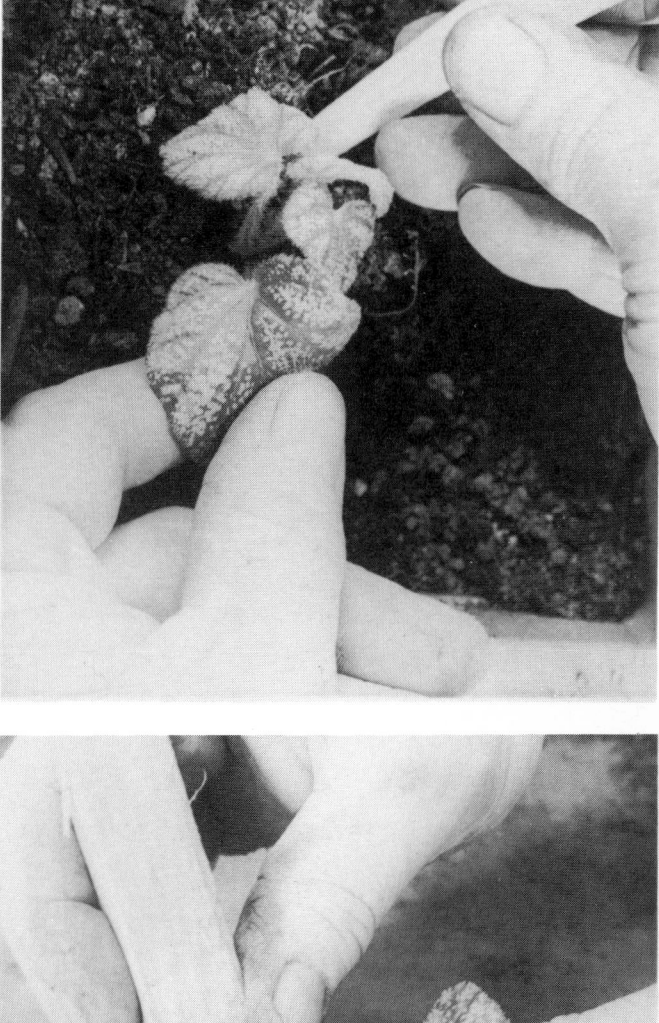

j ▷

PROPAGATION BY GRAFTING

Grafting is the ancient art of attaching the buds or stems of one plant to the roots or stems of another. If the work is successful, the two grow together permanently. The methods used are very varied, but all depend basically on making the cambium layers of the two parts touch closely and link up together.

Grafting is extensively used to change the habit of plants. A slow growing type may be grafted on to rapid growing roots, so speeding up flowering and fruiting. A dwarf root may have buds of fruit grafted in, making a tiny tree, easy to deal with in small gardens.

For propagating though, the reasons for grafting are mainly to produce a large number of plants of some fine variety, within a short time, by grafting stems, or even single buds, on to roots of much cheaper and commoner sorts. The most widespread example is the rose. Most bushes are in fact grafted buds, grown on roots of common wild roses.

You can in fact graft your own rose bushes at home, giving yourself healthy young plants at a fraction of the cost of buying them. Many gardeners fight shy of grafting things because they feel that it is too technical and needs too much attention and skill. You need have no such fears with most bush roses. It is quite simple to get a reasonable number of successes. The cost is very small. All you need are some seedling briars, which you can buy from most

nurserymen, and some leaf buds from your own best rose trees. The seedlings are very similar to the wild roses found in the countryside. In fact you can even collect wild roses, transplant them to your garden and graft them quite successfully! Single buds of your cultivated variety are then grafted to these. In due course the buds develop and the stem of the "stock" bush above the bud is cut away. This results in the cultivated rose stems and leaves growing on roots of the cheap variety. The bud gains something of the vigour and hardiness of wild roots and you can produce large numbers of valuable plants from only one or two cultivated types.

Professional growers have to cover the cost of labour, fertilisers, the land on which they grow the roses, the transport, sale, advertising and of course the printing of their beautiful catalogues! We at home have no such costs and if some of our grafts fail, very little is lost. How then do you set about growing your own rose bushes by graft-budding? Firstly, buy a number of suitable rose stocks. These vary in price according to the thickness of their stems and the exact variety. Several sorts are available, including *rugosa* and *multiflora* stocks. For amateur use there is little to choose between them. Buy your seedling stocks at the very latest in early spring. It is better in fact to get them in autumn, and put them into the ground before the frosts. They will then be well established by the following summer, when the actual budding is done.

The process itself is very simple in theory but requires a little practice. You will need a small sharp knife with a good blade and a supply of raffia. At the base of each rose leaf, you will see a tiny bud. In bud-grafting gardeners lift out the whole of a leaf and its bud and slide it under the bark of the seedling stock. The bud is then tied in place and left to graft itself. The following spring the grafted bud should develop rapidly and then all the wild seedling above it is cut off. By autumn the grafted rose stems, leaves and flowers all of the cultivated sort, is ready for transplanting.

It is best to graft in two buds per plant in case one fails. To make the tall, standard roses you need firm healthy stems of strong wild rose types. These are staked upright and then the bud of your cultivated variety grafted at the desired height. Once it has become established all wild growth above it is cut away. The main stem is kept clear of side growths, because these will be of the wild variety. In exposed places, strong winds can rip the fragile bud free from its developing connection with the stem. To avoid this, stake the new bud stem as it develops.

(a) Clematis are often grafted on wild roots.
(b) *Rosa gallica* "Versicolor" is one variety achieved by grafting.
(c) *Rosa albertine* is a climber/rambler.
(d) *Rosa* "Penelope" – a hybrid musk rose.
(e) The hybrid tea rose, "Pink Parfait."

b ▷

c ▷

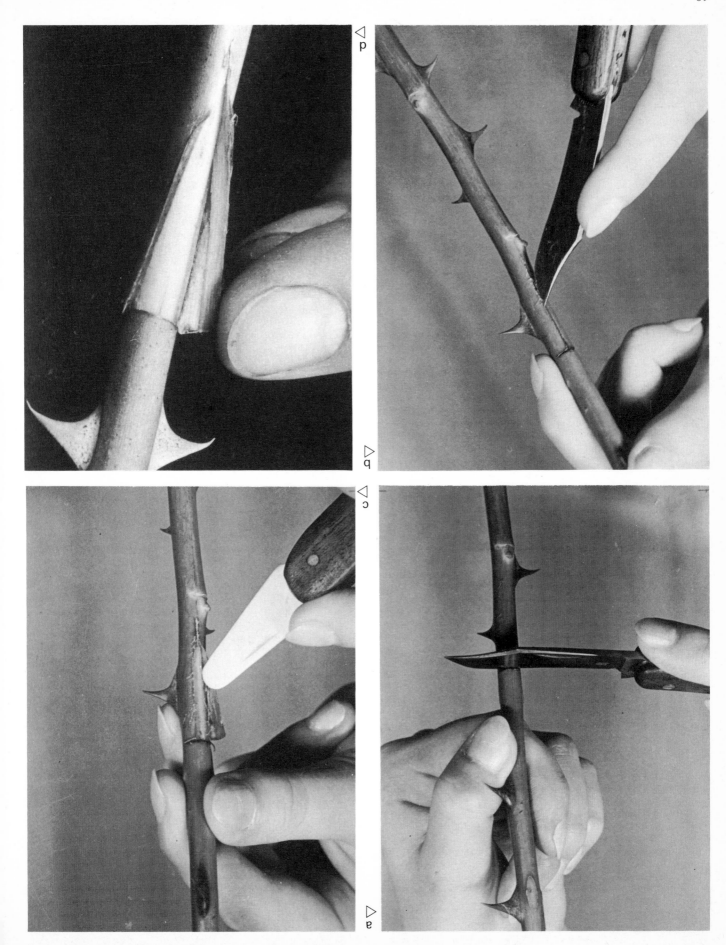

d ◁

b ▷

c ◁

a ▷

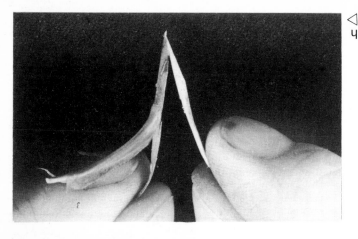

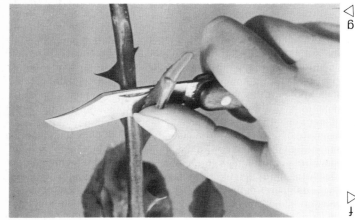

How to Bud Roses

PREPARING THE ROOT STOCK

(a) This wild rose stem will become a standard rose by grafting a bud 4ft (120cm) from the ground. (To make bush roses, clear the ground away to get at the very base of the root stock.) Take a sharp knife and make a thin cross-cut just through the outer green bark of the stem.

(b) Slice vertically down the stem, again just to bark depth and no further.

(c, d) Using the blunt, ivory blade of your budding knife gently lift the bark of the stem away, exposing

the white heartwood beneath. Grafting is only successful if the two parts of the plant are touching at the cambium layer, the soft, sappy layer between the bark and the hard wood beneath.

PREPARING THE BUDS

(e) Your cultivated rose should have a healthy stem which has already borne a flower. Choose the bud from one of the leaves about halfway up the stem. Those at the top will be too soft and sappy and those at the bottom may be too hard and over mature. Cut off the leaf itself, leaving about ½in (12mm) of its stalk intact.

(f) Placing a thumb behind the stem, cut inwards and upwards, starting about ¾in (18mm) below the leaf, cutting through the bark and slightly into the heartwood beneath.

(g) Continuing this cut upwards, pull off the whole of the leaf base and its tiny bud.

(h) At the back of the bud a sliver of heartwood is still attached. This is peeled away with the finger nails. If the whole of the leaf bud is pulled away by mistake, you will see a tiny hole right through the remaining part of the bud. Throw it away and start again! A grafting bud should have none of the heartwood attached but should still have the leaf bud securely fixed in it.

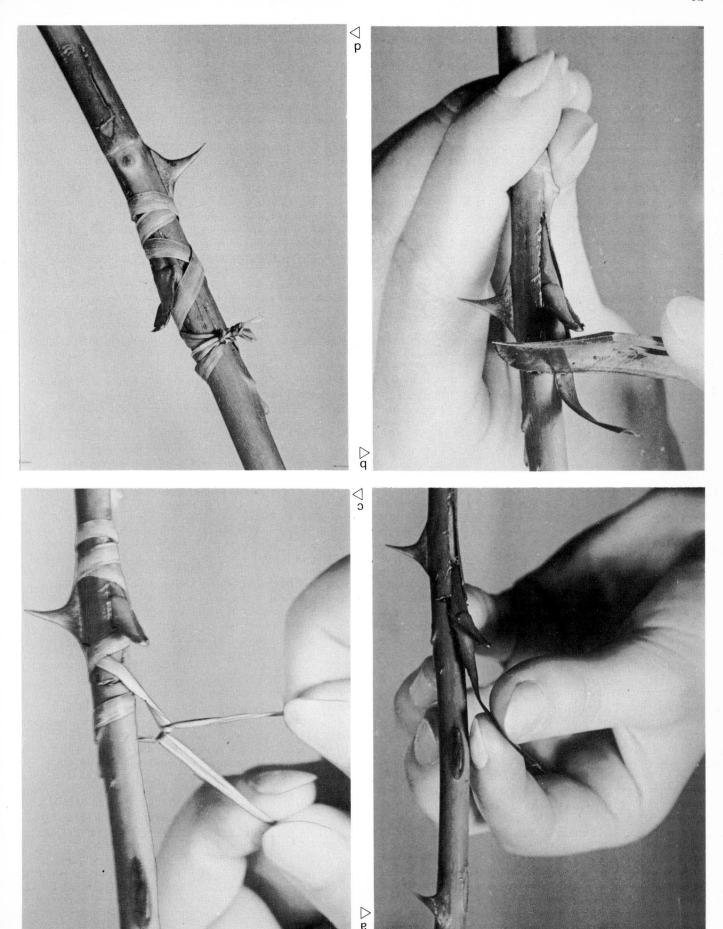

d

b

c

a

e

INSERTING THE BUD

(a) Slide the prepared bud under the loosened bark of your stock rose stem, seating it well down so that the back of the bud is pressed firmly against the cambium layer of the stem.

(b) Then cut away the surplus bark of the bud exactly at the top of the T-cut in the stem.

(c) To hold the bud securely until it is grafted wrap raffia around the stem. When grafting bush roses this raffia will be underground and will rot away in a few months (e). On standard roses, where grafts are made above the ground, the raffia must be cut once the bud has become established, otherwise the growing stem will be strangled.

(d) The completed bud in the stem. Two or more can be put in, spaced 2in (5cm) apart. After a few months shoots will have grown out and the following spring the whole of the wild stem above the bud positions must be cut away. You will have a fine standard rose at a cost of only one wild stem and a few buds from one of your own rose bushes!

Grafting is possible with other plants, but requires rather more skill. Lilac (f) for example, can be grafted on privet stocks; cultivated clematis (fig g) on stocks of wild clematis. Few aspects of gardening are so fascinating as this and once you have succeeded with a few, you can really consider yourself a skilled propagator!

MAKING YOUR OWN PROPAGATING EQUIPMENT

Although the basic equipment needed for propagating is very simple, you can greatly improve your chances of success by buying or making a few extra items of equipment.

A *potting tray* will save mess and damage near the working area especially when used indoors as plants are being pricked out or potted, and composts prepared. It is a deep, three-sided tray which can be laid on a table or greenhouse staging.

Sieves of various kinds are constantly in use where composts are being mixed and seedboxes filled.

A *propagating case* will provide evenly moist and warm conditions for seeds and cuttings. One can be knocked together from a few planks and stood in any convenient light place.

MAKING A POTTING TRAY

(*a*) This is a three-sided tray with a hardboard or plywood base about 2ft (60cm) square and a surround of 3 x 1in (8cm x 2.5cm) timber. Set these sides around the base. Nail the base to the sides from below, using 1in (2.5cm) round headed nails. To secure the corners drive in longer nails. Blunt their tips with a hammer blow before use, to avoid splitting the wood.

c ▷

a ◁

b ◁

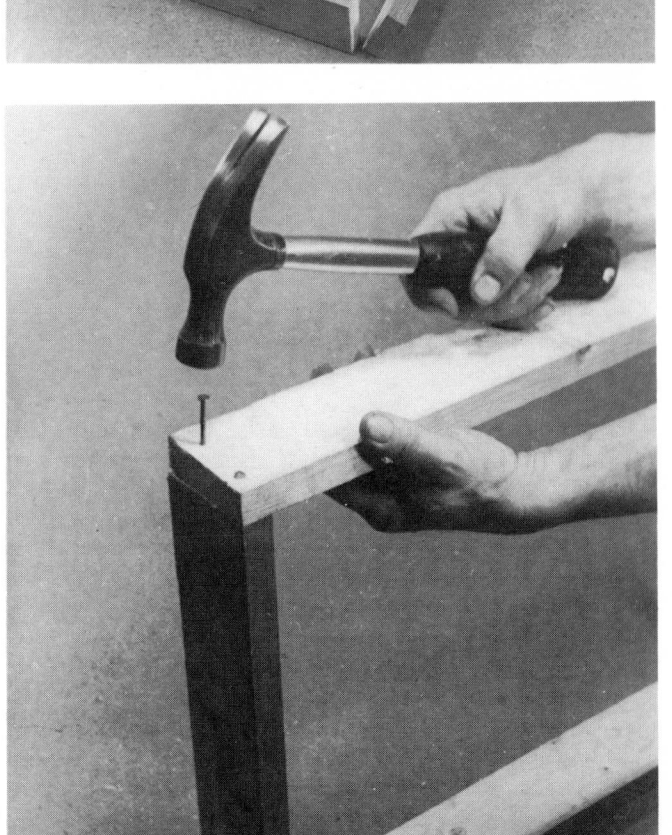

d ▷

MAKING A COMPOST SIEVE

(*b*) Compost for potting must always be sieved down to remove hard lumps or stones. You can make a handy small sieve using only a few lengths of timber and a piece of plastic or wire mesh netting.

(*c*) Make the main frame from planed timber 3 x 1in (8 x 2.5cm) in section. The size can be varied. The one we show has sides 15in (45cm) and 22in (55cm) long. Place the four parts in a rectangle.

(*d*) Secure the corners, angling the nails slightly towards each other.

(*e*) Reinforce corners with 3in (8cm) blocks of wood.

(*f*) Nail them securely whilst resting them over the corner of a vice. (Otherwise, you may split the blocks or force them away from the sides.)

(*g*) Plastic mesh sheeting is easily cut with ordinary scissors. For potting composts choose a mesh of approximately $\frac{1}{4}$in (3mm). Galvanised metal mesh can also be used.

(*h*) Hold the mesh securely in place with strips of wood roughly $\frac{3}{4}$in (18mm) square nailed down to the frame sides. The netting is gripped between.

(*i*) Screw handles made of $1\frac{1}{2}$ x 1in (3cm x 2.5cm) timber, sandpapered smooth, to the sides, a little way in from the ends.

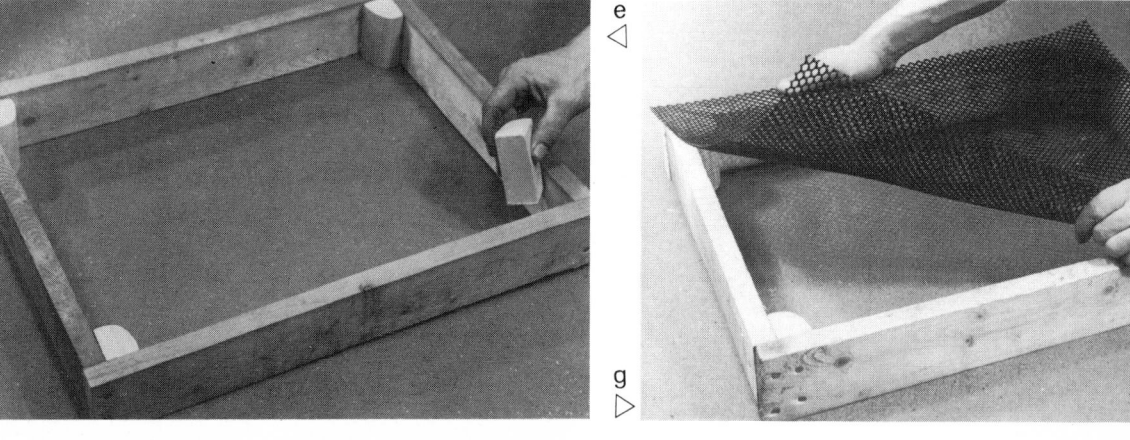

e ◁

g ▷

f ◁

h ▷

i ▷

73

MAKING A PROPAGATING CASE FOR
HOME OR GREENHOUSE

(*a*) A propagating case is basically a box with a glass cover and an outlet for surplus water to drain away. Within this, deep layers of sand and peat are placed to provide well-drained but moist and humid conditions, free from draughts. By providing heating (perhaps by electric cables) underneath the peat, first class germination and rooting conditions will be produced. The case is stood on any flat, well-lit surface such as greenhouse staging. In the home, stand it on a watertight shallow tray within which is scattered a 1in (2.5cm) deep layer

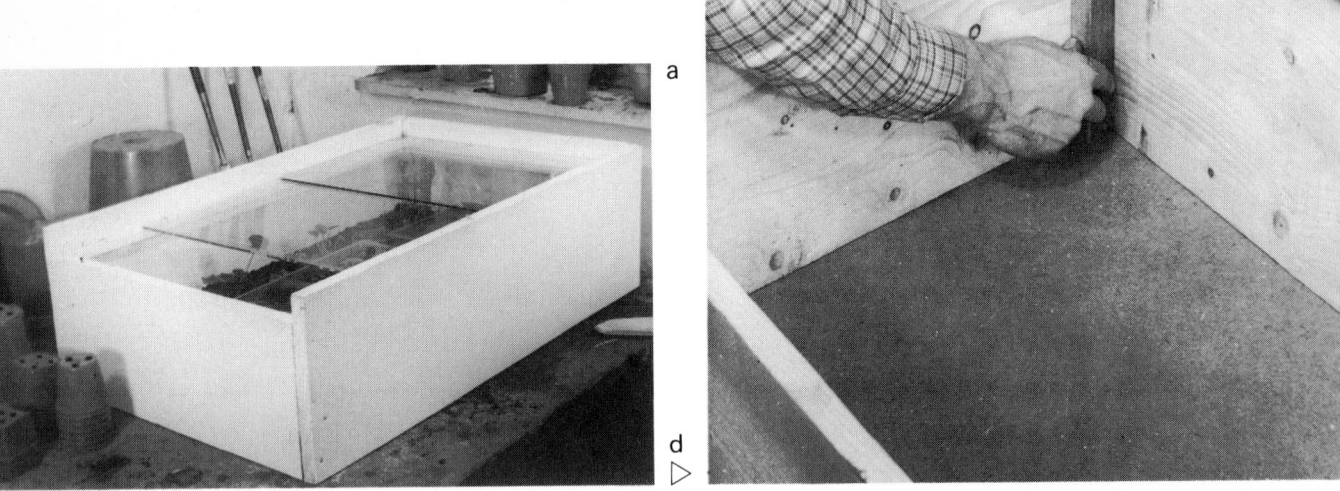

a

d ▷

b ◁

c ◁

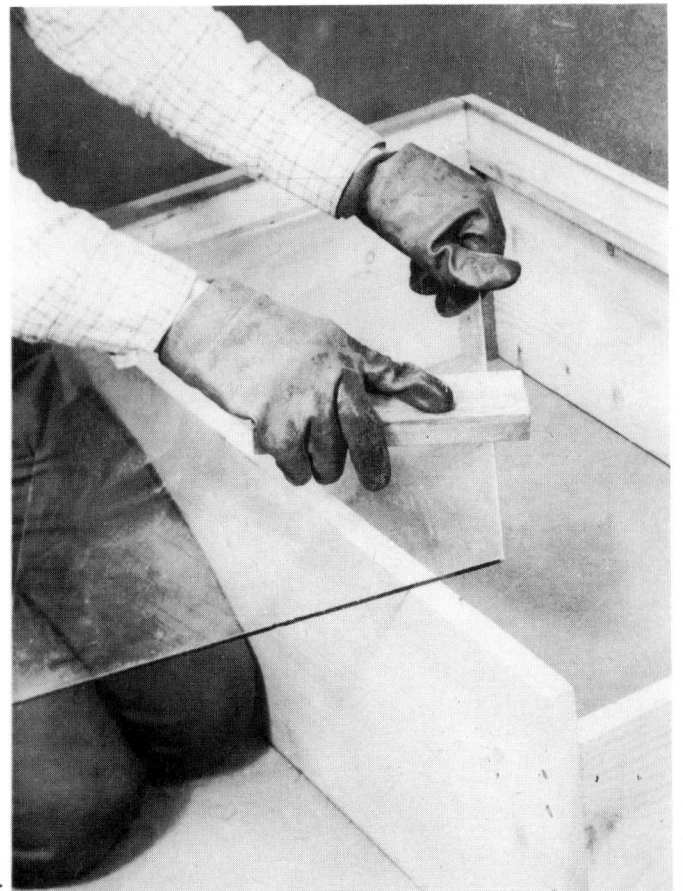

e ▷

74

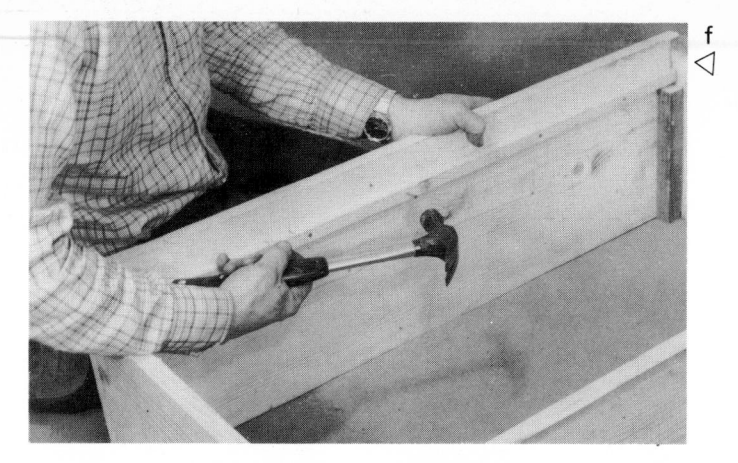

f ◁

of coarse, clean shingle. Properly maintained, drain water will be limited, the aim being to give only enough water to keep the case moist, without surplus.

(*b*) Use wood 8 x 1in (20 x 2.5cm) in section and nail two sides, 3ft (1m) long, to an end 2ft (60cm) long.

(*c*) Make the other end of wood 1in (2.5cm) narrower.

(*d*) Nail blocks of wood at each corner for reinforcement.

(*e*) Cut two pieces of glass, wide enough to be a loose fit between the sides. They should be long enough to overlap about 3in (8cm) in the middle of the case. Carefully round the sharp edges with a flat oilstone. Alternatively ask a glazier to smoothly round the edges and corners.

(*f*) 1in (2.5cm) below the top edge of the sides nail 1in (2.5cm) square strips as runners on which the glass can rest.

(*g*) By using two overlapping pieces, the glass can be slid back, and all parts of the case can then be reached.

(*h*) A 1in (2.5cm) layer of coarse sand, followed by 3in (8cm) of moist granulated peat is about right for propagating cases.

(*i*) Within this peat, pots can be sunk and will be kept in a warm and humid atmosphere.

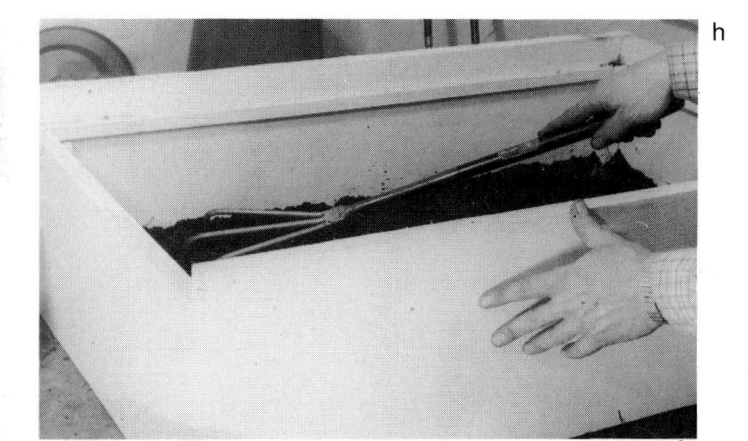

g ◁

h

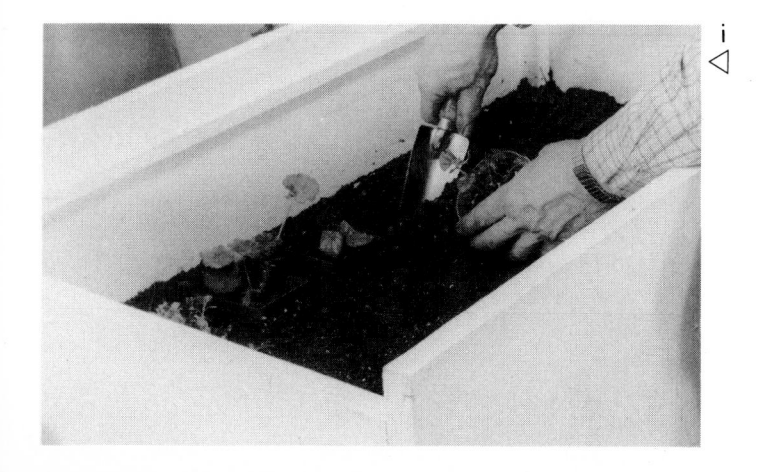

i ◁

GUIDE TO HOUSEPLANT PROPAGATION AND CARE

In recent years there has been a great increase of interest in houseplants, both for flowers and foliage. Many can be propagated without real difficulty, but it is vital, even more than with outdoor plants, that those chosen for propagation be in sound health. For the successful growing on of young plants too, some knowledge of each plant's special needs is essential. There is little point in successfully potting a cutting if it cannot be cared for to maturity.

This brief guide gives simple cultural details for most of the houseplants that may be propagated at home, grouped under seed, root, stem or leaf propagation.

Latin names are almost universal here, and indeed many plants are quite familiar under these titles.

HOUSEPLANTS THAT CAN BE INCREASED BY SEED PROPAGATION

Note: nearly all plants can in the last resort be propagated by seed, but with many of them it may present great difficulties, a heated greenhouse being practically essential. The list below simply gives a number that can be attempted in the home with reasonable hope of success. In most cases, other methods of propagation can also be used and the cultural details may be found in lists given later.

Aglaonema commutatum (see stem propagation)
Anthurium scherzerianum (see root propagation)
Begonia rex (see leaf propagation)
Cyperus diffusus (see root propagation)
Grevillea robusta. This easy, rapid growing plant with its feathery, attractive leaves can be grown in cool to warm conditions (45–60°F). It needs a moderate amount of light but no direct sun and will tolerate dry air. Water freely in summer. It is quite easy to propagate from seeds.
Neanthe elegans. This palm is the one most suited to growing in the home. Large plants can be increased by division but it is also fairly easy to grow from seed. It will live happily in cool conditions (45–60°F) and does not require too much light. It tolerates dry air but needs plenty of water in summer. In direct hot sun the leaves often turn brown.
Philodendron bipinnatifidum. This is a moderately easy plant grown for its large, deeply cut leaves. Unlike other Philodendrons, it is not a climber, the leaves springing from soil level. It needs a warm room (50–60°F) and a moderate amount of light but will tolerate dry air provided it is adequately watered.
Pilea cadierei (see propagation by stems)
Spathiphyllum. Moderately easy to grow, the tall white flowers produced twice a year last a very long time but the plant requires very humid conditions, out of direct sun. Warmth is essential (50–70°F). Large plants can also be increased by division.
Senecio cruentus. Seed sown in summer is used to propagate this gay plant which is usually thrown away after flowering. It needs very frequent watering in a temperature of 45–50°F, away from the sun. Seed is sown from April to June. Mature plants will need feeding fortnightly.
Calceolaria. Seeds sown in June and August will produce plants for flowering in April and July. Watering must be done with care, the plant being kept evenly moist without ever being soaking wet. It needs cool conditions in partial shade.
Cyclamen. Seeds for these plants can be sown in January and February or in August. All Cyclamen need very humid atmospheres but not high temperatures. 50–60° is adequate. Keep out of direct sun and always water from beneath to prevent the corms being wetted. The compost used must be free of lime.
Gloxinia. Seeds sown in February to June can be used to produce these attractive flowering plants. They need a fair amount of warmth and moderately humid atmosphere. (See also leaf propagation.)
Impatiens (see stem propagation)
Primulas. Seeds sown in June or July can be used, covering the fine seed only lightly, if at all. They may prove difficult to grow on after germination. Keep the seed boxes away from sun and never over water. (Also see root propagation.)

HOUSEPLANTS THAT CAN BE INCREASED BY ROOT PROPAGATION

Root Division

Adiantum. This very popular fern is nonetheless quite delicate. It requires a temperature of 50–60°F but will withstand considerable shade. Humid air around it is essential but take care when watering as its roots must not stand in stagnant compost.
Anthurium scherzerianum. This is a delicate plant grown partly for its unusual flowers as well as for its foliage. It needs quite a high temperature (50–70°F) in a position that is partially shaded. The air must be humid, such conditions being best provided by double potting with peat, as shown on page 18. Give plenty of pure rainwater in summer. It can also be propagated by seed.
Asparagus plumosus. An easy plant suited to bright or shady conditions. Tolerant of a wide range of temperature,

provided it is kept well watered in spring and summer. Seed sowing is also used.

Chlorophytum. This well-known and very easy foliage plant will withstand temperatures ranging from cool to quite warm and positions in extreme shade or bright sunlight. It tolerates dry air in central heated rooms and provided it is well watered in summer will continue to grow for many years. It develops young plants at the end of long shoots which may be planted up (see page 35).

Cyperus diffusus. Another very easy plant grown for its unusual umbrella-shaped foliage. It needs a warm room (50–60°F) in a moderately light position. Humidity around its leaves is essential or it will tend to shed them. Watering should be frequent and plentiful in summer. It is also possible to propagate by seed.

Helxine soleirolii. A rapidly growing creeping plant with tiny leaves and very easy to grow. Must be kept moist always. The pots can be stood in water-filled saucers. Partial shade is best.

Maranta leuconeura. A moderately easy plant grown for its unusual leaves, some varieties having semi-transparent "windows" through these. It requires fairly warm conditions (50–70°F) and humid air is essential. Plenty of water in summer will keep it growing well.

Peperomia magnoliaefolia. There are many varieties of *peperomia* which may be increased by division, but most may also be propagated by leaf cuttings (see page 60).

Platycerium alcicorne. An unusual fern very suitable for wall decoration and grown for its very distinctive, irregularly shaped fronds. It does well in fairly cool conditions (45°F) and in partial shade. Since it does not object to dry air it is useful for centrally heated rooms. Do not over water. Division of the roots is usual but many plants also produce offsets which can be planted up individually.

Primula. Flowering well in the home, in any bright spot out of full sun, there are many varieties of Primula. Water freely when flowering and over winter in a bright, cool room. *Primula obconica* can cause skin trouble. Division is easy, see page 35.

Sansevieria. Few plants could be so easy and hardy as this. It will withstand almost any neglect and may be left without water for several months without suffering permanent damage. It is grown for its upright, stiff and succulent leaves and prefers a moderately high temperature (50–70°F). It may be stood either in complete shade or bright sun without ill effect. Sansevieria is tolerant of dry air and fumes. It requires watering only occasionally and it is best to give a thorough soaking and then leave it to dry right out. Although propagation by division and offsets is easiest you can also grow them from cuttings but these tend to lose the variegations which are an attractive feature of many varieties.

Offsets

Aechmea. An easy and distinctive plant somewhat like a cactus which develops attractive flowers from time to time. It needs warm conditions (50–70°F) and will tolerate rather more light than is usual with houseplants. It will also withstand dry air and gas fumes which makes it suitable for centrally heated rooms. Watering should be done only with care. The many offsets which are produced around the base of the plant will grow easily if potted up separately.

Cryptanthus. An easy plant grown largely for its foliage and which will tolerate cool conditions down to 45°F and a moderate degree of shade. Dry air and gas fumes do not affect it. Water only in moderation.

Neoregelia. A fairly difficult plant grown for its foliage and needing warm conditions (from 50°F upwards). It likes a moderately shaded place but will tolerate dry air. All watering should be done with care.

Pandanus sanderi. A rather delicate foliage plant which needs moderately warm conditions (50–60°F) but is another houseplant which can tolerate bright sunshine. However, humidity must be maintained about its leaves which means that double potting should be practiced. Regular, liberal watering is needed, especially in summer.

Saxifraga stolonifera. A very easy plant grown for both flowers and leaves, withstanding quite cool conditions down to less than 45°F. In dry rooms it requires liberal watering. The offsets formed on this plant root very quickly, indeed even when simply resting on the soil.

Vriesia splendens. An easy plant grown for flowers and foliage which needs warmth (50–70°F), and grows best in moderate shade. It tolerates dry air and gas fumes and needs only moderate watering in summer.

HOUSEPLANTS THAT CAN BE INCREASED BY STEM PROPAGATION

Layering

Beloperone. An easy plant with unusual flowers that can be grown in a moderate temperature (50–70°F) and in partial shade. It needs humid air to keep growing well but watering should be restrained, the plant being kept rather dry in winter. It can also be grown from cuttings.

Dracaena. A handsome, reasonably easy plant needing a temperature of at least 50°F, and humid air. Reduce water in winter. Tolerates partial shade. Air layering is useful on older plants.

Ficus elastica decora. This well-known and very easy large leaved plant can be propagated in a number of ways, cuttings, layering and so on. (See page 43.) Its temperature needs are from 50°F upwards and it can tolerate quite heavy shade and dry air. It needs plenty of water in summer but too heavy winter watering can cause leaves to drop.

Hedera helix. Few plants can be so easy as the Ivy to grow in the home and there are many interesting varieties with variegated leaves of differing shapes. It will tolerate even quite cool temperatures but should not be kept long at a temperature higher than 60°F. They will grow even in considerable shade and will tolerate dry air and gas fumes well. Plenty of water in summer. Propagation is easy since the plant develops aerial roots along most of its stems.

Monstera deliciosa. This plant is fairly easy to grow and yet is one of the biggest of houseplants. Its giant leaves are cut into attractive shapes. Although technically it is a climber most houseplants remain as a very large bush and prefer a temperature of 50–60°F. Keep it out of direct, hot sunlight but it will tolerate the dry air of ordinary rooms provided it is adequately watered during the summer. Cuttings will also succeed but it is not always easy to find suitable material.

Philodendron scandens. The climbing varieties of Philodendron are easily propagated by layering. Most are fairly easy and grown for their attractively shaped foliage although there are other kinds grown as bushes. These last are often propagated by seed. Their temperature needs are fairly easy to satisfy, the climbers growing in any temperature (45–70°F) in a moderately shaded position. They will tolerate dry air provided they are well watered in the summer. (See page 38.)

Rhoicissus rhomboidea. An easy plant grown for its glossy, dark green foliage, it is a climber and suited for even quite cool rooms down to 45°F. It prefers coolness to excessive heat and can be grown in either light or shade. It tolerates dry air well, provided it is generously watered. Cuttings, even of single leaves, will also succeed.

Scindapsus aureus. A moderately difficult plant with variegated leaves which may be grown as a climber or trailer. It needs a moderate temperature (50–60°F) in a lightish position. The variegated sorts tend to lose their variegations in shade. It will tolerate dry air quite well but needs plenty of water in the summer. The young growth is sensitive and should not be handled.

Cuttings

Note: those marked "flowering plant" are on a separate list but could be included in the places indicated. (Very many plants can be propagated by cuttings and the list given below is not exhaustive.)

Aglaonema commutatum. A moderately easy plant grown largely for its foliage and requiring a warm room (50–60°F) in partial shade. Variations in temperature in winter can be fatal, and watering then must be reduced. It tolerates dry air well. Can also be propagated by seed.

Aphelandra. This is a fairly delicate plant, rather difficult to keep growing satisfactorily. It tends to shed its lower leaves. These are vigorously striped in white on a dark green background and are if anything more attractive than the curious yellow flowers. It needs quite evenly hot conditions (60–70°F) and prefers partial shade. Humid air around its leaves is vital and plenty of water is required in the summer.

Azalea – flowering plant.

Begonia. Many begonias can be propagated by cuttings although the vast majority of the *rex* type are most interestingly propagated by one form or other of leaf propagation. (See page 62.)

Begonia (fibrous rooted) – flowering plant.

Cissus antarctica. This is a very easy plant, grown for its foliage as a trailer or climber. It will tolerate quite cool conditions in light or shade as well as being indifferent to dry air. Avoid over watering. (See page 56).

Chrysanthemum – flowering plant.

Citrus. Moderately easy plant grown for flowers and foliage. (Fruits may appear if the flowers are pollinated with a small brush, dabbed into them as they open.) They require fairly warm conditions (50–60°F) and will tolerate bright sun and dry air. Water generously.

Coleus. A very easy summer plant, but needing minimum temperature of 55°F in winter. Requires plenty of water and bright conditions. Very easy and quick to propagate by cuttings.

Dieffenbachia – WARNING. This plant is easy to care for but remember, its juice is poisonous. It is therefore *not* advisable to propagate it. It prefers a warm temperature (55–70°F) in moderate shade. Humid air around it is essential but over watering must be avoided. Professionally, cuttings are often made from short sections of stem buried 2in (5cm) deep in a propagating frame. The joints of the stem send up new plants which can be potted on. Air layering can also be used.

Dizygotheca (*Aralia*). A rather delicate plant grown largely for its foliage and requiring warm to hot temperatures (50–70°F) in moderate shade. Humid air is essential and it requires plenty of water in summer. Double potting is helpful to maintain humidity.

Euphorbia – flowering plant.

Fatshedera lizei. A very easy plant grown for its large, attractively shaped leaves and preferring cool conditions down to 45°F. It may be kept in shade or moderate brightness and will tolerate centrally heated rooms quite well provided it is well watered.

Fatsia japonica. Another easy plant grown for its foliage and tolerating cool conditions down to 45°F and any degree of light except direct sun. Dry air does not affect it, provided it is adequately watered. Air layering is used sometimes on big specimens.

Ficus pumila. This plant is quite unlike the usual rubber plant, having small trailing leaves. Its propagation is by small cuttings with two or three leaves. It requires quite low temperatures (45–60°F) and will withstand any degree of light, except direct sunshine, provided humid air is kept circulating around its leaves. Give plenty of water in the summer.

Fittonia. A delicate plant grown for its foliage and requiring warm conditions (50–70°F) in shade. Humid air around its leaves is vital at all times, but watering must only be done with care, especially in winter. This is a handy plant for a darkish corner in a warm room, and is best double potted in wet peat. (See page 18.)

Hoya. The unusual waxy flowers of this plant make it attractive for the home and it is fairly easy to grow. Its temperature requirements are 50–70°F and it needs some shading from direct sun. It will however tolerate dry air provided it is adequately watered in summer, but kept rather dry in winter. (Variegated sorts do not bloom well.)

Hydrangea – flowering plant.

Impatiens – flowering plant.

Kalanchoe – flowering plant.

Pelargonium – flowering plant.

Pilea. This is a moderately easy plant to grow for its unusual leaves with a silvery texture. Its temperature needs are 50–60°F in moderate shade and it must have humid conditions around its leaves or it will tend to shed them. Adequate watering in summer must be given, the pot never drying right out. Seed propagation can also be used.

Setcreasea. A moderately easy trailing plant grown largely for its coloured foliage. It will tolerate quite cool conditions of 45°F provided it is kept fairly light. It does not mind dry air provided it is given plenty of water in the summer. Cuttings root very easily.

Sparrmannia africana. A very easy, rapid developing plant grown for both flowers and foliage and tolerating temperatures 45–60°F. It is a sunshine loving plant that can be kept in bright conditions and in dry air but must have adequate watering during summer. Cuttings root even in water.

Syngonium. Quite a difficult plant, grown for its attractive foliage. It is a climber, producing aerial roots, and requires fairly warm conditions (50–70°F). It will not tolerate the darkest shade. Withstands dry air quite well, provided it is watered frequently.

Tradescantia. The easiest of all houseplants, with its rapidly growing, trailing habit and variegated striped oval leaves. It is ideal for cool rooms and will grow in partial shade but the variegations then tend to be lost. It will withstand dry air or gas fumes and is an ideal plant for centrally heated rooms. Plenty of water in summer is needed to keep it growing well. (See page 50.)

Zebrina. Again another easy plant, very like the Tradescantia, grown for its coloured foliage. It can withstand cool, dry conditions and gas fumes. Give plenty of water in the summer. (See page 50.)

Flowering Plants from Cuttings

Azalea. Cuttings taken in July and August are the usual method of increasing this well-known flowering plant. It requires cool, bright conditions but well out of full sun and the water and soil must be completely free of lime. Spray the foliage frequently, especially in warm conditions. After flowering (which usually takes place in winter) plunge the pot out of doors for the whole of the summer.

Begonia (fibrous rooted). The fibrous rooted begonias bloom from July to September or even later and must be kept fairly moist. They require cool, light conditions out of full sun. Cuttings can be taken in April or May. Over watering must be avoided as it causes developing buds to fall off.

Chrysanthemum. Dwarfed kinds in pots are widely popular and can now be purchased in flower at any time of the year. They are propagated commercially by special methods. At home, cuttings are taken from the fresh shoots arising near the base in February or April, and grown on in light conditions 45–50°F.

Euphorbia poinsettia. Grown mainly for the bright red, white or pink leaves surrounding small yellow flowers. This popular plant comes to its best from December to March. It can be propagated by stem cuttings taken the following April. Light conditions at a temperature of about 65°F are best. The watering is critical, since over or under-watering both cause the buds to fail.

Hydrangea. One of the easiest of the flowering plants to both grow and propagate, it flowers from March to August and requires frequent watering whilst in full bloom. It likes bright conditions and a rich, lime-free compost.

Impatiens. Few plants can be so easy to propagate as this. Even small sections of stem broken off and pushed into the soil will root rapidly. It is also possible to propagate from seed. Kept in a temperature of 50°F it will withstand quite bright and warm, sunny conditions.

Kalanchoe. Flowering in December to April, this popular houseplant can be stood in a frame outdoors throughout the summer and brought in again for the subsequent winter. Stem cuttings are taken in spring and the mature plants kept in bright sunny conditions.

Pelargonium. Pelargoniums (often known as Geraniums) are one of the most popular of all houseplants and are frequently used out of doors. Over watering must be avoided, but otherwise they are very easy to maintain. Stem cuttings taken between March and September will grow rapidly and the mature plants can withstand full sun. (See page 52.)

HOUSEPLANTS THAT CAN BE INCREASED BY LEAF PROPAGATION

Begonia rex. These plants are grown almost entirely for their beautifully coloured leaves and are relatively easy to care for. They need warmish rooms (50–60°F) and tolerate partial or fairly deep shade. Humid air around them is essential but care is needed in watering. Water from below, keeping splashes off the leaves as these may cause rot. Besides leaf propagation, large plants can also be divided, cuttings often succeed and layering can be practised. Seeds are used in a few cases. (See page 62.)

Calathea mackoyana. This is a moderately easy plant, grown for its attractive foliage but needing very warm conditions (60–75°F). Shade is desirable and a humid atmosphere is essential for good growth. It needs plenty of water in the summer. Large plants can also be increased by division.

Peperomia. There are many varieties of *peperomia*, none of them really difficult. A few have attractive flowers but most are grown for their leaves. They need warm conditions (50–60°F) in a moderately light place. Humid air is essential around their leaves but care is needed in watering. As with many succulents too much water around their bases may cause stems to rot. Division is another method used with larger plants. (See page 34)

Tolmiea menziesii. An easy plant, grown for its foliage and needing only cool conditions (45–50°F). It will tolerate a certain amount of shade and dry air but needs plenty of water. In summer its main interest lies in its habit of developing small young plantlets on top of the mature leaves. These can be easily grown on as described under leaf layering.

Saintpaulia. These attractive flowering plants with velvety leaves are most popular but not specially easy to grow. They need steady warmth and a very humid atmosphere. Watering from below must be done as splashes of water around the bases of the stems can cause them to rot. Rooting of leaf cuttings is rapid, the difficulty mainly arising in growing on young plants into flower. (See page 61)

Streptocarpus. A moderately delicate plant with attractive trumpet-shaped flowers and requiring fairly warm conditions. The leaves are somewhat similar in appearance to those of primulas, with a strong central vein. In propagating leaves, the base of this vein is inserted in the soil and a young plant will develop from it. Keep the main part of the leaf itself above the soil.

Gloxinia. The leaves of *gloxinia* can be treated similarly to those of Streptocarpus although propagation of these plants is usually done commercially by seed. They need moderately warm conditions out of direct sun and regular feeding and watering when in growth. Watering must be reduced as the leaves turn yellow.

HOUSEPLANTS FROM FRUIT PIPS

Most of our common fruits can be successfully grown in the home. Orange, lemon, tangerine, and grapefruit pips all germinate quite quickly. Plant them in small pots in moderately warm, bright conditions.

All are hardy and after germination must be kept fairly cool. Growth may be slow but many will flower and even fruit in due course. Pot them gradually into larger and larger pots and then into tubs to be placed in a bright conservatory or (in summer) stood outdoors on a warm terrace, preferably against a south-facing wall.

The avocado pear stone will also grow if planted blunt end downwards in a 5in (13cm) pot in warm conditions. In appearance this is not unlike a rubber plant and requires a fair degree of warmth. Its leaves tend to brown at the tips and sides.

Pomegranate seeds can also be grown to produce a plant but one which loses its leaves in winter.

A GUIDE TO LATIN NAMES OF COMMON PERENNIALS, CLIMBERS, TREES AND SHRUBS WHICH MAY BE INCREASED AT HOME.

Find the Latin name of any plant before consulting the methods guide that follows. Column 1 indicates plant type. Where more than one classification is given this means that certain varieties within the same group have different habits.

B	Bulb	R	Rock plant
C	Climber	S	Shrub
P	Perennial	T	Tree
	W Water or bog plant		

	ENGLISH	LATIN		ENGLISH	LATIN
S	Abelia	*Abelia*	S	Bear berry	*Arctostaphylos*
B	Aconite winter	*Eranthis*	P	Beard tongue	*Penstemon*
S	Alder	*Alnus*	P	Bear's breeches	*Acanthus*
S	Allspice	*Calycanthus*	B	Begonia, tuberous	*Begonia*
P	Alum root	*Heuchera*	B	Belladonna lilies	*Amaryllis*
R	American cowslip	*Dodecatheon*	P	Bellflower	*Campanula*
P	American wood lily	*Trillium*	P	Bergamot	*Monarda*
S	Aralia	*Aralia*	S	Bilberry	*Vaccinium*
W	Arrow head	*Sagittaria*	S	Bittersweet	*Celastrus*
P	Avens	*Geum*	C.S.	Blackberry	*Rubus*
S	Azalea	*Rhododendron*	S	Bladder nut	*Staphylea*
P	Balloon flower	*Platycodon*	S	Bladder senna	*Colutea*
W	Bamboo	*Arundinaria*	P	Bleeding heart	*Dielytra*
S	Barberry	*Berberis*	P	Bleeding heart	*Dicentra*
P	Barrenwort	*Epimedium*	B	Bluebell	*Scilla*
S	Bay laurel	*Laurus*	R	Bluets	*Houstonia*

	ENGLISH	LATIN		ENGLISH	LATIN
W	Bog arum	*Calla palustris*	S	Elder	*Sambucus*
S	Box	*Buxus*	P	Evening primrose	*Oenothera*
S	Broom	*Cytisus*	S	Evergreen laburnum	*Piptanthus*
S.R.	Broom	*Genista*	P	Everlasting pea	*Lathyrus*
S	Buddleia	*Buddleia*	S.T.	False cypress	*Chamaecyparis*
P	Bugbane	*Cimicifuga*	P	False dragon's head	*Physostegia*
P	Bugloss	*Anchusa*	P	Feverfew	*Pyrethrum*
P	Burning bush	*Dictamnus*	C	Firethorn	*Pyracantha*
S	Bush honeysuckle	*Diervilla*	W	Flag	*Iris*
S	Butcher's broom	*Ruscus*	P	Flea-bane	*Erigeron*
W	Buttercup, water	*Ranunculus*	P	Foam flower	*Tiarella*
R	Calamint	*Calamintha*	S	Formosa	*Leycesteria*
S	Calico bush	*Kalmia*	S	Forsythia	*Forsythia*
R	California fuchsia	*Zauschneria*	P	Foxtail lily	*Eremurus*
B	Californian hyacinth	*Brodiaea*	B	Fritillary	*Fritillaria*
P.S.	Californian tree		S	Fuchsia	*Fuchsia*
	poppy	*Romneya*	P	Funkia	*Hosta*
S	Camellia	*Camellia*	S	Furze	*Ulex*
P	Cape figwort	*Phygelius*	S.C.	Garrya	*Garrya*
P	Carnation	*Dianthus*	R	Germander	*Teucrium*
S	Caryopteris	*Caryopteris*	P	Giant asphodel	*Eremurus*
S	Castor oil plant	*Fatsia*	W	Giant rhubarb	*Gunnera*
R	Catch-fly	*Silene*	B	Gladioli	*Gladiolus*
P	Cat mint	*Nepeta*	P	Globe flower	*Trollius*
R	Cats-ear	*Antennaria*	P	Globe thistle	*Echinops*
C.S.	Ceanothus	*Ceanothus*	B	Glory of the snow	*Chionodoxa*
P.R.	Chalk plant	*Gypsophila*	P	Goat's beard	*Astilbe*
P	Chamomile	*Anthemis*	P	Goat's rue	*Galega*
P	Chelone	*Penstemon*	R	Gold dust	*Alyssum*
S	Chilean glory flower	*Eccremocarpus*	R	Golden drop	*Onosma*
R	Chinese bellflower	*Platycodon*	S	Golden heath	*Cassinia*
S	Chinese hawthorn	*Photinia*	P	Golden rod	*Solidago*
P	Chinese lantern	*Physalis*	S	Gorse	*Ulex*
P	Christmas rose	*Helleborus*	B	Grape hyacinth	*Muscari*
R.S.	Cinquefoil	*Potentilla*	R	Gromwell	*Lithospermum*
R	Clover	*Trifolium*	S.T.	Hawthorn	*Crataegus*
P	Columbine	*Aquilegia*	S	Hazel	*Corylus*
P	Cone flower	*Rudbeckia*	S	Heath	*Erica*
C	Convolvulus	*Convolvulus*	R	Hedgehog broom	*Erinacea*
R	Cotoneaster	*Cotoneaster*	P	Hellebore	*Helleborus*
P.R.	Crane's bill	*Geranium*	R	Herons bill	*Erodium*
B	Crocus	*Crocus*	S	Hibiscus	*Hibiscus*
S	Crown vetch	*Coronilla*	S	Holly	*Ilex*
R	Cup flower	*Nierembergia*	S.C.	Honeysuckle	*Lonicera*
P	Cupid's dart	*Catananche*	R	House-leek	*Sempervivum*
S	Currant	*Ribes*	B	Hyacinth	*Hyacinthus*
S.T.	Cypress	*Cupressus*	S.C.	Hydrangea	*Hydrangea*
B	Daffodil	*Narcissus*	R	Immortelle	*Anaphalis*
R	Daisy	*Bellis*	S.T.	Indian bean tree	*Catalpa*
S	Daisy tree	*Olearia*	B	Iris, bulbous	*Iris, bulbous*
S	Daphne	*Daphne*	S	Irish heath	*Daboecia*
P	Day lily	*Hemerocallis*	C	Ivy	*Hedera*
S	Deutzia	*Deutzia*	P	Jacob's ladder	*Polemonium*
B	Dogs tooth violet	*Erythronium*	C	Japanese allspice	*Chimonanthus*
S.T.	Dogwood	*Cornus*	J	Japanese cedar	*Cryptomeria japonica*
C	Dutchman's pipe	*Aristolochia*	S.C.	Japanese quince	*Chaenomeles*

81

	ENGLISH	LATIN		ENGLISH	LATIN
C	Jasmine	*Jasminum*	C	Myrtle	*Myrtus*
S	Jasmine box	*Phillyrea*	B	Narcissus	*Narcissus*
S	Jerusalem sage	*Phlomis*	R	New Zealand burr	*Acaena*
C	Jessamine	*Jasminum*	S	New Zealand flax	*Phormium*
S	Jews mallow	*Kerria*	P	Obedient plants	*Physostegia*
R.S.	Juniper	*Juniperus*	P	Old man	*Artemisia*
P	Kansas gay feather	*Liatris*	S	Oleaster	*Elaeagnus*
R	Kidney vetch	*Anthyllis*	P	Orange sunflower	*Heliopsis*
P.W.	Kingcup	*Caltha*	P	Oriental poppy	*Papaver orientale*
P	Knapweed	*Centaurea*	P	Paeony	*Paeonia*
R	Knot wort	*Polygonum*	C	Passion flower	*Passiflora*
T	Laburnum	*Laburnum*	S	Pearl bush	*Exochorda*
R	Lady's slipper	*Cyananthus*	R	Perennial candytuft	*Iberis*
S	Laurel, variegated	*Aucuba japonica*	P	Perennial larkspur	*Delphinium*
S	Lavender	*Lavandula*	S	Periwinkle	*Vinca*
S	Lavender cotton	*Santolina chamaecyparissus*	R	Persian candytuft	*Aethionema*
R	Leadwort	*Certatosigma*	P	Peruvian lily	*Alstroemeria*
P	Leopard's bane	*Doronicum*	P	Phlox	*Phlox*
S	Lilac	*Syringa*	R	Pimpernel	*Anagallis*
B	Lilies, bulbous	*Lilium*	P	Pincushion flower	*Scabiosa*
B	Lily of the field	*Sternbergia*	P	Pinks	*Dianthus*
P	Lily of the Nile	*Agapanthus*	T	Plane	*Platanus*
P	Lily of the valley	*Convallaria majalis*	P	Plantation lily	*Hosta*
T	Lime	*Tilia*	S	Plumbago	*Ceratostigma*
T	Linden tree	*Tilia*	P	Plume poppy	*Bocconia*
S	Ling	*Calluna*	W	Pond-lily	*Nuphar*
T	London plane	*Platanus*	T	Poplar	*Populus*
P	Loosestrife	*Lysimacha* or *Lythrum*	P	Prairie mallow	*Sidalcea*
W	Loosestrife, water	*Lysimachia*	W	Prickly rhubarb	*Gunnera*
R	Lung-wort	*Pulmonaria*	S	Privet	*Ligustrum*
P	Lyre flower	*Dicentra*	P	Purple cone flower	*Echinacea*
S	Magnolia	*Magnolia*	P	"Purple" loosestrife	*Lythrum*
S	Mahonia	*Mahonia*	S	Ragwort	*Senecio*
T	Maple	*Acer*	R	Rampion	*Phyteuma*
W	Marigold, March	*Caltha*	P	Red hot poker	*Kniphofia*
W	Marsh marigold	*Caltha*	R	Rock Cress	*Arabis*
S	Marsh rosemary	*Andromeda*	R	Rock Cress	*Aubrietia*
S.T.	May	*Crataegus*	R	Rock forget-me-not	*Omphalodes*
P	Meadow rue	*Thalictrum*	R	Rock jasmine	*Androsace*
B	Meadow safron	*Colchicum*	R	Rock pinks	*Dianthus*
S.P.	Meadow sweet	*Spiraea*	S	Rock rose	*Cistus*
S	Medlar	*Mespilus*	T	Rose acacia	*Robinia*
S	Mexican orange blossom	*Choisya*	S	Rosemary	*Rosmarinus*
P	Michaelmas daisy	*Aster* (perennial)	S	Rose of Sharon	*Hypericum*
R	Mint	*Mentha*	T	Rowan	*Sorbus*
S	Mock orange	*Philadelphus*	W	Rush	*Juncus*
W	Monkey flower	*Mimulus*	W	Rush, flowering	*Butomus*
P	Monk's hood	*Aconitum*	C	Russian vine	*Polygonum baldschuanicum*
B	Montbretia	*Crocosmia*			
R	Moon-wort	*Soldanella*	P	Sage	*Salvia*
P	Moses in the bullrushes	*Tradescantia*	P	St Bernard's lily	*Anthericum*
T	Mountain ash	*Sorbus*	S	St Peter's wort	*Symphoricarpus*
R	Mountain avens	*Dryas*	R	Sand-wort	*Arenaria*
S	Mulberry	*Morus*	R	Satin-flower	*Sisyrhinchium*
P	Mullein	*Verbascum*	P	Satin-leaf	*Heuchera*
			R	Saxifrage	*Saxifraga*

	ENGLISH	LATIN		ENGLISH	LATIN
P	Saxifrage, giant	*Bergenia*	T	Sweet gum	*Liquidambar*
P	Scabious	*Scabiosa*	S.T.	Tamarisk	*Tamarix*
S.T.	Sea buckthorn	*Hippophae*	P	Thrift	*Armeria*
P	Sea holly	*Eryngium*	R	Thyme	*Thymus*
R	Sea lavender	*Acantholimon*	B	Tiger flower	*Tigridia*
P	Sea lavender	*Limonium*	P	Torch lily	*Kniphofia*
P	Sea pink	*Armeria*	T	Tree of heaven	*Ailanthus*
P	Shasta daisy	*Chrysanthemum maximum*	P	Trumpet flower	*Incarvillea*
			B	Tulip	*Tulipa*
R	Skull cap	*Scutellaria*	P	Valerian	*Centranthus*
S.T.	Smoke tree	*Rhus cotinus*	S	Veronica	*Veronica (Hebe)*
P	Sneeze weed	*Helenium*	P	Vervain	*Verbena*
S	Snowberry	*Symphoricarpos*	S.T.	Viburnum	*Viburnum*
B	Snowdrop	*Galanthus*	C	Vine	*Vitis*
S.T.	Snowdrop tree	*Halesia*	C	Virginia creeper	*Vitis*
B	Snowflake	*Leucojum*	R	Virginian cowslip	*Mertensia*
R	Snow-in-summer	*Cerastium*	W	Water buttercup	*Ranunculus*
R	Soap wort	*Saponaria*	W	Water forget-me-not	*Myosotis*
P	Solomons seal	*Polygonatum*	W	Water hawthorn	*Aponogeton*
S	Southernwood	*Artemisia*	W	Water-lily	*Nymphaea*
S	Spanish broom	*Spartium*	W	Water violet	*Hottonia palustris*
P	Speedwell	*Veronica*	P	Welsh poppy	*Meconopsis*
P	Spiderwort	*Tradescantia*	S	Whin	*Ulex*
S	Spindle tree	*Euonymus*	T	White beam	*Sorbus*
B	Spire lily	*Galtonia*	R	Whitlow grass	*Draba*
B	Spring meadow saffron	*Bulbocodium*	S	Whortleberry	*Vaccinium*
P	Spurge	*Euphorbia*	S.T.	Willow	*Salix*
B	Squill, blue	*Scilla*	W	Willow herb	*Epilobium*
B	Squill, striped	*Pushkinia*	P	Windflower	*Anemone*
B	Star of Bethlehem	*Ornithogalum*	B	Winter aconite	*Eranthus*
B	Star tulip	*Calochortus*	P	Winter cherry	*Physalis*
R	Starwort	*Aster*	C	Wisteria	*Wistaria*
R	Stone-crop	*Sedum*	S	Witch hazel	*Hammaelis*
T	Strawberry tree	*Arbutus*	R	Wood-ruff	*Asperula*
S.T.	Sumach	*Rhus typhina*	P	Wormwood	*Artemisia*
S	Summer fir	*Artemisia*	P	Yarrow	*Achillea*
R	Summer starwort	*Erinus*	T.S.	Yew	*Taxus*
P	Sunflower	*Helianthus*	S	Yucca	*Yucca*
R	Sun rose	*Helianthemum*	P	Zebra grass	*Miscanthus*
W	Sweet flag	*Acorus*	B	Zephyr flower	*Zephyranthes*

A GUIDE TO PROPAGATING METHODS OF PERENNIALS, CLIMBERS, TREES AND SHRUBS WHICH MAY BE INCREASED AT HOME.

How to use the Guide

First   Find the Latin name of the plant you wish to increase. (See index.)

Second   The first column tells you whether it is a tree, shrub, climber, etc, as a check that you have the right plant.

Third   Read the methods recommended in the code given below, and note the recommended season and special details.

The Code of Methods

S	–	Seed in protection	CL –	Cuttings, leafy or half-ripe
SH	–	Seed, hardy, sown outdoors	CS –	Cuttings, soft under cover
SU	–	Suckers	CR –	Cuttings of roots
D	–	Division	L –	Layers
CH	–	Cuttings, hard or ripe, outdoors	AL –	Air layering

NOTE: The methods recommended are those we feel most likely to be successful with a newcomer. However, almost any plant can be increased by more than one method. Conditions vary from place to place and garden to garden – not to mention from gardener to gardener! If the first does not succeed, by all means try another method. Almost any plant in the list can be increased, for example, by cuttings of some sort. The season recommended too is only a guide. Much will depend on your local climate. The work may well need doing later in the north, earlier in the south or west.

In the notes, FRAME means a heated one. COLD FRAME means glass protection without heat. KEEP CLOSE means to give a humid atmosphere.

	LATIN	ENGLISH	METHOD		SEASON	REMARKS
S	*Abelia*		CLh		July	Frame
			CH		November	Cold frame
C	*Abutilon*		CL		June-July	Frame
			CH		November	Cold Frame
T	*Acacia*		CL			After flowering
R	*Acaena*	New Zealand burr	D		Summer or Autumn	Easy
R	*Acantholimon*	sea lavender	L			Sandy soil
			CLh		Summer	In Pots
P	*Acanthus*	bear's breeches	D		Spring	
			CR		October	
T	*Acer*	maple	L		October	Not easy
P	*Achillea*	yarrow	D		Spring or Autumn	
			CS		Spring	
P	*Aconitum*	monk's hood	D		Spring or Autumn	
W	*Acorus*	sweet flag	D		Spring	
R	*Aethionema*	persian candytuft	CL		Summer	
P	*Agapanthus*	lily of the Nile	D		Spring	
			SU		Spring	
T	*Ailanthus*	tree of heaven	CR		Spring	
			SU		Spring	
S	*Akebia*		CL		Summer	Cold frame
B	*Allium*		D		Autumn	
P	*Alstroemeria*	peruvian lily	D		Spring or Autumn	

	LATIN	ENGLISH	METHOD	SEASON	REMARKS
R	*Alyssum*	gold dust	CS	Spring	
			SH	March	
B	*Amaryllis*	belladonna lily		July	Offsets (see text) not more frequently than four years.
S	*Amelanchier*		L	Autumn	Older wood
			SU		As available
C	*Ampelopsis*		CL	July	Frame
			CH	October	Cold frame
R	*Anagallis*	pimpernel	D	May	Into pots
			CS	May	
P	*Anaphalis*	immortelle	D	May	
			CS	May-June	
P	*Anchusa*	bugloss	D	Autumn or Spring	
			CR	Spring	4in in open or cold frame
S	*Andromeda*	Marsh rosemary	L	August-September	
			D	Autumn	
R	*Androsace*	rock jasmine	D	Spring	
			SU	August	Into pots
PB	*Anemone*	windflower	CR	Spring	(or into pots. Autumn)
			D	Spring	Tubers lift July or plant Autumn
R	*Antennaria*	cat's ear	D	Spring or late summer	
P	*Anthemis*	chamomile	D	Spring or Autumn	
			CS	Spring	
P	*Anthericum*	St Bernard's lily	D	Spring	
B	*Antholyza*		D	August	Replant October
R	*Anthyllis*	kidney vetch	CLh	Summer	
W	*Aponogeton*	water hawthorn	D	Spring	
P	*Aquilegia*	columbine	D	Spring	
			SH	Spring	
R	*Arabis*	rock cress	D	Autumn or late Spring	
			CL	Summer	
S	*Aralia*		SU	Spring usually	Depends on varieties
			CR	Autumn	
T	*Arbutus*	strawberry tree	L	Autumn	Difficult
S	*Arctostaphylos*	bear berry	CL	Autumn	Tip cuttings in frame
R	*Arenaria*	sand wort	D	Autumn	
			CL	Summer	
C	*Aristolochia*	dutchman's pipe	CLh	Summer	In frame
P	*Armeria*	thrift: sea pink	D	Autumn or Spring	
			CL	Summer	
S	*Aronia*		CH	Late Autumn	
P	*Artemisia*	wormwood: old man	D	Autumn	After flowering
W	*Aruncus*		D	Spring	
W	*Arundinaria*	bamboo	D	April-May	
R	*Asperula*	wood-ruff	D	Spring	
			CL		After flowering
P	*Aster, perennial*	michaelmas daisy	D	Spring	
			CS	Spring	
R	*Aster*	starwort	D	Spring and Autumn	
			CS	Summer	
P	*Astilbe*	goat's beard	D	Spring	Whilst dormant, every 5 years

	LATIN	ENGLISH	METHOD	SEASON	REMARKS
R	*Aubrietia*	rock cress	D	Autumn	
			CL	Summer	Sandy soil
			S	May	
S	*Aucuba*	variegated laurel	L		Easy
			CH	Autumn	Easy
S	*Azaleas*		L		As rhododendron
S	*Azara*		L	Spring or Summer	Especially variegated sorts.
			CH	Autumn	Cold frame
B	*Begonia, tuberous*	begonia, tuberous	D	When dormant	Cut bulbs in pieces, each with a bud, at planting
R	*Bellium*		D	August	
C	*Berberidopsis*		L	Autumn	
			CS	Spring	
S	*Berberis*	barberries	CLh	Autumn	6in long cold frame
			D	Spring	in some sorts
			SU	Spring	in some sorts
			L	Autumn	in some sorts
P	*Bergenia*	giant saxifrage	D	Autumn	or after flowering
C	*Bignonia*		L	Spring	
P	*Bocconia*	plume poppy	D	Spring	
B	*Brodiaea*	californian hyacinth	D	Autumn	
S	*Buddleia*		CH	October-November	8in long
			CS	July	3in long
			CL	Late Summer	
B	*Bulbocodium*	spring meadow saffron	D	Autumn	When dormant by offsets
W	*Butomus*	flowering rush	D	Spring	
S	*Buxus*	box	CHh	September	Cold frame
			D	Autumn or late Spring	(of dwarf forms for edging)
R	*Calamintha*	calamint	D	Autumn or early Spring	
			CL	Summer	
W	*Calla palustris*	bog arum	D	Spring	
S	*Calluna*	ling	L		
			CL		Short cuttings
B	*Calochortus*	star tulip	D	Autumn	
PW	*Caltha*	marsh marigold kingcup	D	Spring	
P	*Calycanthus*	allspice	L	Spring and Autumn	
C	*Camellia*	camellia	L	September	Or leaf cuttings with stalk and a bud around pot of sandy peat.
R	*Campanula*	bell flower	D	Spring or Autumn	
			CS	Spring	Before flowering
S	*Caryopteris*		CS	Summer	
			CH	November	Cold frame
P	*Catananche*	cupid's dart	D	Spring	
			S	Spring	
S	*Cassinia*	golden heath	CLh	Autumn	4in Cold frame
S	*Cassiope*		CLh	August	in shade: peaty soil
ST	*Catalpa*	indian bean tree	CL	Summer	
			CR	Autumn	$1\frac{1}{2}$in horizontal in pots
			L	Autumn	
C	*Ceanothus*		L		some sorts

	LATIN	ENGLISH	METHOD	SEASON	REMARKS
			CL	Summer	cold frame
			CH	October	cold frame
	Celastrus		CL	Late Summer	
P	*Centaurea*	knapweed	D	Spring or Autumn	easy
			S	Spring	easy
P	*Centranthus*	valerian	D		
R	*Cerastium*	snow -in-summer	D	September	
			CL	Summer	
S	*Ceratostigma*	plumbago	CL	Late Spring	Sandy soil: Cold frame
SC	*Chaeonomeles*	japanese quince	L	September	
			CHh	October	cold frame
ST	*Chamaecyparis*	false cypress	CLh	October	cold frame
			CS	Summer	keep close
C	*Chimonanthus*	Japanese allspice	L	Autumn	
B	*Chionodoxa*	glory of the snow	S		in frames, but take 2 years to develop.
S	*Choisya*	mexican orange blossom	CHh	October	4in cold frame
P	*Chrysanthemum maximum*	shasta daisy	D	Spring or Autumn	easy
			SU		as available
P	*Cimicifuga*	bug bane	D	Spring	
S	*Cistus*	rock rose	CLh	Summer	
			CH	October	cold frame
C	*Clematis*		LC	Spring	
SC	*Clerodendron*		CR	Winter	Thick pieces 2½in in sand planted outdoors April-May
S	*Clethra*		L	Autumn	
			CL	Summer	
B	*Colchicum*	meadow saffron	D	June	
S	*Colutea*	bladder senna	CL	Summer	Cold frame
P	*Convallaria majalis*	lily of the valley	D	September	
CR	*Convolvulus*		D	April	
			CL	Summer	
P	*Coreopsis*	tick-seed sunflower	D	Spring or Autumn	
			SU		as available
			CS	Early Summer	
ST	*Cornus*	dogwood	L	July	
S	*Corokia*		L	Autumn	
R	*Coronilla*	crown vetch	CL	Summer	
			L	Summer	Occasionally
R	*Cortusa*		D	March	
S	*Corylus*	hazel	L	Autumn	
			SU		If available
C	*Cotoneaster*		CH	November	
			CL	Summer	Some sorts in frame
R	*Cotula*		D	Autumn	
R	*Cotyledon*		CL	Summer	Easy
ST	*Crataegus*	thorn, May			Varieties are grafted on common hawthorn by buds – July as roses.
B	*Crinum*		D	March	
P	*Crocosmia*	montbretia	D	Autumn	In cold parts replant in spring.
B	*Crocus*	crocus	D	Summer	After foliage has died
T	*Cryptomeria*	japonica	CHh	Autumn	in frame
ST	*Cupressus*	cypress	CS	Spring & Summer	

	LATIN	ENGLISH	METHOD	SEASON	REMARKS
			CL	Autumn	
			D	Spring or Autumn	
R	*Cyananthus*	lady's slipper			
C	*Cydonia*	(see chaenomeles)			
C	*Cytisus*	broom	CLh	August	Side shoots
S	*Daboecia*	irish heath	CH	October	under light
S	*Daphne*		CL	Summer	
			L	Autumn	
P	*Delphinium*	perennial larkspur	CS	Spring	
			D	Summer	After flowering but difficult.
S	*Desfontainea*		CH	October	Cold frame
S	*Deutzia*		SU	Spring	If available
			CH	November	8in long, protect in cold areas.
R	*Dianthus*	rock pinks	L	Summer	
		carnation	CL	Spring	Cutting made from growing tips.
		pinks	D	Autumn	
P	*Dicentra*	bleeding heart:	D	Spring	
		lyre flower	CR	Spring	
P	*Dictamnus*	burning bush	S	Spring	
P	*Dielytra*	(see dicentra)			
S	*Diervilla*	bush honeysuckle	CL	July	Cold frame
			CH	November	6in long in open
R	*Dodecatheon*	american cowslip	D	April or August	
P	*Doronicum*	leopard's bane	D	Autumn or Winter	
R	*Draba*	whitlow grass	SU	Spring	
R	*Dryas*	mountain avens	SU		As available
			CLh	August	
S	*Eccremocarpus*	chilian glory flower	S	Autumn	Sow outdoors in boxes exposed to cold. Bring into warmth Feb. to start germination. Easy.
P	*Echinacea*	purple cone flower	D	Spring	*Rudbeckya* is alternative name.
P	*Echinops*	globe thistle	D	Spring or Autumn	
			CR	Spring	Cold frame
S	*Elaeagnus*	oleaster	L		
			CH	Autumn	4in long cold frame
S	*Enkianthus*		L	Spring	Young shoots
W	*Epilobium*	willow herb	D and Su	Spring	
P	*Epimedium*	barren wort	D	Autumn	After flowering
B	*Eranthis*	winter aconite	D	October-November	
P	*Eremurus*	giant asphodel: foxtail lily	SH	Autumn	Very slow, not easy
S	*Erica*	heath	L	Spring	Methods vary with varieties.
			D	Spring	
			CL	Spring	1½in long in pots kept close.
P	*Erigeron*	fleabane	D	Autumn to Spring	
R	*Erinacea*	hedgehog broom	CL	Autumn	After flowering, into pots
R	*Erinus*	summer starwort	D	Late Summer	Easy
R	*Eriogonum*		D	Spring	
			CLh	Summer	
R	*Erodium*	heron's bill	S	Spring	
P	*Eryngium*	sea holly	D	Spring or Autumn	

	LATIN	ENGLISH	METHOD	SEASON	REMARKS
			CR	Spring	
B	Erythronium	dogs-tooth violet	D	August	Offsets
C	Escallonia		CH	October	Protect in cold areas.
C	Euonymus	spindle-tree	L	Summer	Younger wood
			CL	Summer	Keep close
			D	Spring	some varieties
P	Euphorbia	spurge			A large group propagated by different means according to habit of growth.
S	Exochorda	pearl bush	L	Autumn	
			SU		As available.
S	Fatsia		CL	Spring or Summer	
S	Forsythia		CH	October	6in in open easy
			CL	Summer	Keep close at first
S	Fothergilla		L	Autumn	Slow
			SU		As available
B	Fritillaria	fritillary	D	August	Offsets
S	Fuchsia		CS	Any season	Easy
P	Funkia	see hosta			
P	Gaillardia	blanket flower	D	Spring	
			SU	Spring	
			S		Variable from seed
B	Galanthus	snowdrop	D	Autumn	
P	Galega	goat's rue	D	Spring or Autumn	
B	Galtonia	spire lily	D	Autumn	Offsets
CS	Garrya		L	Autumn	
			CHL	Autumn	Protect in frame
S	Gaultheria	wintergreen	D	Spring or Autumn	Varies with variety
			L	Summer	
SR	Genista	broom, furze	CL	September	2–4in sandy soil cold frame
R	Gentiana	gentians	D		After flowering. Not easy.
			SH		As soon as ripe Expose to frost after sowing.
PR	Geranium	crane's bill	D	Spring or Autumn	
			SU	Summer or Autumn	
P	Geum	avens	D	Spring	
			SH	Summer	In frame (depends on varieties, try both)
B	Gladiolus	gladioli	D		Offsets when dormant
S	Griselinia		L	Summer	
			CHL	October	In frame, easy
			CS	Summer	
W	Gunnera	prickly rhubarb giant rhubarb	D	Spring	
PR	Gypsophila	chalk plant	CS		Before flowering
			D	Spring or Autumn	Not easy
			CR	Autumn	3in long $\frac{1}{4}$in thick, sandy soil. Not easy.
T	Halesia	snowdrop tree	L	Autumn	
			CR	Spring	In frame
S	Hamamelis	witch hazel	L	Spring	
C	Hedera	ivy	CL	Spring or Summer	Easy
			CH	Autumn	Easy

	LATIN	ENGLISH	METHOD	SEASON	REMARKS
P	*Helenium*	sneeze weed	D	Late Autumn or Winter	
			SU		As available
R	*Helianthemum*	sun rose	CL	After May	Easy: cold frame if available
P	*Helianthus*	sunflower	D	Autumn or Winter	
			CS	Spring	Easy
P	*Heliopsis*	orange sunflower	D	Spring	
P	*Helleborus*	hellebore: christmas rose	D	Spring or Autumn	
P	*Hemerocallis*	day lily	D	Spring or Autumn	
P	*Heuchera*	alum root: satin leaf	D	Spring	
S	*Hibiscus*		CL	Summer	Not easy
ST	*Hippophae*	sea buckthorn	L	Summer	
			SU		As available
P	*Hosta*	Funkia: plantain lily	D	Spring	Slice clumps with spade
W	*Hottonia palustris*	water violet	D	Spring	
R	*Houstonia*	bluets	D	Autumn or Spring	
R	*Hutchinsia*		D	Spring	
B	*Hyacinthus*	hyacinths			Not easy. Special treatment to produce Offsets is given by lifting dormant bulbs and cutting base away to expose the scale bases. Packed *dry*, cut end up, and dipped in dry lime, in box, the ends callus over. Then, warm, moist conditions produce bulblets for planting out.
C	*Hydrangea*		CS	Spring	
			CL	Spring and Summer	See text.
S	*Hypericum*	rose of Sharon old man's beard	D		After flowering. Easy
R	*Iberis*	perennial candytuft candytuft	S	Spring	Easy
S	*Ilex*	holly	L	Autumn	Slow
			CL	Summer	
P	*Incarvillea*	trumpet flower	D		After flowering
W	*Iris*	flag	D	Summer	
B	*Iris, bulbous*	iris, bulbous	D	July-August	Offsets: divide early sorts when dormant
P	*Irises (bearded)*		D	Autumn or Spring	Immediately after flowering
C	*Jasminum*	jasmine: Jessamine	CH	November	
			CL	Summer	Close: new season's growth
W	*Juncus*	rushes	D	Spring	
RS	*Juniperus*	juniper	CH	August to October	In frame
S	*Kalmia*		L	Autumn	
S	*Kerria*	jew's mallow	D	Autumn	
			CS	July	Tips of growth
P	*Kniphofia*	red hot poker: torch lily	D	Spring	
T	*Koelreuteria*		L	Autumn	

	LATIN	ENGLISH	METHOD	SEASON	REMARKS
			CL	Spring	Sandy soil
S	Kolkwitzia		CL	Autumn	Young growth. In frame
T	Laburnum		L	October-November	
			CHh	Autumn	9–12in long
P	Lathyrus	everlasting pea	D	Spring	
S	Laurus	bay laurel	L	Spring & Summer	
			CHh	Early Winter	In frame
S	Laurustinus				See Viburnum
P	Lavandula	lavender	CS	All growing Season	
			CL		
S	Leptospermum		CL	Summer	2–3in long in frame
B	Leucojum	snowflake	D	September	Offsets
S	Leucothoe		D	Autumn	
			L	Autumn	
R	Lewisia		D	Spring	If growth allows
S	Leycesteria	formosa	CH	Autumn	Not easy
			D	Spring	When growth allows
P	Liatris	Kansas gay feather	D	Spring	
			SU	Spring	
S	• Ligustrum	privet	CLh	Spring or Summer	
			CH	Autumn or Winter	3in long. Close: easy 10in long in open
B	Lilium	lilies, bulbous			Scale division, see text
P	Limonium	sea lavender: statice	D	Spring	
R	Linaria	toad flax	D	March	
			SH		
P	Linum	flax	D	Spring	
			CLh	Summer	
T	Liquidambar	sweet gum	L	Autumn	Slow
R	Lithospermum	gromwell	CL	Summer	Use peat/sand mix
SC•	Lonicera	honeysuckle	CS	June	Tips of new growth
			CH	Autumn	6in cold frame. (Nitida variety in the open)
P	Lupinus	lupins, perennial	D		With care
			CS	April	Into pots
			SH	Early Summer	Not true from seed
R	Lychnis	campion	D	Spring or Autumn	Some sorts
			SH	Spring	Some sorts
W	Lysimacha	loosestrife, water	D	Spring	Easy
P	Lythrum	purple loosestrife	D	Spring	Easy
C	• Magnolia		AL	Autumn	Not quick
			L	Autumn	Not quick
S	Mahonia		CH	Autumn	Firmly in pots sandy soil
			CL	June	6in keep close
T	Malus	apple			Most varieties are grafted, a process not covered in this volume. Flowering sorts are difficult for this reason
R	Mazus		D	Spring or Autumn	
P	Meconopsis	welsh poppy	SH	Autumn	As available
			S	Spring	With heat if possible
R	Mentha	mint	D	Spring	
R	Mertensia	Virginian cowslip	D	Autumn	
S	Mespilus	medlar	L	Spring or Autumn	
P	Mimulus	monkey flower, musk	D	Spring	Easy

	LATIN	ENGLISH	METHOD	SEASON	REMARKS
			CL	Summer	Keep moist
P	Miscanthus	zebra grass	D	Early Spring	
P	Monarda	bergamot	D	Spring	
			CS	Spring	Basal shoots
R	Morisia		CR	June-August	1in long in moist sand
S	Morus	mulberry	CHh	Autumn	
B	Muscari	grape hyacinth	D	Autumn	
W	Myosotis	water forget-me-not	D	Spring	Easy
			S	Spring	
L	Myrtus	myrtle	CL	Early Summer	Cold frame
B	Narcissus	daffodil narcissus	D		
P	Nepeta	cat mint	D	Spring	
			CS	Summer or Autumn	3in tips of new growth
R	Nierembergia	cup flower	D	August	(N. repens & N. rivularis)
			CL	Summer	(N. caerulea & N. frutescens)
W	Nuphar	pond lily	D	Spring	Into pots submerged in shallow water
W	Nymphaea	water lily	D	May	Each part having one 'eye' in pots plunged into shallow water
R	Oenothera	evening primrose	D	Spring	
			CLh	June-August	Easy
S	Olearia	daisy tree	CHh	October	In frame
			CL	Summer	Easy
R	Omphalodes	rock forget-me-not	D	Spring or Autumn	
R	Onosma	golden drop	CL	July	After flowering
B	Ornithogalum	star of Bethlehem	D	Autumn	
S	Osmanthus		L	Autumn	
			CHh	October	Into sandy peat
P	Paeonia	paeony	D	Spring	Detach young rooted "crowns"
			D	Autumn	Divide leaving each piece with an "eye"
			CL	Summer	
P	Papaver orientale	oriental poppy	CR	September	4in long in sandy soil Protect in winter
C	Passiflora	passion flower	CL	Summer	Young shoots in frame
P	Penstemon	beard tongue: chelone	S	June	
			CL	Autumn	In frame for winter
S	Pernettya		L	Autumn	Young shoots
			D	Spring	
			CL	Summer	
P	Perowskia		CL	Summer	Keep close: sandy soil
S	Philadelphus	mock orange	CS	June-July	3in long
			CL	Summer	Side shoots
			CHh	Summer-Autumn	9in light soil in open
S	Phillyrea	jasmine box	CL		Easy. Light sandy soil
S	Phlomis	Jerusalem sage	CS	Summer	Easy
			CH	October	3in long in frame
PR	Phlox	phlox	D	Spring or Autumn	
			SU		As available
S	Phormium	New Zealand flax	D	Spring	
S	Photinia	chinese hawthorn	CL	Summer	
S	Phygelius	Cape fig-wort	CS	Spring-early Summer	

	LATIN	ENGLISH	METHOD	SEASON	REMARKS
P	*Physalis*	winter cherry:	D	Spring	Natural increase is rapid
		chinese lantern	SU	Spring	
P	*Physostegia*	false dragon's head:	D	Autumn	
		obedient plants	SU	Spring	
R	*Phyteuma*	rampion	D	Autumn	
S	*Piptanthus*	evergreen laburnum	CLh	August	Do not expose pith at cutting's base
S	*Pittosporum*		CLh	July	Keep close
T	*Platanus*	plane	L	Autumn	(*P. orientalis*)
			CHh	Autumn	10in long. Slow
P	*Platycodon*	balloon flower	D	Spring	
P	*Polemonium*	jacob's ladder	D	Spring or Autumn	
P	*Polygonatum*	Solomon's seal	D	Spring or Autumn	
R	*Polygonum*	knot wort	D	Spring	
		russian vine	CL	Summer	
			CHh	Summer	In pots in cold frame for *Baldschuanicum* (Russian vine)
W	*Pontederia*	cordata	D	Spring	
T	*Populus*	poplar	CH	Autumn	8in long light soil, easy
			CLh	Summer	
S	• *Potentilla*	cinquefoil	D	Spring or Autumn	
			CLh	July	Small cuttings kept close
			CH	October	Cold frame
PRBI	*Primula*		D	Summer	After flowering
ST	*Prunus*				This group includes many varieties of plum, cherry, etc, many of which are grafted on to wild cherry stocks which is not easy. CH sometimes succeed.
R	*Pulmonaria*	lung-wort	D	Spring or Autumn	Easy
B	*Pushkinia*	striped squill	D	October	Offsets
C	*Pyracantha*	firethorn	CLh	August	Small cuttings prefer cold frame
			CH	Autumn	
P	*Pyrethrum*	feverfew	D	Spring	Or after flowering
			SU		After flowering
T	*Pyrus*	pear			This group include many that are propagated by grafting, a technique not covered in this volume
R	*Ramonda*		SU	March	Leaf cuttings Spring 1in long with bud at base in sandy compost
R	*Ranunculus*		SH	May	
W	*Ranunculus*	water buttercup	D	Spring	
R	*Raoulia*		D	Late Summer or Autumn	
S	*Raphiolepis*	indian hawthorn	CLh	September	Keep close
S	*Rhododendron*	rhododrendron, azalea	L		See text
			AL		Of some varieties, especially the evergreen azalea group. Always use compost containing soil from near roots of a growing specimen
			CH		

	LATIN	ENGLISH	METHOD	SEASON	REMARKS
ST	*Rhus cotinus*	smoke tree	L	Autumn	
ST	*Rhus typhina*	sumach	CR	Winter	1½in long in pots
S	*Ribes*	currant	CH	Autumn	6in long, easy
T	*Robinia*	rose acacia	SU		When available
			CR	Spring	3in long in open ground not easy
W	*Rodgersia*		D	Spring	
			CR	Spring	
PS	*Romneya*	californian tree poppy	CR	Winter	1in long horizontal in sandy compost
B	*Romulea*		D	Autumn	
S	*Rosa*	roses			See text
PS	*Rosmarinus*	rosemary	L	Spring	
			CL	Spring	
CS	*Rubus*	blackberry family	D		Where growth allows
			L	Spring, Summer	of tips of arching stems Easy
P	*Rudbeckia*	cone flower	D	Spring or Autumn	
S	*Ruscus*	butcher's broom	D	Spring	
W	*Sagittaria*	arrow-head	D	Spring	Makes natural runners
ST	*Salix*	willow	L	Spring or Summer	
			CL		All very easy
			CH		
P	*Salvia*	sage	D	Spring	
			SU	Spring	
S	*Sambucus*	elder	CHh	Late Summer	Do not expose pith at the heel
			CL	July	Keep close
S	*Santolina chamaecyparissus*	lavender cotton	CS	July	
R	*Saponaria*	soap-wort	D	Autumn	
			CS	Summer	
R	*Saxifraga*	saxifrage	D	March	A wide variety of plants in this group
			CL	Summer	Division is usual
P	*Scabiosa*	scabious: pincushion flower	D	Spring	
B	*Scilla*	bluebell, squill	D	Autumn	
R	*Scutellaria*	skull cap	CS	Summer	
PR	*Sedum*	stone-crop	D	Spring or Autumn April	Very easy some even grow by being broken and scattered over ground surface open ground. Easy.
			SH		
R	*Sempervivum*	house-leek	SU	May	
			SU	September	Into pots, replanted out in following Spring
S	*Senecio*	groundsel, ragwort	CL	August	
			CH	Autumn	Cold frame
P	*Sidalcea*	prairie mallow	D	Spring or Autumn	Easy
R	*Silene*	catch-fly	D	July-August	
			CL	Summer	
R	*Sisyrhinchium*	satin-flower	D	Spring or Autumn	
S	*Skimmia*		L	Autumn	
			CH	Autumn	Cold frame slow
R	*Soldanella*	moon-wort	D	June	

	LATIN	ENGLISH	METHOD	SEASON	REMARKS
P	*Solidago*	golden rod	D	Autumn to Spring	Easy
B	*Sparaxis*		D	Autumn	
S	*Spartium*	spanish broom	CH	Autumn	Cold frame
R	*Spiraea*	meadowsweet	D	Spring	Herbaceous kinds. Make certain one leaf bud left on each portion
			SU	Spring	Shrubby sort
			CH	Autumn	8in long in open
S	*Staphylea*	bladder nut	CL	Summer	
			L	Autumn	
S	*Stephanandra*		SU	Spring	
B	*Sternbergia*	lily of the field	D	Autumn	
S	*Stransvaezia*		CL	Summer	Sandy soil
S	*Styrax*		L	Autumn	
S	*Symphoricarpos*	snowberry, St. Peter's wort	D	Autumn	Where growth allows
			SU	Autumn	When available
			CH	October	In open
S	*Symplocos*		CH	Autumn	Light soil
S	*Syringa*	lilac	L	Spring	
			CL	July	Cold frame. Not all varieties succeed. Many are grafted
ST	*Tamarix*	tamarisk	CH	Autumn	8in long in permanent site in open. Shelter during first winter
T	*Taxodium*	swamp cypress	CL	Summer	Slow. Keep moist at all times
TS	*Taxus*	yew	CL	Spring	Terminal shoots. Keep close
C	*Tecoma*		L	Spring	
SP	*Teucrium*	germander	CS	June	
P	*Thalictrum*	meadow rue	D	Spring	As growth starts
			SU	March	
ST	*Thuia (Thuja)*		CLh	September	Side shoots. Cold frame
R	*Thymus*	thyme	D	Spring or Autumn	
			CS	Summer	
P	*Tiarella*	foam flower	D	Spring	
B	*Tigridia*	tiger flower	D	October	But replant in spring
T	*Tilia*	lime: linden tree	L	Summer	But not usually easy
			AL	Summer	
P	*Tradescantia*	spiderwort: Moses in the bullrushes	D	Spring or Autumn	
			CS	Summer	
S	*Tricuspidaria*		CL	Summer	Sandy soil, cold frame
R	*Trifolium*	clover	D	Spring or Autumn	
P	*Trillium*	american wood lily	S		When seed is ripe
P	*Trollius*	globe flower	D	Autumn	
B	*Tulipa*	tulips	D	Summer	Offsets but these arise only from large bulbs. Offsets increased by removing flower HEADS (not stalks) as soon as they open.
R	*Tunica*		CL	July-August	
S	*Ulex*	gorse: furze: whin	CL	July	Sunny position but spray overhead frequently
S	*Vaccinium*	bilberry, cranberry,	L	Autumn	

	LATIN	ENGLISH	METHOD	SEASON	REMARKS
P	*Verbascum*	mullein	D	March	
			CR	Spring	
RP	*Verbena*	vervain	CL	Autumn	Winter indoors
R	*Veronica*	speedwell, hebe	D	Spring or Autumn	
			CS	Summer	
			CLh	Summer	3in long, cold frame
S	*Viburnum*	snowball tree, guelder rose (many varieties)	CL	July-August	Keep close. There are many varieties, some may be increased by CH Autumn or L spring usually easy but slow
S	*Vinca*	periwinkle	D	early Spring	
			CS	Summer	Tips of growth
			CH	Autumn	Cold frame
P	*Viola*	pansy	S	Spring	Treat as Biennial
			D	Autumn	Some varieties
			CS	Summer	
C	*Vitis*	vine: virginia creeper	L	Spring or early Summer	
			CH	October	6in long, cold frame
R	*Wahlenbergia*		D	March	
S	*Weigela*	diervilla	CL	Spring	Light soil
			SU	Spring	As available
C	*Wistaria*		L	Summer	Young shoots
S	*Yucca*		CR	Autumn	2in long, in boxes. Place in heat for winter
			SU		As available
R	*Zauschneria*	californian fuchsia	D	Spring	
			CL	Summer or Autumn	
L	*Zenobia*		L	Autumn	
B	*Zephyranthes*	zephyr flower	D	Autumn	

Power Frequency Magnetic Fields and Public Health

William F. Horton
Saul Goldberg

Electrical Engineering Department
California Polytechnic State University

CRC Press

Boca Raton   New York   London   Tokyo

Library of Congress Cataloging-in-Publication Data

Catalog record is available from the Library of Congress

No claim to original U.S. Government works
International Standard Book Number 0-8493-9420-1
Printed in the United States of America 1 2 3 4 5 6 7 8 9 0
Printed on acid-free paper

This book is dedicated
to the memories of
Bill's father
Arch W. Horton
and
Saul's father
Morris Goldberg
and
Brother-in-law Dr. Joseph G. Smith

Contents

Preface

Our interest in power frequency electromagnetic fields began in 1991 when we carried out a study of the electric and magnetic fields produced by a 500 kV transmission line of the Pacific Intertie. This study led to a sequence of seminars on the characteristics of the magnetic fields emanating from a variety of sources in the electric power delivery system. In preparing and giving these seminars, and responding to questions during and after the seminars, we realized that there is a need for a reference that comprehensively combines:

- significant information on the magnetic field health effects issue

- a description of the principal sources of power frequency magnetic fields

- characterizations of the fields from the principal sources

- principles of magnetic field management

- positions taken by regulatory agencies relative to the magnetic field health effects issue.

We wrote this book with the concept that the material in it should be accessible to readers who are not highly conversant with electromagnetics, field theory or electric utility practices. Rather, our aim was to provide a book that would be available to a non-technical reader but would be equally of interest to an engineer, scientist or technologist. To accomplish this we placed highly technical and/or mathematical developments in appendices, allowing the main body of the book to be somewhat descriptive. This, we believe, permits the non-technical reader to understand the nature of power frequency magnetic fields and their human health consequences without being bogged down with the underlying (and very technical) developments of epidemiological science and electromagnetic field theory. For example, Chapter Two on Magnetic Fields and Human Health Issues is supported by Appendix D—The Science of Epidemiology. Chapter Three, The Nature and Measurement of Power Frequency Magnetic Fields, is supported by Appendix A—The Physics of Magnetic Fields. This pattern continues throughout the book. Our goal in designing the layout of the book was to provide satisfying coverage for both non-technical and technical readers. Wherever appropriate we have used examples to illustrate concepts and exercises at the end of chapters to allow readers to test their understanding of important points. Answers are given for most of the exercises.

In preparing this book it became very clear to us that we all live in a kind of ubiquitous soup of magnetic fields. This soup becomes a little stronger every year. It is only natural to question whether our immersion in this medium, which has existed in the present form for less than 100 years, may produce or promote disease. To date there has been no convincing argument that it does or does not. This opinion may change as more and more epidemiological and biological studies are carried out. If, in fact, it is found that power frequency magnetic fields are bound up with human disease, our society will be forced to make some difficult decisions. As we show in Chapter Eight of this book, field mitigation is both difficult and very costly to achieve. Mitigation on a large scale would be an almost unthinkable economic burden using today's technology. This is reflected in the decisions made by public utility commissions across the United States. Generally, public agencies have chosen to invoke the principle of *prudent avoidance* in dealing with the magnetic field-health effects issue. We trace this development in Chapter Nine and follow with an appendix dealing with prudent avoidance on a personal basis.

We would like to acknowledge the principal sources of the information that we have used in writing this book.

- The transactions of the Institute of Electronic and Electrical Engineers (IEEE).

- The publications of the Electric Power Research Institute (EPRI).

- The publications of a number of public utility commissions but particularly the report of the Electro-magnetic Health Effects Committee, Public Utility Commission of Texas, Austin, Texas, March 1992.

- Pacific Gas and Electric Company, EMF Distribution Guidelines.

We utilized the Southern California Edison Company computer program, FIELDS, to compute magnetic field profiles of transmission and distribution lines. This program is available free of charge from Southern California Edison Company, and a coupon for it is included in the book.

Substation fields were computed using the EPRI computer program SUBCALC.

We also acknowledge the help and encouragement we received from our good friend Larrie Ciano. Mrs. Ciano, a principal of TechWrite, located in San Luis Obispo, California, was invaluable in organizing, editing and correcting the material in this book.

Finally, we hope that this book will serve as a useful reference to a wide audience of readers involved in the magnetic field-health effects issue.

William F. Horton
Saul Goldberg

About the Authors

Dr. William F. Horton and Dr. Saul Goldberg are professors in the Electrical Engineering Department, College of Engineering, California Polytechnic State University, San Luis Obispo, California. Dr. Goldberg is also chairman of the Electrical Engineering Department. Dr. Horton and Dr. Goldberg have published more than 40 technical papers in the subject area of electric power systems. Both are senior members of the IEEE. As principals of the consulting company Power Systems Consultants, located in San Luis Obispo, they have served as consultants to a wide variety of clients in the power industry.

William F. Horton received the Bachelor of Science and Master of Science Degrees (Electrical Engineering) from the California Institute of Technology and the Doctor of Philosophy Degree (Engineering) from the University of California-Los Angeles. His industrial experience includes assignments at Westinghouse Electric Company, Lear-Siegler, Hughes Aircraft, Pacific Gas and Electric Company and San Diego Gas and Electric Company. He has been a professor of Electrical Engineering for 25 years and teaches in the areas of control systems, power systems, electromagnetics and random processes.

Saul Goldberg received the Bachelor of Science Degree (Electrical Engineering) from Fairleigh Dickinson University and the Master of Engineering and Doctor of Philosophy Degree (Engineering) from the University of Florida-Gainesville. His industrial experience includes assignments at Bendix-Eclipse Pioneer Division, Mount Sinai Hospital (Department of Radiology, Miami Beach, Florida), Southern California Edison Company and Pacific Gas and Electric Company. His university teaching spans 26 years, with three years as department chairman; he teaches in the areas of control systems, power systems, biomedical engineering and computer software and hardware.

1 Introduction

If you're reading this book, you're already involved in the raging debate over the possible adverse health effects of electromagnetic fields (EMFs). Electromagnetic fields are produced by the electric power delivery system and the electrical devices (appliances, motors, computing equipment, to name a few) we use on a daily basis. Thus, the prospect that magnetic fields may foster disease is an alarming thought—one that raises such immediate questions in most of us as: What are these invisible forces? How are they produced? Is there conclusive evidence that they are harmful to the human body? How do we protect ourselves against possible harm from them?

1.1 Goal in Writing This Book

Our goal in writing this book is to address the above and other questions about magnetic fields, bringing together in a single reference volume the scientific background, current status of health research, and means to reduce the impact of EMFs in our environment.

1.2 Intended Audiences

For non-technical readers, we believe this book can be a useful tool in developing an understanding of the relationship between EMFs and life in an industrialized society. (The extensive glossary is intended to aid in bridging the gap between the non-technical and technical reader.)

For the engineer or scientist, we have provided the in depth mathematical explanations and exercises necessary for developing a professional level of knowledge of the characterization and measurement of power frequency magnetic fields, including further rigorous development in the Appendices of the physics of magnetic fields and modeling and calculating of transmission line fields.

1.3 Possible Applications for This Book

Designed originally as the basis for a short course at the university level in power frequency magnetic fields, the material in this book is also applicable as a reference for courses in environmental engineering, city and regional planning, and electrical engineering.

Safety officers, public utility commissioners and staff, planners, and public health officials will find resources herein for developing policies and procedures related to human interaction with magnetic fields.

Architects, building contractors, appraisers, and real estate professionals will find the concept of prudent avoidance and the understanding of its basis a valuable tool in the design, construction, and valuation of homes, offices, retail facilities, and factories.

For utility engineers this book provides theory, computer simulation results, measured data and recommendations for magnetic field management, all of which are useful as a power system design and construction resource.

1.4 Using the Book's Features

Reference Numbers and Reference Lists. The numbers in square brackets, for example, [1], shown in the text of each chapter correspond to the references listed at the end of that chapter.

Examples and Exercises. Whenever appropriate we provide examples to illustrate concepts and, at the end of each chapter, an additional set of exercises (with solutions) to help readers solidify their understanding of the material presented.

Appendices. In the Appendices we have placed materials targeted for specific reader groups and perhaps not of interest to our audience as a whole.

Appendices A, B, and C provide a technical, academic examination of magnetic fields, focusing on their nature and measurement. Note especially that Appendix B contains a registration form to be filled out for obtaining a **free** copy of the FIELDS personal computer program (DOS format) for calculating magnetic fields resulting from current flow in transmission or distribution lines (both overhead and underground).

In Appendix D we present an overview of the science of epidemiology, the study of disease in population groups.

Appendix F provides a series of personal "prudent avoidance" strategies, reasonable, practical, and relatively simple and inexpensive actions to take to reduce personal exposure to magnetic fields.

With the extensive glossary in Appendix E we hope to provide background for the non-technical reader.

1.5 Overview of Chapters Two Through Nine

Chapter Two: Magnetic Fields and Human Health Issues. Despite the failure of more than forty studies of disease in the populations of industrialized countries to conclusively prove a connection between magnetic field exposure and adverse health effects, particularly cancer, the magnetic field issue has in recent years become an area of increasing public concern. That concern is in response to studies "linking" magnetic field exposure to cancer. (A prime example is the 1992 Swedish study which found that children living near power lines had about four times the leukemia rate of other children.) To provide a chronological and historical view of health research, in Chapter Two we have compiled a summary of the literature available to date on studies, findings, and reports done in the past 15 years in the areas of human health effects which might be attributed to exposure to power frequency magnetic fields.

Chapter Three: The Nature and Measurement of Power Frequency Magnetic Fields. To provide a basis for understanding measurement, in Chapter

Three we examine the nature of magnetic fields as they are produced by single and multiphase currents. Here the *nature* of the field involves frequency, polarization and harmonic content. Since the magnetic field is a vector field, its measurement involves not only magnitude but also its direction in space. The appropriate measure of a magnetic field is considered in Chapter Three, in connection with commercially available measurement systems. Very sophisticated instrument systems have been developed, and are commercially available, to measure and map these fields.

Chapter Four: Sources of Power Frequency Magnetic Fields in the Electric Energy Delivery System. As the sources of power frequency magnetic fields are power frequency currents, magnetic fields result from the generation, transmission, transformation, distribution, and utilization of electric energy. In Chapter Four we introduce the components of a typical electric power delivery system and describe the magnetic fields associated with each stage of energy delivery. (We have omitted discussion of power generation because the general public is usually isolated from generating stations.) The delivery system may be divided into the four parts described below:

- The transmission system transmits electric power from a generating station to a distribution substation.

- The distribution substation provides for voltage transformations from transmission or subtransmission voltages to primary distribution level voltages.

- The primary distribution system distributes electric power from the distribution substation to distribution transformers adjacent to customers.

- The secondary and services system distributes electric power from distribution transformers to customer service entrances.

Each of these parts of the delivery system produces magnetic fields which we characterize in Chapters Five and Six.

Chapter Five: Characterization of Fields from Transmission and Distribution Lines. In Chapter Five we discuss in detail the magnetic fields produced by electric power transmission and distribution lines, addressing overhead and underground lines and the effects of construction (configuration of the physical apparatus—towers, conductors) on both. Transmission lines provide an overhead electrical path for transmission of large amounts of power at high voltage (115,000 to 765,000 volts and higher). Distribution lines, which may be overhead or underground, provide an electrical path for distribution of power from distribution substations throughout residential, commercial, and industrial areas. We present an extensive parametric study of the fields from lines. Study parameters include line configuration, line height or depth, and grounding return methods. The modeling process that leads to the field characterization is fully developed in Appendix B.

Chapter Six: Characterization of Fields from Distribution Substations and Transformers. In Chapter Six we focus on the magnetic fields produced by distribution substations, with their transformation and switching equipment and by distribution transformers. Distribution substations provide for voltage transformations from transmission or sub transmission voltages (66,000 to 765,000 volts) to distribution level voltages (4,000 to 35,000 volts). Distribution transformers provide voltage transformation from distribution voltages (4,000 to 35,000 volts) to secondary or utilization voltages (120/240 to 277/480 volts). Normally located in close proximity to residential, commercial, or industrial power consumers, distribution transformers may be overhead, on the surface (pad-mounted at ground level), or sub-surface (below ground).

Chapter Seven: Power Frequency Magnetic Fields in the Home. In the home (or at any location at which electric power is consumed) power frequency magnetic fields are produced by both the apparatus of delivery of electric energy **and** by the appliances, equipment, and machines that utilize that energy. In Chapter Seven we look at household appliances, wiring, lighting and office equipment as sources of magnetic fields, addressing, in addition, the measurement of residential power frequency magnetic fields.

Chapter Eight: Management of Magnetic Fields by Shielding. Magnetic fields are subject to management—reduction of the magnitude of the field—under appropriate conditions by shielding, a relatively undeveloped mitigation technique. In Chapter Eight we present the principles of shielding and a study of applications of passive shielding (magnetic shunt and eddy current) and active shielding (cancellation).

Chapter Nine: Regulatory Issues and Prudent Avoidance. The absence of a definition of a dose (amount and duration of exposure) is a handicap in formulating protective policies—for example, it is as yet unknown whether such an exposure as to two milligauss for one hour is more or less risky than exposure to 0.5 milligauss for two hours. In Chapter Nine we survey the present status of EMF regulations in the United States, other industrialized nations, and an international association summarizing and analyzing both proposed and existing regulations for consistency and practicality. In the United States a number of states have formulated, or are close to formulating, regulations which limit the magnitudes of magnetic fields produced in public areas. In addition, other countries, most notably Sweden, have announced that future regulations will be based on the assumption that there is a connection between exposure to power frequency magnetic fields and cancer.

In conjunction with our survey of regulations, we discuss the available means to protect ourselves as individuals from the possible harmful effects of EMFs, the foremost of which is prudent avoidance. A common-sense response to the EMF problem, prudent avoidance means taking such actions as using an electric razor only in its cordless (or battery) mode and **not** placing your baby's crib on the opposite side of the wall from your home's electrical service panel.

2 Magnetic Fields and Human Health Issues

2.0 Overview

Chapter Two contains a compendium of information obtained from representative studies, findings and reports published during the past 15 years in the area of human health effects associated with exposure to power frequency magnetic fields.

Most of the efforts to date have concentrated on examining the evidence relating prolonged magnetic field exposure to several forms of cancer, including leukemia, in human populations.

The effects of power frequency magnetic fields on human health have been difficult to uncover for several reasons:

- early epidemiological studies were subject to a number of confounding factors

- some epidemiological studies were contradictory

- no definitive interaction mechanism has been found

- there is no accepted measure of exposure.

A preponderance of the evidence, however, points to a link between exposure to power frequency magnetic fields and health risks. This has led the Swedish National Board for Industrial and Technical Development to announce that "it will act on the assumption that there is a connection between exposure to power frequency magnetic fields and cancer, in particular childhood cancer" [14].

2.1 Introduction to Human Health Issues

Although exposure to power frequency magnetic fields is widespread, the risk to an individual due to this exposure is apparently very low. Consequently, the health impact cannot be related to exposure without elaborate statistical studies (epidemiological studies). The concern over power frequency magnetic fields began with a case controlled epidemiological study of childhood leukemia, carried out by Wertheimer and Leeper in Denver, Colorado, in 1979 [1]. Until that study, there had been no evidence of a relationship between very low energy fields and human health. While the Wertheimer and Leeper study has been amply criticized for its design and methodology, it must also be recognized as an historical turning point.

Since 1979, the magnetic field issue has become an area of increasing public concern. The public has shown heightened sensitivity to the siting of new and the redesign of existing electric facilities. Much of this sensitivity centers on the exposure to magnetic fields and their potential impact on human health. The introduction of nearly every new transmission line is being challenged by the public because of concerns over possible health effects from the exposure to the magnetic fields associated with such lines. There also exists a new awareness of the magnetic fields associated with many of the tools and appliances of life in an electrified society.

The proposition that power frequency (50 hertz in Europe and 60 hertz in North America) magnetic fields can affect human health is hardly credible when viewed from an energy perspective. Figure 2.1-1 shows the frequency spectrum of electromagnetic radiation. From this figure we see that the frequency spectrum is divided between ionizing and non-ionizing radiation.

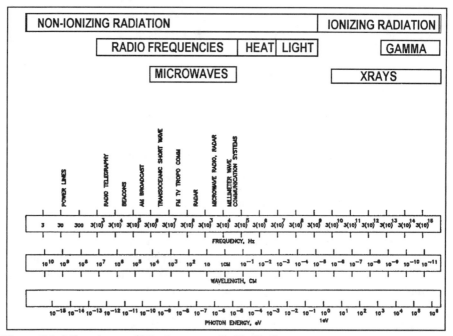

Figure 2.1-1 Spectrum of Electromagnetic Radiation

Ionizing radiation, which has high energy, is known to affect biological tissue, influence reproduction, cause birth defects and increase cancer rates. While nonionizing radiation is not completely harmless (it includes microwaves), low levels of non-ionizing radiation have been considered to be benign. Power frequencies are in the lowest energy end of the non-ionizing radiation spectrum. The photon energy associated with power frequencies is of the order of 10-15 electron volts as opposed to 103 electron volts for x-rays, an astounding ratio of 10-18. The energy comparison leads us to the conclusion that the effect of power frequency magnetic fields on human health must be subtle. Thus, the magnitude of the effect can probably be detected only by large-scale population studies such as those described in Section 2.4, Epidemiological Studies.

2.2 Historical Development

We now turn our attention to a chronology of events leading to the present state of understanding of the effects of power frequency magnetic fields on human health. While most of these events involve the publication of scientific studies, some are publications of significant articles and books. The articles and the books, *Currents of Death* and *The Great Power Line Cover Up*, by Paul Brodeau, have had an enormous influence on the public perception of power frequency magnetic fields. Public interest and anxiety, in turn, have led to a significant increase in the number and size of scientific studies initiated.

A chronology of events beginning with the 1979 study by Wertheimer and Leeper is shown in Figure 2.2-1.

EVENT	REFERENCE NUMBER	YEAR
Wertheimer & Leeper Study [1]		1979
State of Florida Proposes EMF Regulations [2]		1985
Savitz Case Study [3]		1988
New Yorker Magazine-Annals of Radiation by Brodeur [4]		1989
Telephone Lineman Study (Johns Hopkins) [10]		1989
Interim Guidelines on Limits of Exposure [5]		1990
Currents of Death book by Brodeur [6]		1990
IEEE EMF Seminar [7]		1991
Currents of Death Rectified Article by E. R. Adair [8]		1991
USC/EPRI Study [9]		1991
New Yorker Magazine-Article on Slater School by Brodeur [11]		1992
Swedish Study #1 [12]		1992
Swedish Study #2 [13]		1992
Swedish Board [14]		1992
Danish Study [15]		1992
SCE/UCLA Occupational Study [16]		1993
Mount Tamalpais Article [17]		1993
EPRI Home Survey [18]		1993
The Great Power-Line Cover-Up [27]		1993
Canadian/French EMF Occupational Study [28]		1994
North Carolina Study [29]		1994
Alzheimer's Disease Study [30]		1994

Figure 2.2-1 A Time Line of Significant Events

The large-scale studies carried out to establish relationships between magnetic fields and human health are reported in references [1], [3], [9], [12], [13], [15], [16], [28], [29] and [30]. Termed epidemiological studies, these studies are concerned with groups or populations rather than individuals and are discussed in detail in Section 2.4. Epidemiology as a science is the subject of Appendix D. Here the epidemiologic methods identified as prevalence studies, cohort studies, case-control studies, proportional mortality studies and cluster studies are defined and discussed.

The events reported in references [11] and [17] fall into a category that could be termed episodal. They address cancer clusters at an elementary school in Fresno, California, [11] and in a neighborhood in the San Francisco Bay area [17]. The episodes, while unusual and tragic in themselves, do not provide a basis for scientific conclusions regarding magnetic fields and human health. They do help to energize further investigations of the issue.

The book, ***Currents of Death: Power Lines, Computer Terminals, and the Attempt to Cover Up Their Threat***, by Paul Brodeur [6] and the resulting reaction were largely responsible for the organization of an Electromagnetic Fields Seminar by the Industry Applications Society (IAS) of the Institute of Electrical and Electronics Engineers (IEEE) in 1991. The Proceedings [7] of this seminar provide a technical position from which to view the human health studies. Of particular interest is the paper by E. R. Adair, *Currents of Death Rectified* [8], a direct challenge to Brodeur's conclusions. Dr. Adair argues that there is no evidence of a link between low-level magnetic fields and human health. She states that it is the consensus of the IEEE-USA Committee On Man And Radiation (COMAR) that the book *Currents of Death* authored by Paul Brodeur does not provide a complete or balanced account of the research conducted to date on health effects that may be associated with exposure to low frequency, low intensity, electric and magnetic fields. This imbalance is evidenced by a disregard of scientific principles that guide the research and analysis process and by the omission of any discussion of established, prudent regulatory practices.

The Swedish National Board for Industrial and Technical Development [14] spoke to the issue on September 30, 1992. *It formally announced that from then on it will act on the assumption that there is a connection between exposure to power frequency magnetic fields and cancer, in particular childhood cancer.*

2.3 Human Exposure to Magnetic Fields

The exposure of the public to power frequency magnetic fields is usually viewed in terms of the strength of the field and the time duration of exposure. The field strength is measured in units of tesla, gauss, microtesla or milligauss, a sometimes confusing array of different units but related as follows:

1 tesla = 10^4 gauss = 10^6 microtesla = 10^7 milligauss.

(A one-tesla field is a very large field usually only found internally in electrical equipment such as machines and transformers). An indication of the field strengths associated with several types of electrical appliances and power lines is shown in Figure 2.3-1.

As shown in Figure 2.3-1, the power frequency fields emanating from electrical appliances are probably the largest encountered by the public. Possible exceptions would be fields encountered by those who work on energized power lines, in power plants or in substations (electrical workers).

While the field strengths associated with operating appliances are relatively large, the time spent close to the appliance may be short. The field strengths drop off rapidly with distance from the appliance. For example, from Figure 2.3-2, a man's exposure as a result of daily shaving with an electric shaver might be of the order of 60,000 milligauss-hours per year.

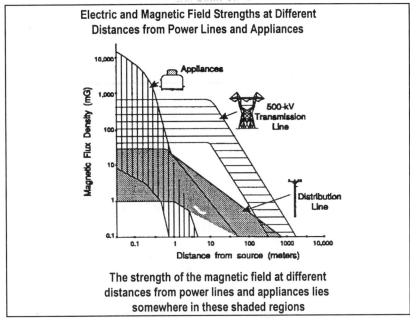

Figure 2.3-1 Magnetic Flux Density Profiles [18]

The fields measured in the vicinity of some specific appliances are shown in Figure 2.3-2.

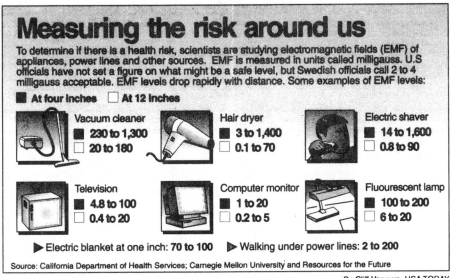

Figure 2.3-2 Magnetic Fields for Common Household Appliances

Electric transmission and distribution lines produce significant magnetic field strengths over relatively wide corridors, as seen in Figure 2.3-1. The actual field levels depend on such factors as the magnitudes of currents in the lines and ground return paths for the currents. For those of the public living close to a line, time spent in these fields may be long or almost all of the time, resulting in a large exposure. For example, a person spending 12 hours a day in a ten milligauss environment accumulates an exposure of about 44,000 milligauss-hours per year.

The background (ambient) field strength in residences not close to power lines is usually below 0.5 milligauss. In the field contour for a residential living room shown in Figure 2.3-3, field strengths rising well above the background are found in the vicinity of the service drop, a clock, entertainment center and fish tank.

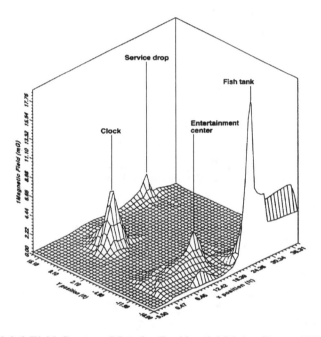

Figure 2.3-3 Field Contour Map for Residential Living Room [18]

Sometimes measured fields inside residence spaces reach values as high as 20-30 milligauss because of old fashioned wiring known as **knob and tube** wiring. Such wiring produces higher fields than newer cable wiring because the wires are spaced more widely apart, with the result that fields do not self-cancel as effectively. A bank of lights, for example, can cause readings of 20 milligauss when such a wiring system is utilized. Knob and tube wiring was found in about seven percent of residences during a recent nationwide EMF survey of 1000 residences by the Electric Power Research Institute, EPRI [18]. Other factors such as stray currents in water pipes and radiant heating in floors or ceilings can cause the **ambient** magnetic field value in a residence to be as high as 30 milligauss.

It would be speculative at this time to suggest that exposure should be measured in terms of a sum or integral of the product of field strength and time duration, that is, a cumulative effect. Such a measure would equate, from a human health viewpoint, a short time spent in a strong field with a long time spent in a weak field. Though some recent epidemiological studies [12, 13] do lend credence to this speculation by concluding that the risk of developing cancer increases with field strength, over long periods of time, it should be pointed out that there is no direct evidence of a cumulative effect. In fact there may exist definite thresholds for specific biological endpoints [8]. In any case, a large fraction of the public in an urban community experiences exposures consisting of field strengths from one to ten milligauss for long periods, even continuously. For reference, an environment of two milligauss, experienced 24 hours a day, results in an exposure of 17,720 milligauss-hours per year.

It is of interest to note that the public is also subject to a constant (not time varying) magnetic field due to the earth. The earth's magnetic field and the power frequency magnetic field combine to produce a net field strength. This is shown in Figure 2.3-4.

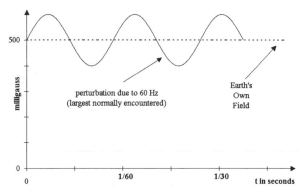

Figure 2.3-4 Largest Magnetic Field Usually Encountered by Humans

It has been speculated that health effects are a result of the Alternating Current (AC) modulation of the earth's Direct Current (DC) field. There is no evidence, however, to confirm this speculation.

2.4   Epidemiological Studies

The branch of medical science that studies the spread of disease in human populations and the factors influencing that spread is called Epidemiology. (An exposition of this branch of science is provided in Appendix D.) Unlike other medical disciplines, epidemiology concerns itself with groups of people rather than individuals. Developed in the 19th century out of the search for causes of human disease-especially of epidemics, one of its chief functions remains the identification of populations at high risk for a given disease so that the cause may be identified and preventive measures begun. Epidemiological studies correlate historical data for a large population of people; they may be classified as descriptive or analytic. The results of an epidemiological study can only show an association with a stimulus (electromagnetic fields, for example) since there are many unresolved variables associated with each individual. As detailed below, risk ratios greater than 1.0 indicate increasing associations between the stimulus and the disease.

In descriptive epidemiology, surveys are used to determine the nature of the population affected by a particular disease, noting such factors as age, sex, ethnic group, and occupation among those afflicted. Other descriptive studies may examine the occurrence of a disease over several years to determine changes or variations in incidence or death rates; geographic variations may also be noted. Descriptive studies also help to identify new disease syndromes or suggest previously unrecognized associations between risk factors and disease.

Line Configuration		Category
VHCC	Very High Current	Transmission line within 50 feet "Thick" primary within 50 feet "Thin" primary within 25 feet
OHCC	Ordinary High Current	Transmission line within 50-150 feet "Thick" primary within 50-130 feet "Thin" primary within 25-65 feet First span secondary within 50 feet
OLCC	Ordinary Low Current	"Thin" primary within 65-130 feet First span secondary within 50-130 feet Any non-end pole secondary within 130 feet
VLCC	Very Low Current (Endpole)	Endpoles
	Buried Lines	Underground

Figure 2.4-1a Wertheimer-Leeper Wire Code

In analytic epidemiology, studies are carried out to test the conclusions drawn from descriptive surveys or laboratory observations. These studies divide a sample population into two or more groups selected on the basis of suspected cause of the disease-for example, cigarette smoking-and then monitor differences in incidence, death rates, or other variables. In one form of analytic study, called the prospective-cohort study, members of a population are examined over time to observe differences in disease incidence. Statistics are used to analyze the incidence of diseases and their prevalence. If, for example, a disease has an incidence rate of 100 cases per year in a given region, and, on the average, the affected persons live three years with the disease, the prevalence of the disease is 300. Statistical classification is another important tool in the study of possible causes of disease. Statistical classification studies-as well as epidemiologic, nutritional, and other analyses—have made it clear, for example, that diet is an important consideration in the causes of atherosclerosis (the buildup of fatty deposits on the walls of arteries). The statistical analyses drew attention to the role of high levels of animal fats in the diet as the possible causes of atherosclerosis.

The analyses further drew attention to the fact that certain populations that do not eat large quantities of animal fats but instead live largely on vegetables, vegetable oils, and fish have a much lower incidence of atherosclerosis. Thus statistical surveys are of great importance in the study of human disease.

Epidemiological studies of the health effects of magnetic fields to date have focused on cancers (such as leukemia and tumors). Over the past ten years more than 30 such studies have been carried out both in and outside of the United States. In general, the results of the studies have been inconclusive. Epidemiologists have been unable to confirm or refute, conclusively, the negative effect on human health due to exposure of power frequency magnetic fields. However, three residential studies, two conducted in the United States and one in Sweden [3, 9, 12], have given support to associating an increased incidence of cancer as a result of exposure to power frequency magnetic fields at levels as low as two milligauss. The combined impact of the three studies on the scientific community (and especially the public) has been much stronger than might be warranted by the results of any one of the studies taken alone.

Residential Studies. The first widely publicized epidemiological study proposing the possibility of adverse health effects due to low frequency magnetic fields was done in Denver, Colorado, by an epidemiologist, Nancy Wertheimer, and a physicist, Ed Leeper [1]. The results of the study were published in 1979 with Wertheimer and Leeper concluding that a statistical correlation existed between childhood leukemia and proximity to ambient magnetic fields. No magnetic field measurements of any kind were used in their study. Rather measuring actual magnetic fields, the authors depended on a concept based upon a wire code system that they defined. Exposure to power line frequency magnetic fields was assumed to be related in some way to the wire code assigned to the power lines in the vicinity. The Wertheimer-Leeper Wire Code itself is based on the thickness and configuration of the electricity distribution line in the vicinity of the child's residence (refer to Figure 2.4-1a). For example, a residence within 50 feet of a transmission line or a three phase distribution line with thick primary wires was categorized as having very high current configuration (VHCC). Note that the 1993 EPRI EMF survey [18] found that measured fields corresponding to the VHCC code were substantially higher than those corresponding to other wire code assignments. Measured fields corresponding to the lower three current code assignments and underground lines were largely overlapping. (The survey results are summarized in Figure 2.4-1b.) EPRI's Environment Division is currently using data from the 1000 home survey to develop refinements in the procedure for assigning wire codes.

This first epidemiological study was met with a great deal of disbelief. Epidemiological studies by their very nature are often inaccurate, and in this instance, no measured field data was obtained. No convincing scientific evidence was found that such weak magnetic fields could cause biological effects.

Expecting to discredit Wertheimer's results, a group set up by the New York State Public Utilities Commission hired David Savitz, an epidemiologist, to conduct another study in the same general area (Denver) [3]. This study was conducted more rigorously than Wertheimer's and was done with a different group of children. The research group led by Savitz included a group of epidemiologists and engineers. In contrast to the first study, as well as using the wire codes defined by Wertheimer [1], the Savitz group measured actual field values in children's homes. In 1988 the Savitz results verified correlation between the Wertheimer-Leeper wire codes and the fields in the homes, but not close correspondence. In addition, the study found correlation between wire codes and the incidence of leukemia and brain cancer, but very small correlation between these same diseases and any measured fields. A strong correlation was also found between wire codes and heavy automobile traffic near the homes, a possible confounding factor. Savitz reported a doubling of risk for leukemia and a 50 percent increase in all cancers among children who lived in homes where power frequency magnetic fields, presumably from the nearby distribution wires, measured above two milligauss. He found that level in about 20 percent of the homes in the study.

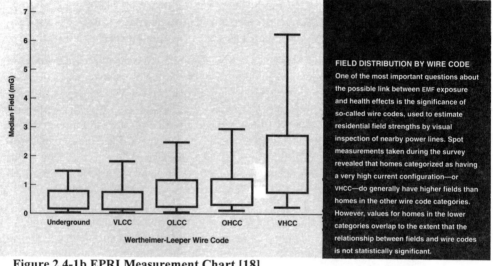

Figure 2.4-1b EPRI Measurement Chart [18]

A large and relatively well controlled study of childhood leukemia was completed in 1991 in the Los Angeles area [9]. As in the second Denver study, a much stronger correlation was found between wire code and cancer than between measured fields and cancer. This time a new result was found, a correlation between certain electrical appliances and cancer occurrence.

While the Wertheimer-Leeper study [1] was criticized as flawed in a number of ways, it has triggered some 30 to 50 additional epidemiological investigations related to power frequency electric and magnetic fields as a possible human carcinogen. The number (30-50) used depends on the definition of a valid study. Some cited studies have been published simply as a letter to the editor. Others have been based on a carefully selected, statistically valid population and published in a refereed scientific journal.

While most studies have been concerned either with childhood leukemia, brain cancer, or other types of cancers or cancer rates among workers exposed to higher than average doses of power frequency magnetic fields, the watershed 1992 Swedish residential study (Feychting and Ahlbom, Stockholm) [12] provided evidence that the stronger the field, the greater the incidence of leukemia. The study of nearly 500,000 Swedes between 1960-1985 found that children living near power lines had nearly four times the rate of leukemia as others. In addition, Feychting and Ahlbom found the cumulative effects of exposure to fields as low as two milligauss to be significant.

As reported at the IEEE 1991 EMF Seminar [7], although a number of biological effects have been observed, no health hazards have been proved to be directly linked to magnetic field exposure. Overall, the epidemiological data for both children and adults show relatively low risk ratios of approximately 1.5 to 3.0 with a large uncertainty. (A risk ratio of 1.0 is the same risk as the control group.) These results make the epidemiological studies suggestive, but not reliable. With such low risk ratio values and large uncertainty, positive statistical inferences are not possible.

Occupational Studies. Studies based on the effects of occupation on cancer rates seem to show a stronger correlation or risk ratio than residential studies. For example, of twelve U. S. and eight international studies, only two Swedish studies have found no association. A Johns Hopkins study was conducted by Genevieve Matanoski, an epidemiologist with the University's School of Hygiene and Public Health, and two colleagues of 50,000 telephone company employees [10]. The study found that the incidence of brain tumors among 4500 telephone line cable splicers was almost twice as high as that of office workers and that their leukemia rate was seven times as high. An increased risk ratio of ten was found for brain cancer among East Texas electric utility workers, and an increase in incidence of leukemia was found in ten out of eleven men working in close association with electric or magnetic fields.

The telephone line cable splicer and electric utility worker studies, however, are far from conclusive in their findings. For example, other factors present when telephone splices are made include possibly carcinogenic vapors in the manholes as a result of the solvents used. Inappropriate conclusion are easily made [20]. (The fact that dogs bark at burglars does not mean that barking dogs are responsible in any way for crime.)

More than 220,000 men, employees of Ontario Hydro and Hydro-Quebec in Canada and Electricite de France-Gaz de France in France, were covered in an occupational study reported in 1994 [28]. Lead author Dr. Gilles Theriault of McGill University, Montreal, Canada summarized the study results-"our study does not provide a final answer to the question" of whether electromagnetic fields are linked to cancer. The study does indicate that utility workers with heightened exposure to magnetic fields had three times the risk of acute myeloid leukemia than those with less exposure. Weaker evidence was found for a link with the brain cancer astocytoma. The study is to be published in the American Journal of Epidemiology. Dr. David Savitz of the University of North Carolina School of Public Health characterized the study as "one of the best if not the best study of this issue ever done; still the results are not definitive or conclusive".

For the first time, electromagnetic fields were implicated in breast cancer, in a study involving some 140,000 death certificates of women from 24 states. The study [29], published in 1994 in the Journal of the National Cancer Institute found that women in electrical occupations had a 38 percent higher rate of breast cancer mortality than those in other occupations. The lead author, Dana P. Loomis of the University of North Carolina School of Public Health, acknowledged that the study has "important limitations." But he said it does add new information on the suggestion from other studies that low-level electromagnetic energy can cause cancer. "I don't think we've proven it," Loomis said in an interview. "But we have taken it one step closer."

Loomis said all of the data used in the study were taken from death certificates that are susceptible to clerical error. Also, he said, the data do not take into account other exposures (confounding factors), such as diet or smoking, that could contribute to the cancer risk of women.

Three studies carried out by researchers in Finland and Los Angeles have indicated an increased incidence of Alzheimer's disease among people with a high occupational exposure to magnetic fields. The research results were reported at the 1994 Fourth International Conference on Alzheimer's Disease and Related Disorders, Minneapolis, Minnesota, by Dr. Eugene Sobel of the University of Southern California (USC). The two Finnish studies conducted by USC with the Universities of Kuopro and Helsinki involved a total of 386 Alzheimer's patients and 475 healthy subjects. Sobel reported that people with higher occupational exposure to magnetic fields were about three times as likely to develop Alzheimer's as those with low exposure. The relative risks found in the three studies were quite consistent at 2.9, 3.1 and 3.0. Women were more susceptible than men to the magnetic field effects. In the combined studies, women with high exposure had a relative risk ration of 3.8 compared to women with the lowest exposure.

High exposure to magnetic fields was associated with the use of sewing machines, both industrial and domestic. People who use electric sewing machines regularly have a greater exposure to magnetic fields than those in other occupations. Accordingly, dressmakers and tailors were found to be overrepresented among Alzheimer's victims.

Alzheimer's disease afflicts several million Americans each year, most over the age of 65. It is characterized by memory loss, disorientation, and deterioration of bodily functions and is ultimately fatal. The cause of Alzheimer's is unknown.

A mechanism by which magnetic fields may be associated with the development of Alzheimer's has been suggested by one of Sobel's co-authors, neurologist Zorek Davanipour of Loma Linda University, Loma Linda, California. This mechanism involving the production of calcium ions in nerve cells is a subject of Section 2.5.

2.5 Interaction Mechanisms

Laboratory studies have shown that certain temporary changes in biological rhythms, such as routine sleep-wake patterns, may be initiated through magnetic field exposure. There has been no indication that this type of effect is permanent or that it adversely affects health. These experiments do show, however, that magnetic field exposure can change biological function. This finding shows that non-ionizing radiation, such as that created by 60 hertz AC, can have an effect on biological function. Prior to 1980 this was thought not to be true.

Reasons given that power frequency magnetic fields could not be responsible for biological or health effects are that non-ionizing fields do not break chemical bonds or heat body tissue. Recent cellular research now suggests alternative mechanism processes may be involved.

Several authoritative reviews of this subject have been published [22-24]. Although scientific consensus has grown over the existence of biological changes resulting from magnetic field exposure, the effects of these changes on human health and safety remain unclear. A number of ongoing studies are focused on developing knowledge of mechanisms that can explain how magnetic fields interact with biological systems.

Recent experiments with rats have shown that their melatonin production is reduced when they are subjected to large doses of magnetic field radiation. The rats were exposed nightly to pulsed DC magnetic fields of 200 - 400 milligauss, at one-minute intervals for about an hour. Some researchers feel that the suppression of melatonin as a result of magnetic field exposure may be a possible link between magnetic fields and health effects such as cancer. Melatonin is believed, among other things, to aid in the body's defense against cancer. Laboratory studies to investigate whether magnetic fields suppress melatonin production in humans will be started in the near future.

One established mechanism of action of power frequency magnetic fields is via induced current. Cells may use electrical signals to communicate with each other, to *whisper* to each other, as one researcher says [21]. When these currents are large enough, nerve and muscle cells can be stimulated. A vertical electric field tends to induce predominantly vertical fields and currents, while a horizontal magnetic field tends to induce circulatory or eddy currents as predicted by Faraday's law. This effect is illustrated in Figure 2.5-1. The average currents as well as currents in various body locations depend on the size and shape of the animal. The effects of cell stimulation by induced currents, however, are not well understood. Currents induced from external magnetic fields are very small even when compared to those currents present inside the body that are due to internal processes. Examples of naturally occurring low frequency (below 100 hertz) currents are found in the brain, the heart and nerve synapses. Current densities in the brain are approximately one milliamp per meter-squared and in the heart about ten milliamps per meter-squared.

Medical researchers do not yet fully understand how cells can be affected by externally induced currents that are so much smaller than currents that flow in these same cells due to natural body functions. However, there does exist a large body of evidence about how this might occur.

18

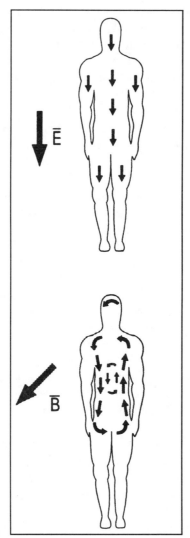

Figure 2.5-1 Simplified Patterns of Induced Electric Fields and Currents From Electric and Magnetic Fields [21]

As shown in Figure 2.5-2, cells are surrounded by a phospholipid bi-layer membrane. Embedded in this layer are numerous proteins that act as receptors for various molecules including enzymes. This assembly of receptors on the membrane surface translates signals on the cell's exterior into its interior. The signals can then trigger various biological processes. Therefore, the receptors can bring a small or weak signal to a cell's interior and essentially, at the same time, amplify the signal.

A number of biological experiments have shown that low-frequency magnetic fields can affect the calcium ions present across the cell membrane. Calcium ions regulate many important functions of the cell; thus, the interaction between calcium ions and magnetic fields may be an important mechanism of reaction relating to power frequency magnetic fields. No specific results relating to this type of mechanism had been known until neurologist Zorek Davanipour of Loma Linda University reported research showing that magnetic fields can increase the number of calcium ions in nerve cells. It is speculated that the increased levels of calcium ions kill nerve cells. Through this mechanism magnetic fields could affect neurological diseases such as Alzheimer's disease.

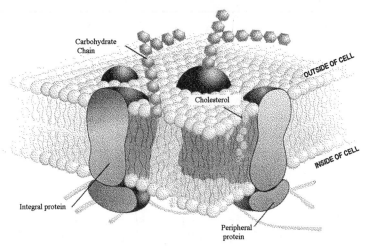

Figure 2.5-2 Simplified Illustration of a Cell Membrane [31]

In 1992 a group of scientists at Cal Tech discovered a possible mechanism based on the finding of crystals of the iron-containing mineral magnetite in the human brain [25]. The team, whose findings were accepted for publication in the National Academy of Sciences, was made up of Joseph Kirschvink, Associate Professor of Geobiology, Atsuko Kobayashi-Kirschvink, Associate Research Engineer and Barbara J. Woodford, formerly a Cal Tech Research Fellow now at the University of Southern California.

This discovery of magnetite in the brain may provide an explanation of how exposure to power frequency electromagnetic fields might cause increased rates of certain types of cancers.

Kirschvink, however, asserts that no conclusions about any possible connection between his team's findings and magnetic fields can be drawn until scientists are able to pinpoint the magnetite's location in the brain tissue and determine what purpose, if any, it serves there. "Because this study has the potential to be misunderstood," continued Kirschvink, "I want to be absolutely clear about what I am saying and what I am not saying. Yes, there is magnetite in human brain tissue. Yes, it is possible that the presence of magnetite may mediate any health effects of electromagnetic fields. But in my opinion, the jury is definitely still out on whether electromagnetic fields actually do have health effects. I'm a geobiologist, and the extremely difficult job of determining whether electromagnetic fields have health effects belongs to the field of epidemiology. We also don't know what normal function magnetite might serve in humans, and in particular I want to be explicit about the fact that we have no evidence at this time that humans have a magnetic sense."

2.6 Studies Under Way

The United States Environmental Protection Agency in a 1990 Review Draft provides an evaluation of the possible carcenogenicity of magnetic fields: "With our current understanding, we can identify 60-hertz magnetic fields from power lines and perhaps other sources in the home as a possible, but not proven, cause of cancer in humans. The absence of key information....makes it difficult to make quantitative estimates of risk."

A large bloc of the world's industrial countries are involved in magnetic field research projects to identify and quantify the links, if they exist, between magnetic fields and human health. Approximately 25 countries are now sponsoring research projects with a total annual budget of about $8.9 million in contract research in 1992. World-wide, as of December 1992, there were more than 230 research projects under way, including epidemiological studies, laboratory studies on biological effects, and exposure and measurement studies. The number of countries, research projects and dollar budget amounts has grown steadily since 1990. A 1991 listing of international Federal Agency Research Projects is shown in Figure 2.6-1.

The federal government of the United States has expended over $60 million in the past on EMF research. The rate of expenditure decreased in the early and mid-1980s, but increased to over $10 million in fiscal year 1992. Congressional efforts to better coordinate federally sponsored magnetic field health effects research led to the Department of Energy's designation as the lead agency. Other agencies currently conducting or planning studies related to the health impacts of magnetic field exposure are the Environmental Protection Agency, the National Cancer Institute, the National Institute of Environmental Health Sciences, the National Institute of Occupational Safety and Health, and the Food and Drug Administration.

Many other private and public groups, as well the government agencies listed above, are conducting research of one form or another on their own. In particular, the electric utility industry, through its research arm-the Electric Power Research Institute (EPRI), headquartered in Palo Alto, California-is very actively engaged in various research projects related to power frequency magnetic fields. Funding support for EPRI's EMF health-related Health Studies Program increased from about $5.5 million in 1991 to $8.9 million in contract research in 1992. The increase in funding was primarily to fund new studies in wire codes as reliable indicators for residential magnetic field exposures and to support an accelerated effort in laboratory experiments on melatonin suppression in humans as a result of magnetic field exposure. The experiments concentrate on three important issues: (1) the specific conditions under which the effect occurs, (2) whether such effects occur in humans under realistic exposure, and (3) whether such effects have any relevance to cancer or other suspected magnetic field induced illnesses. In addition, EPRI spent more than $4 million in 1993 evaluating ways to manage magnetic fields produced by the power delivery system, including grounding. In 1993, research sponsors, including EPRI, allocated more than $40 million to EMF studies [31].

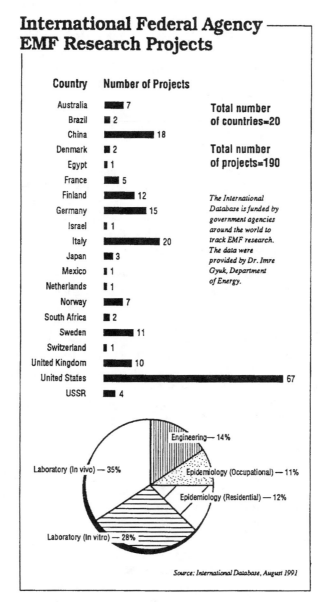

International Federal Agency EMF Research Projects

Country	Number of Projects
Australia	7
Brazil	2
China	18
Denmark	2
Egypt	1
France	5
Finland	12
Germany	15
Israel	1
Italy	20
Japan	3
Mexico	1
Netherlands	1
Norway	7
South Africa	2
Sweden	11
Switzerland	1
United Kingdom	10
United States	67
USSR	4

Total number of countries=20

Total number of projects=190

The International Database is funded by government agencies around the world to track EMF research. The data were provided by Dr. Imre Gyuk, Department of Energy.

Engineering— 14%
Laboratory (In vivo) — 35%
Epidemiology (Occupational) — 11%
Epidemiology (Residential) — 12%
Laboratory (In vitro) — 28%

Source: International Database, August 1991

Figure 2.6-1 International Federal Agency Research Projects [19]

2.7 Summary

Over the past 20 years a body of evidence has been developed suggesting that there may be health risks associated with exposure to power frequency electromagnetic fields. Opinions on this vary across the full range of the spectrum, from those who firmly believe that power frequency electromagnetic fields pose absolutely no health threat, to those who are completely convinced that these same fields are causing a number of serious health problems. As is often the case, perhaps it will be shown that the truth lies somewhere between these two diverse opinions. Although biological effects due to low frequencies have been demonstrated, and generally accepted, health effects are still controversial and will likely remain so for some time to come. The term *electrophobia*-the irrational fear of electromagnetic fields-, has been coined by Dr. Eleanor Adair, a senior research associate and lecturer at Yale University, to describe this fairly new concern [26].

Exposure to magnetic fields can occur in a variety of ways. Typically the largest fields contained inside a residence or office are due to electrical equipment, appliances or wiring, rather than outside electrical lines. Field intensities are normally given in milligauss or microtesla; one tesla = 10,000 gauss. Typical *ambient* values of exposure in the home or office range from .1 to 2 milligauss although it is not unusual to routinely find fields up to ten milligauss. Particularly high values of fields, as much as 20-30 milligauss, are sometimes found in older houses using *knob and tube* wiring. A value as high as 90 milligauss has been reported as a result of a corroded aluminum wire at the service drop on a house roof.

A knowledge of cellular mechanisms permitting magnetic fields to interact with human cells and possibly alter biological function is the key to determining the impact, if any, that magnetic fields have on human health. Results of a number of recent cellular biological studies now suggest there may be appropriate mechanisms to link power frequency magnetic fields with biological processes. Under certain conditions, magnetic field exposures can induce a range of cellular responses, including the production of chemical agents associated with known cancer promoters. Cancer promoters induce a cancer cell in its first or *latent* stage of development to become tumorous or cystic. Research to date indicates that magnetic fields may act like a cancer promoter [26] rather than a cancer initiator. It appears that the research, while not yet conclusive, suggests that magnetic field exposure may reduce the cell's ability to protect itself from cancer.

Epidemiology is the branch of medical science that studies the spread of disease in human populations and the factors influencing that spread. Over the past 10 years more than 30 such studies have been carried out both in and outside of the United States. In general, the results of these studies have been inconclusive. Epidemiologists have been unable to conclusively confirm or refute the negative effect on human health due to exposure to power frequency magnetic fields. However, three studies, two conducted in the United States and one in Sweden, have given support to associating an increased incidence of cancer as a result of exposure to power frequency magnetic fields at levels as low as two milligauss. The joint impact of the three studies on the scientific community (and especially the public) has been much stronger than might be warranted by the results of any one of the studies taken alone.

Power frequency magnetic fields are ubiquitous in our environment and are clearly a part of a high standard of living as we know it in an industrialized country. The interactions and resultant biological effects depend on many characteristics of the exposure and the physiological state of the biological system. The conclusions and subsequent decisions concerning health effects of power frequency magnetic fields should be based on technical and scientific evidence drawn from valid, reproducible studies and experiments (when possible). It should not become, as Dr. Asher Sheppard says, *"part of the media circus in which the public is alternately scared to death and then passivated"*.

The evidence to date is not totally convincing that magnetic fields are in fact harmful to human health. Still, researchers have published enough suggestive data to raise many unanswered questions. Research into possible health effects as a result of exposure to power frequency magnetic fields should continue. We especially should support biological, neurological and cellular research in order to gain key information on the relationship between magnetic fields and the human body.

2.8 References

[1] N. Wertheimer and E. Leeper, *Electrical Wiring Configurations and Childhood Cancer*, American Journal of Epidemiology, 109, 273 (1979)

[2] Florida Electric and Magnetic Fields science Advisory Commission, 1985, *Biological Effects of 60 Hertz Power Transmission Lines*. Florida Department of Environmental Regulation, Tallahassee, FL: NTIS No. P8-85200871.

[3] Savitz 1988; Savitz, D., Wachtel, H., Barnes, F., John, E., and Tvrdik, J., *Case-Control Study of Childhood Cancer and 60-Hertz Magnetic Fields*, American Journal of Epidemiology, Vol. 128, pp. 21-38, 1988.

[4] Paul Brodeur, *Annals of Radiation, the Hazards of Electromagnetic Fields*, in three parts, The New Yorker, pp 51-88, June 12, pp 51-88, June 19, pp 39-68, June 26, 1989.

[5] *Interim Guidelines on Limits of Exposure ... Electric and Magnetic*, USA Health Physics, Vol. 58, No.1, pp 113-122, January, 1990.

[6] Paul Brodeur, *Currents of Death: Power Lines, Computer Terminals, and the Attempt to Cover Up Their Threat*, Simon and Schuster, New York, NY, 1989.

[7] Proceedings of the Industry Applications Society, Electromagnetic Fields Seminar, The Institute of Electrical and Electronic Engineers, 345 East 47th Street, New York, NY, 10017-2394, 1991.

[8] Eleanor R. Adair, *Currents of Death Rectified*, Proceedings of the Industry Applications Society, Electromagnetic Fields Seminar, The Institute of Electrical and Electronic Engineers, 345 East 47th Street, New York, NY 10017-2394, 1991.

[9] *Exposure to Residential Electric and Magnetic Fields and the Risk of Childhood Leukemia*, prepared by the University of Southern California, November 1991, Electric Power Research Institute Report, EPRI EN-7464.

[10] Matanoski G.M., Breysse P., and Elliott E.A., 1989, *Cancer Incidence in New York Telephone Workers*, Poster at DOE/EPRI Contractors Review Meeting, (Portland, Oregon, November 15).

[11] Paul Brodeur, Annals of Radiation, *The Cancer at the Slater School*, The New Yorker, pp 86-119, December 7, 1992

[12] Maria Feychting, and Anders Ahlbom, *Magnetic Fields and Cancer in People Residing Near Swedish High Voltage Power Lines*, IMM-rapport 6/92, Institutet för miljömedicin, Karolinska institutet, Stockholm, 1992

[13] B. Floderus, T. Persson and others, *Occupational Exposure to Electromagnetic Fields in Relation to Leukemia and Brain Tumors*. A Case-Control Study, PM edition: National Institute of Occupational Health, Solina, Sweden, 1992.

[14] IEEE Power Engineering Review, Vol. 13, Number 10, October 1993, Stig Goethe, *EMFs and Health Risks: Research and Reactions in Sweden*.

[15] J.H. Olsen, A. Nielsen, G. Schulgen, A. Bautz and V. Larsen, A scientific translation from the Danish: *Residence Near High-Voltage Facilities and the Risk of Cancer in Children*, Sponsored by EMF Information Project, P. O. Box 10149, Minneapolis, Minnesota 55414, October 1992.

[16] Jack D. Sahl, Michael A. Kelsh and Sander Greenland, *Cohort and Nested Case-Control Studies of Hematopoietic Cancers and Brain Cancer Among Electric Utility Workers*, Epidemiology, Vol. 4, Number 2, pp 104-114, March 1993.

[17] Peter White, *Poison In Paradise*, IMAGE Magazine, pp 19-29, March 14, 1993.

[18] Electric Power Research Institute, EPRI Journal, April/May 1993.

[19] *Electric and Magnetic Field Effects,* A Program Description of the DOE Office of Utility Technology, Imre Gyuk, U. S. Department of Energy, Office of Energy Management, CE-141, 1000 Independence Avenue, Washington, DC 20585.

[20] Montoya, Paul, *Communicating Potential Health Risks Due to EMF*, Co-op Professional Report, California Polytechnic State University, EE Department, June 4, 1993.

[21] Stuchly, Maria A., *Electromagnetic Fields and Health*, IEEE Paper 0278-6648/93, appeared in Institute of Electrical and Electronic Engineers Potentials Magazine, pp 34-39, April 1993.

[22] WHO 1984: Environmental Health Criteria 35, Extremely Low Frequency (ELF) Fields, WHO, Geneva, 1984.

[23] WHO 1987: Environmental Health Criteria 89, Magnetic Fields, WHO, Geneva, 1987.

[24] WHO 1989: Non-ionizing Radiation Protection (Second Edition) Eds. M. J. Seuss and D. A. Benwell-Morison WHO Regional Publication. European Series No. 25. WHO Regional Office for Europe. Copenhagen, 1989.

[25] *Of Magnetite and Men*, Caltech News, Vol. 26, No. 3, California Institute of Technology, Pasadena, California 91125, (818) 356-6256, June 1992.

[26] Eleanor R. Adair, *Speakout* (Letters to the editor), *Nurturing Electrophobia*, IEEE Spectrum, Vol. 27 No. 8, p11-14, August 1990

[27] Paul Brodeur, *The Great Power-Line Cover Up*, Little Brown and Company, 1993.

[28] Theriault, G. et al., *Cancer Risks Associated With Occupational Exposure to Magnetic Fields Among Utility Workers in Ontario and Quebec, Canada, and France*:, 1970-1989. American Journal of Epidemiology 1994; 139(6): 550-72.

[29] Dana P. Loomis, et al., North Carolina Study. Journal of the National Cancer Institute, 1994.

[30] Sobel E, et. al., *Increased Risk of Alzheimer's Disease Due to EMF Exposure*, Fourth International Conference on Alzheimer's Disease and Related Disorders, Minneapolis, Minnesota, 1994.

[31] *EMF Research - BR-102833*, Electric Power Research Institute, 3412 Hillview Avenue, Palo Alto, CA 94303, 1994.

3   The Nature and Measurement of Power Frequency Magnetic Fields

3.0 Overview

Power frequency magnetic fields are as pervasive as the air we breathe. Everyone, especially citizens of industrialized nations, lives in a medium of such fields. In Chapter Three we discuss the power frequency field in the context of the entire electromagnetic spectrum. The field concept is explored and field magnitudes which are of particular interest in the public health debate are defined.

To provide a basis for measurement, Chapter Three addresses the nature of magnetic fields as produced by single phase and multi phase currents. In the absence of harmonic currents, such fields are respectively linearly polarized or elliptically polarized. Typically, they arise from currents in single phase and three phase transmission lines. Several measures of a field are described.

The description of a single phase field involves a single magnitude along the axis of the field, whereas the description of a multi phase field involves magnitudes along three orthogonal axes. Measurements of magnitudes are carried out using single axis or three axis field meters. In either case, field sensing is based on Faraday's Law. Typically a field meter may measure 10 milligauss fields with a resolution of 0.1 milligauss or larger fields to an accuracy of about one percent.

3.1 The Electromagnetic Spectrum and Sinewaves

A magnetic field is produced by the flow of electrical current through a conducting medium. This relationship was first discovered by the French physicist André Marie Ampere (1775-1836). Later it became clear to the Scottish physicist James Clerk Maxwell (1831-1879) that the magnetic field is one of the two principal components of the electromagnetic field (EMF). The other component is the electric field. Electromagnetic fields account for a myriad of phenomena in our lives, from nerve impulses in our bodies through television signals and light itself.

Magnetic fields exist over a wide range of frequencies. Electrical frequency is used to divide the electromagnetic spectrum into several regions as shown in Figure 3.1-1.

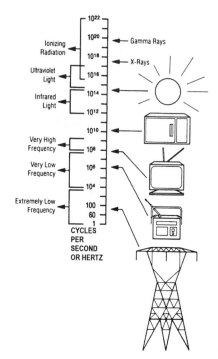

Figure 3.1-1 Electromagnetic Spectrum

Power frequency fields are in the Extremely Low Frequency (ELF) region. The frequencies of the field and the electrical current that produces it are the same, as shown in Figure 3.1-2. In the United States this frequency is almost universally 60 hertz (Hz), or cycles per second, and in Europe 50 hertz, both in the ELF region. As a result, the electrical power that is generated, transmitted, distributed, and utilized all over the world produces power frequency or ELF magnetic fields.

Figure 3.1-2 shows a sinusoidal current alternating at 60 cycles per second or 60 hertz (Hz). The magnetic field flux density, a measure of the magnetic field produced by the current, is shown in the same figure. We see that the magnetic flux density wave is a sinusoid of the same frequency and phase as the current that produced it.

3.2 Fields—A Definition

Thus far, we have used the term *field* liberally without defining it. In the sense of this book, the concept of field is used to explain *action at a distance*. A current carrying conductor produces a magnetic field in the space surrounding it. This may be thought of as conditioning the space. This field may interact with another conductor in the space to induce a voltage in the second conductor. The second conductor may in fact be conducting living tissue in a biological organism (as opposed to a wire) as depicted in cartoon form in Figure 3.2-1. The induced voltage can be thought of as the result of the direct interaction between the current of the first conductor with the second conductor (action at a distance) or as the result of the interaction between the field of the first conductor with the second conductor. The field concept, while a mathematical abstraction, has been found to lead to a better understanding of this and many other complex interactions. For example, the force interaction between bodies is often described in terms of a gravitational field. The field concept is used almost universally in describing EMF phenomena.

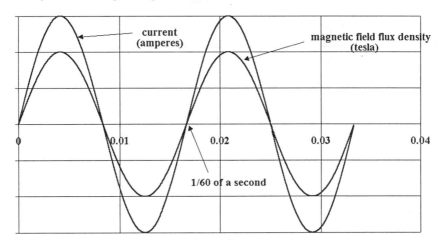

time in seconds

Figure 3.1-2 Current and Magnetic Field Flux Density for a 60 Hertz Alternating Current

The magnetic field is an example of a **vector** field, which means one that is described at every point in space by a magnitude and direction. The magnitude is expressed in units of tesla, gauss or milligauss, depending on the system of units and strength of the field. Appendix A, The Physics of Magnetic Fields, provides a detailed and rigorous discussion with examples. Magnetic fields as small in magnitude as a fraction of a milligauss are easily measured with readily available instrumentation. As a point of interest, the earth's magnetic field, which is a zero frequency naturally occurring magnetic field, is approximately 1000 milligauss (1 gauss), the actual value depending on location. This field is the result of direct currents that flow in the earth's core. For comparison, power frequency magnetic fields as high as 15,000,000 milligauss are commonly found internally in equipment such as power transformers, motors and generators. Although these fields are almost entirely contained within the equipment, some leakage will occur. This book concentrates on magnetic fields in the range of zero to several hundred milligauss. This level of field may be experienced by the general public as a result of operation of electrical equipment, current carrying lines, house wiring, electrical appliances, lighting, office equipment, or any other device energized by power frequency sources—a virtual supermarket of sources.

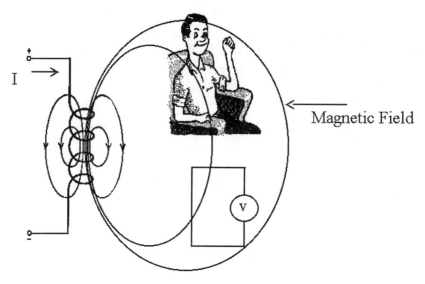

Figure 3.2-1 Interaction With Magnetic Fields

3.3 The Nature of Fields from Single Phase Alternating Current Sources

The magnetic flux density field ($\overline{B}$) produced by a long, straight, isolated conductor carrying a current of 100 amperes has a magnitude of about four milligauss at a distance of 50 meters from the conductor. The flux density field increases in magnitude as we approach the conductor as shown in Figure 3.3-1. At a distance of 20 meters, the field has a magnitude of about ten milligauss and, at a distance of five meters, about 40 milligauss. These field values are obtained using calculations based on the theory developed in Appendix A, The Physics of Magnetic Fields.

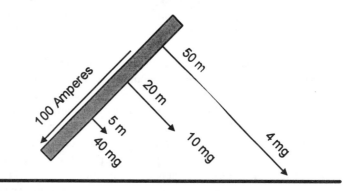

Figure 3.3-1 Long Straight Conductor Carrying 100 Amperes

The field lies in a plane orthogonal to the conductor. In fact, the magnetic flux lines form closed circles in this plane [1]. The geometry is shown in Figure 3.3-2, where the current flows along the z axis and the flux lines lie in the x-y plane.

The direction of the magnetic field (the magnetic flux density vector, $\overline{B}$) at any point is tangent to a magnetic flux line. Measurement of the field at a specified point means determining the magnitude and direction of the vector $\overline{B}$. Clearly, the magnitude and direction of $\overline{B}$ will depend on the current (I) and the x, y coordinates of the point of interest, for example (x_1, y_1), as shown in Figure 3.3-2. If the current I in Figure 3.3-2 varies sinusoidally at a power frequency, the magnitude of $\overline{B}$ will likewise vary sinusoidally at the same frequency.

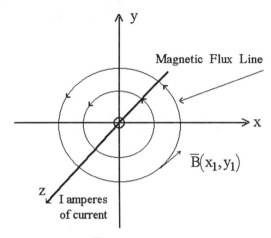

Figure 3.3-2 Coordinate System for $\overline{B}$ Field Around a Long Straight Conductor

The calculation of the magnetic field directly below a two conductor, single phase transmission line is given in Example 3.3-1.

Example 3.3-1. Calculation of a field from a single phase current.

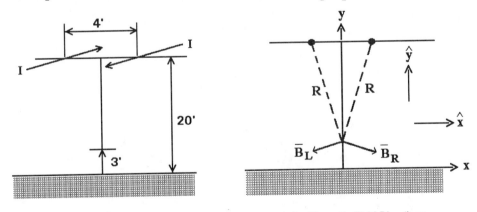

a. Single Phase Transmission Line

b. Magnetic Field Directions

Figure 3.3-3

The current I flows on an overhead transmission line into our page on the left conductor and out of our page on the right conductor, as shown in Figure 3.3-3a. We will find the magnetic field at a point of observation, three feet above the ground, directly under the center of the line.

Solution:

The distance from each conductor to the point of observation is:

$$R = \sqrt{(20-3)^2 + \left(\frac{4}{2}\right)^2} = 17.12 \text{ feet} = 5.22 \text{ meters}$$

As shown in Figure 3.3-3b, $\overline{B}_R$, the magnetic field vector produced by the right conductor current, is pointed down to the right, and B_L is pointed down to the left. The magnitudes of the two vectors are the same and are equal to:

$$B_R = B_L = \mu_o \frac{I}{2\pi R}, \text{ tesla} \qquad = \frac{2I}{R}, \text{ milligauss (see Appendix A)}$$

Note that $\mu_o = 4\pi \times 10^{-7}$ henry / meter and 1 tesla $= 10^7$ milligauss .

If we let $I = 100\sqrt{2} \sin 377t$, amperes (a sinusoidal current at a frequency of 60 hertz with a root mean square (rms) magnitude of 100 amperes),

$$B_R = B_L = \frac{2 \times 100\sqrt{2} \sin 377t}{5.22}$$
$$= 38.3\sqrt{2} \sin 377t \text{ milligauss}$$
$$= 54.2 \sin 377t \text{ milligauss}$$

From Figure 3.3-3b, the horizontal (x) components of $\overline{B}_R$ and $\overline{B}_L$ cancel, but their vertical components (y) add. The vector sum has the magnitude:

$$B = 2 \times 54.2 \times \frac{2}{17.12} \sin 377t$$
$$= 12.7 \sin 377t \text{ milligauss}$$
$$= 8.9\sqrt{2} \sin 377t \text{ milligauss}$$

which acts in the direction of the negative y-axis. Several values of t and their corresponding values of B are summarized in Table 3.3-1.

Value of t in Seconds	Peak Value of B in Milligauss	
t = 0	B = 0	
$\dfrac{1}{240}$	B = 12.7, Down	**Table 3.3-1 Peak Value of B versus time**
$\dfrac{2}{240}$	B = 0	
$\dfrac{3}{240}$	B = 12.7, Up	

Measurement of the magnitude of $\overline{B}$ is relatively easy using simple instrumentation, as will be shown. This is largely due to the fact that the field itself is simple. As it is produced by a single phase current, its direction in space at the point (x_1, y_1) is along a fixed line (linear polarization). As the current I varies sinusoidally, the magnitude of $\overline{B}$ also varies sinusoidally, but its direction is fixed. This is true even if the field is produced by the currents in a number of conductors, providing that the currents are all *in phase*[1] electrically. A common occurrence of this is when the conductors are all carrying the same, single phase current.

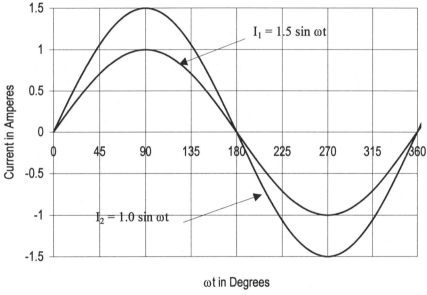

Figure 3.3-4 Currents in Phase

Examples of in phase and out of phase currents are shown in Figure 3.3-4 and 3.3-5 respectively. In Figure 3.3-4, I_1 and I_2 are in phase, and in Figure 3.3-5, I_1 leads

---

[1] Two sinusoidally varying currents are in phase if they have the same electrical frequency, and they reach maximum values at the same instant of time.

I_2 by 30°. In other words, I_1 reaches a maximum value 30° before I_2, and I_1 is said to have a positive phase angle of 30° relative to I_1.

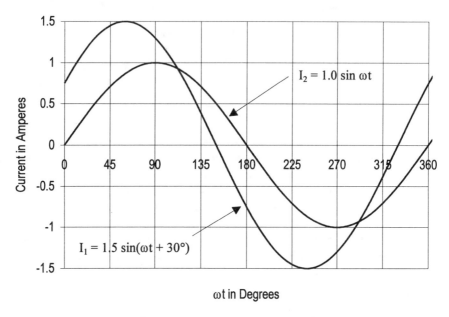

Figure 3.3-5 Currents 30° Out of Phase

3.4 The Nature of Fields from Multiphase Alternating Current Sources

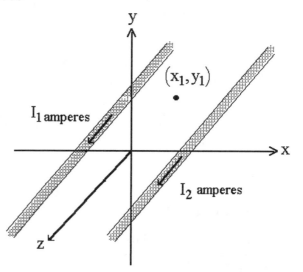

Figure 3.4-1 Coordinate System With Two Long Straight Conductors

Consider that two current carrying conductors are present as shown in Figure 3.4-1. Suppose that the currents have the same frequency but that the current in the first conductor is not in phase with the current in the second (as shown in Figure 3.3-5).

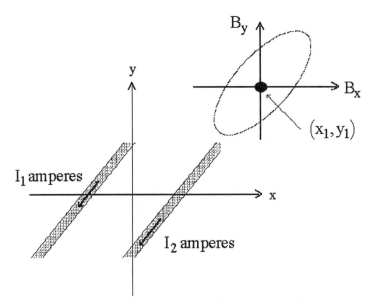

Figure 3.4-2 Field Ellipse Showing the Rotating Magnetic Field Vector $\overline{B}$

The $\overline{B}$ vector is perpendicular to the z axis and is equal to the sum of the $\overline{B}$ vectors produced by each of the currents I_1 and I_2. However, since I_1 and I_2 are not in phase, the locus of $\overline{B}$ is not a straight line. Rather, in general, it is an ellipse as shown in Figure 3.4-2 (elliptical polarization), and the direction of the $\overline{B}$ vector rotates 360 degrees in space for every cycle of the electrical frequency. (At certain points in the space surrounding the conductor, depending on the conductor configuration and currents, the ellipse may degenerate into a line.)

Measurement of the field vector $\overline{B}$ is now more complicated than in the case of single phase currents. A description of the $\overline{B}$ vector involves the angles made by the axes of the ellipse (say the angle between the major axis and the x axis) and the magnitudes of the semi-major and semi-minor axes of the ellipse [2].

Looking at the field ellipse of Figure 3.4-3, in detail, we see that:

- the angle between the x axis and the major axis of the ellipse is given by θ_m

- the maximum magnitude of $\overline{B}$ is $B_{s\ major}$

- the maximum magnitude of $\overline{B}$ in the x direction is B_{xm}

- the maximum magnitude of $\overline{B}$ in the y direction is B_{ym}

- the minimum magnitude of $\overline{B}$ is $B_{s\,minor}$

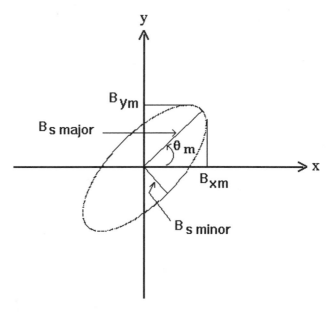

Figure 3.4-3 Expanded Field Ellipse

A commonly used measure of the magnitude of $\overline{B}$ is the resultant, given by:

$$B_{resultant} = \sqrt{B_{s\,major}^2 + B_{s\,minor}^2}$$
$$= \sqrt{\left(B_{xm}\right)^2 + \left(B_{ym}\right)^2}$$

More generally, $B_{resultant} = \sqrt{\left(B_{xm}\right)^2 + \left(B_{ym}\right)^2 + \left(B_{zm}\right)^2}$.

where B_{zm} is the maximum magnitude of $\overline{B}$ in the z direction, which is orthogonal to the x-y plane.

We will see that while $B_{resultant}$ can be calculated as shown above, it cannot be directly measured. When the magnetic field is elliptically polarized, the resultant magnetic field will always be greater than the maximum magnetic field. That is, $B_{resultant}$ is greater than $B_{s\,major}$. When $B_{s\,minor}$ is zero, the magnetic field is linearly polarized, and the maximum and resultant fields are equal.

The $\overline{B}$ field may be produced by currents in a number of conductors, all of different phases, that is, multiphase currents. If these currents are all of the same frequency, the locus of $\overline{B}$ will be an ellipse, as shown (although not necessarily in the x-y plane). The calculation of the parameters of a field ellipse is illustrated in Example 3.4-1.

Example 3.4-1 Calculation of a field from multiphase currents.

The current I_L flows on a transmission line into our page on the left conductor, and I_R flows out of our page on the right conductor as shown in Figure 3.3-3a in Example 3.3-1.

$$I_R = 100\sqrt{2}\sin 377t, \text{amperes and } I_L = 100\sqrt{2}\sin(377t - 90°)\text{amperes.}$$ We will find the magnetic field at a point of observation, 3 feet above the ground, directly under the center of the line.

Solution:

As shown in Figure 3.3-3b, $\overline{B}_R$, the magnetic field vector due to the right conductor current is pointed down to the right and $\overline{B}_L$ is pointed down to the left. The magnitudes of the two vectors are:

$$B_R = \frac{2I_R}{R} \text{ milligauss}$$

$$\text{and} \quad B_L = \frac{2I_L}{R} \text{ milligauss}$$

Since $I_R = 100\sqrt{2}\sin 377t$, amperes and $I_L = 100\sqrt{2}\sin(377t - 90°)$, amperes, we have multiphase currents; I_L lags I_R by 90°. (The return paths for these currents are not considered here.)

As shown in Example 3.3-1, R = 5.22 meters and, at the point of observation:

$$B_R = 54.2\sin 377t \text{ milligauss}$$
$$B_L = 54.2\sin(377t - 90°) \text{ milligauss}$$

Expressed as a vector field

$$\overline{B} = \overline{B}_R + \overline{B}_L$$
$$= 76.11\sin(377t + 45°)\hat{x} - 8.99\sin(377t - 45°)\hat{y} \text{ milligauss.}$$

It can be seen that the field ellipse has a semi-major axis of 76.11 mg in the $\hat{x}$ direction and a semi-minor axis of 8.99 mg in the $\hat{y}$ direction. These are peak values, and the corresponding root mean square (rms) values are 53.82 and 6.35 milligauss rms respectively.

The resultant value of the magnetic field is:

$$B_{\text{resultant}} = \sqrt{53.82^2 + 6.35^2}$$
$$= 54.19 \text{ milligauss, root mean square (rms)}$$

3.5 The Measures and Units of the $\overline{\text{B}}$ Field

The measures and units of the $\overline{\text{B}}$ field as described above,

$$B_{s\,major} \; ; B_{s\,minor} \; ; B_{xm} \; ; B_{ym} \; ; B_{zm} \, ,$$

represent measures of the field along specified directions in space. These measures themselves may be expressed in peak, average, or root-mean-square (rms) units with respect to time. For a sinusoidally varying $\overline{\text{B}}$ field:

$B = B_m \sin\omega t$

B_m is the peak value of B

$B_m \cdot \dfrac{2}{\pi}$ is the average value of B

and $\dfrac{B_m}{\sqrt{2}}$ is the rms value of B

$\left.\begin{array}{c} \\ \\ \\ \\ \\ \end{array}\right\}$ with respect to time

Thus $B_{s\,major}$, $B_{s\,minor}$, B_{xm}, B_{ym}, and B_{zm} can be expressed in terms of peak, average, or rms values or units. The units of $B_{\text{resultant}}$ follow from the units of its components. The calculation of $B_{\text{resultant}}$ from spatial components is the subject of Example 3.5-1.

Example 3.5-1. Calculation of the resultant value of B.

The power frequency magnetic field is 3 milligauss rms in the East-West direction, 4 milligauss rms in the vertical direction, and 0 milligauss in the North-South direction. What is the resultant value of B?

Solution:

Since there is zero field in the N-S direction, the field may be described generally by an ellipse in the (E-W)-Vertical Plane, as shown in Figure 3.5-1. The ellipse may in fact be no more than a straight line depending on the values of $B_{s\,major}$ and $B_{s\,minor}$.

$$B_{resultant} = \sqrt{3^2 + 4^2} = 5 \text{ milligauss, rms}$$

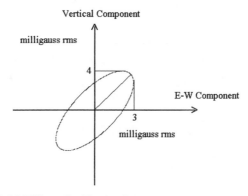

Figure 3.5-1 Field Ellipse for Example

The values of $B_{s\,major}$ and $B_{s\,minor}$ cannot be determined from the data provided above. However, if the value of the resultant and either $B_{s\,major}$ or $B_{s\,minor}$ is known, then the other can be calculated. This is shown in Example 3.5-2.

Example 3.5-2. Calculation of the parameters of the field ellipse.

The maximum value of B, as described in Example 3.5-1, is found to be 4.5 milligauss, rms. What are $B_{s\,major}$ and $B_{s\,minor}$?

Solution:

$B_{s\,major}$ = 4.5 milligauss, since $B_{s\,major}$ is defined as the maximum value of $\overline{B}$ and may be described in peak, average or rms terms as desired.

$$B_{s\,minor} = \sqrt{\left(B_{resultant}\right)^2 - \left(B_{s\,major}\right)^2}$$
$$= \sqrt{5^2 - 4.5^2} = 2.18 \text{ milligauss, rms.}$$

We note that $B_{resultant} = \sqrt{B_{xm}^2 + B_{ym}^2 + B_{zm}^2}$ is defined as a rms unit with respect to space. If B_{xm}, B_{ym}, and B_{zm} are expressed in rms units with respect to time, then $B_{resultant}$ is expressed in rms units with respect to both space and time.

We may ask the question - what is the most appropriate single measure of the $\overline{B}$ field ?

There are two reasonable candidates:

1. $B_{s\,major}$ in rms units.

    This is the rms value of the maximum component of $\overline{B}$. No other space value is larger. $B_{s\,major}$ can be measured directly by a single axis meter.

2. $B_{resultant}$ in rms units.

    This is the rms value of the field in both space and time. As opposed to $B_{s\,major}$, it includes, in an rms sense, $B_{s\,minor}$. Measurement of $B_{resultant}$ is indirect, requiring a three axis field meter and subsequent processing, as described below. In the case of a single phase field, $B_{resultant} = B_{s\,major}$.

Both candidates are in rms units (with respect to time). RMS values are indicative of power capability, whether $\overline{B}$ is sinusoidal or non-sinusoidal. Most electrical units are rms based for this reason.

3.6 Principles of Measurement of Power Frequency Magnetic Fields

Instrumentation for measuring power frequency magnetic fields use a property described by Faraday's Law. The relationship is illustrated in Figure 3.6-1, which shows a coil linked by magnetic flux. Faraday's law, as applied to this configuration, states that:

$$v_c = n\frac{d\phi}{dt}, \text{ volts}$$

where n is the number of turns of a coil linked by ϕ lines of magnetic flux (webers), t is the time in seconds and v_c is the potential across the terminals of the coil (volts).

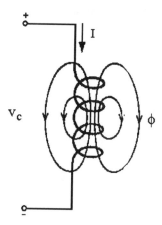

Figure 3.6-1 Illustration for Description of Faraday's Law

The lines of flux linking the coil in Figure 3.6-2 are given by:

$$\phi = BA\cos\theta \text{ webers}$$

where A is the area of the coil in meter^2 and θ is the angle between the vector $\overline{B}$ and a normal to the coil. In this instance,

$$v_c = nA \cos\theta \, \frac{dB}{dt} \text{, volts, since A and } \theta \text{ are not changing with time.}$$

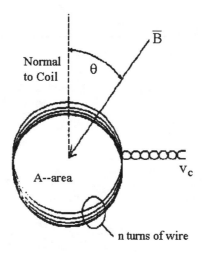

Figure 3.6-2 Coil in a Uniform Field $\overline{B}$

The coil voltage, v_c, may be then processed in one of two ways as follows:

1. Simple Scaling

By the use of simple scaling, an AC voltmeter may be calibrated for use as a detector. Suppose $B = B_m \sin\omega t$ is the field from a single phase source and ω is $2\pi \cdot$ power frequency.

$$\frac{dB}{dt} = \omega B_m \cos\omega t$$

and $v_c = nA\omega \cos\theta \, B_m \cos\omega t$, volts.

The rms value of v_c is:

$$V_{c\,rms} = \frac{nA\omega B_m}{\sqrt{2}} \cos\theta$$

$$= nA\omega B_{rms} \cos\theta$$

where $B_{rms} = \dfrac{B_m}{\sqrt{2}}$.

When the normal to the coil is aligned with $\overline{B}$, $V_{c\,rms} = nA\omega B_{rms}$.

If an rms reading voltmeter is used to measure v_c, $B_{rms} = \dfrac{V_{c\,rms}}{nA\omega}$, *tesla*. In other words, if the voltmeter reading in volts is divided by the scale factor ($nA\omega$), the resulting value is B_{rms}. Of course this result is only true when the $\overline{B}$ field is sinusoidal with the frequency ω. An Alternating Current voltmeter can be labeled in terms of field flux density, based on the scale factor as shown in Example 3.6-1.

Example 3.6-1. Determining the scale factor.

Suppose the coil area $A = 10$ cm^2, and the number of turns of the coil n = 265 turns. Determine the scale factor at a frequency of 60 hertz and label the voltage detector range appropriately.

Solution:

At a frequency of 60 hertz, the scale factor $= N \times A \times \omega = 265 \times 10 \times 10^{-4} \times 2\pi \times 60 = 1000$ volts/tesla or 10^{-3} tesla/volt.

On a 100 mv (millivolt) scale, full scale is 10^{-4} tesla or 1,000 milligauss. In other words, a voltmeter with a 100 mv scale connected to the coil could be marked 1000 milligauss full scale. We note that the scale factor is correct only if the frequency of the magnetic field is 60 hertz.

Simple scaling is applicable when the magnetic field varies sinusoidally at the power frequency.

2.   Processing and Scaling

If the magnetic field does not vary sinusoidally, simple scaling may not be adequate to determine the rms value of the field to the desired accuracy [2]. In this case a processor stage is commonly inserted between the coil and voltage detector as shown in Figure 3.6-3. By this means v_c is processed electronically to produce the voltage v_I [3].

Note that:

$$v_c = nA\cos\theta\,\frac{dB}{dt}.$$

The processor samples v_c, N times per period (T). The sampled voltage is integrated for a time Δt, seconds, to produce v_I.

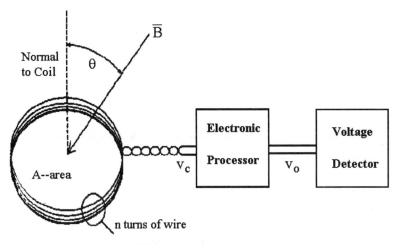

Figure 3.6-3 Typical Measurement Configuration

The change in $B\cos\theta$ during the ith integration interval is:

$$\left(\Delta\, B\cos\theta\right)_i = \frac{1}{nA}\left(v_c\right)_i\Delta t = \frac{1}{nA}\left(v_I\right)_i \text{, where}\left(v_I\right)_i = \left(v_c\right)_i\Delta t$$

v_I is squared, electronically, and the squared value is added in a register and scaled to produce the voltage v_o .

$$v_o = \frac{1}{nA}\left[\frac{1}{T}\sum_{i=1}^{N}\left(v_I\right)_i^2\right]^{1/2} \quad \text{volts, rms}$$

and

$$N\Delta t = T = \frac{2\pi}{\omega} \quad \text{seconds.}$$

v_o , in volts rms, is equal to $\left(B\cos\theta\right)$ in tesla rms.

If B contains frequencies higher than the power system frequency, due to harmonic currents, for example, these frequency components will be correctly processed provided the sampling frequency $\left(\frac{N}{T}\right)$ is sufficiently high. N should be of the order of 100 times the frequency of the highest harmonic component to be included in the computation of $\left(B\cos\theta\right)$.

Magnetic field measurements are routinely made in the region of 1 to 100 milligauss using an instrument called a field meter or gauss meter. Two types of field meters are commercially available.

- single axis field meter

- three axis field meter

The single axis field meter measures the magnetic field magnitude along its principal axis which is orthogonal to the plane of the search coil.

The three axis field meter contains three search coils. The three orthogonal axes of the instrument are perpendicular to the planes of the three search coils. The instrument contains electronics for operating on the voltages from the three search coils. Thus the instrument can determine such quantities as:

$$B_{s\,major} \; ; \; B_{s\,minor} \; ; \; \text{and} \; B_{resultant} \; .$$

3.7   Measurement of Single Phase Fields

By single phase fields, we mean fields produced by single phase currents. As noted above, the locus of the $\overline{B}$ vector at any point is along a line (a linearly polarized field). To measure this field with a single axis field meter we orient the axis of the meter in such a way as to obtain a maximum reading ($\theta = 0°$). The reading is $\dfrac{B_m}{\sqrt{2}}$ tesla rms (or milligauss depending on the instrument being used) when the orientation of the instrument is along the line of the field. On the other hand, if the single axis meter is oriented such that its axis is orthogonal to the line of the field ($\theta = 90°$), its reading will be a null.

Single phase fields may also be measured with a three axis meter. In this case,

$$B_{s\,minor} = 0 \; \text{and} \; B_{resultant} = B_{s\,major} \; .$$

3.8   Measurement of Multiphase Fields

By multiphase fields we mean fields produced by multiphase currents, usually three phase currents. If these currents are at the power system frequency, the locus of the $\overline{B}$ vector, at any point, is generally an ellipse in a plane (an elliptically polarized field). The resultant value of B is now quite difficult to obtain using a single axis field meter. It is almost mandatory that a three axis meter or the equivalent be used for determining the resultant value of the field. Determination of the resultant value, given by

$$B_{resultant} = \sqrt{\left(\frac{B_{xm}}{\sqrt{2}}\right)^2 + \left(\frac{B_{ym}}{\sqrt{2}}\right)^2 + \left(\frac{B_{zm}}{\sqrt{2}}\right)^2} \; \text{tesla rms,}$$

requires simultaneous (or almost simultaneous) measurement of the x, y, and z components of the field as is readily accomplished with a three axis meter.

3.9 The Effects of Harmonics

Use of non-linear devices, such as solid state electronics, produces non-sinusoidal currents and, as a result, non-sinusoidal magnetic fields. These fields can be analyzed as a sum of a field at the power system frequency plus fields at harmonics of the power system frequency. If the power system frequency is 60 hertz, its harmonics, from the second to the seventh, are tabulated as follows:

Harmonic Number	Frequency Hertz
1	60
2	120
3	180
4	240
5	300
6	360
7	420

Table 3.9-1 Harmonic Numbers and Frequencies

Harmonic Number 1 is the fundamental power system frequency. Usually, but not always, if harmonics are present, only the harmonic currents and fields corresponding to the odd harmonic numbers will be present; and, of these, the significant harmonics are the 3rd, the 5th and the 7th. These harmonics are significant because they are commonly produced by equipment typically found in the power system itself or in power system loads. Higher harmonic currents are usually so small in magnitude that they can be ignored. Harmonic currents produce magnetic fields at the frequencies of the currents, just as in the case of the fundamental (60 hertz) power frequency field. Consequently, we may have to consider the effect of the harmonic fields on the field measurements. The concepts of linear polarization and elliptical polarization, developed for the power frequency field, must be modified in the presence of harmonic frequency fields.

As an illustration of this, we compare a field ellipse for sinusoidal fields with a similar figure in which a 10 percent third harmonic component was added to B_y as shown in Figures 3.9-1a and b, respectively. Since B_y is non-sinusoidal in the case of Figure 3.9-1b, the plot of B_y versus B_x is not an ellipse, and the notions of major and minor axes no longer apply. However, the concept of a resultant field is still valid.

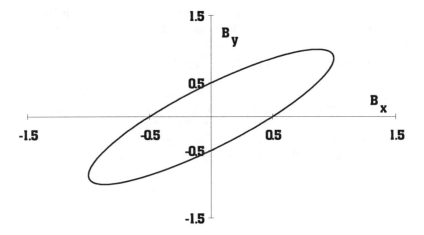

Figure 3.9-1a Field Ellipse With Sinusoidal Fields

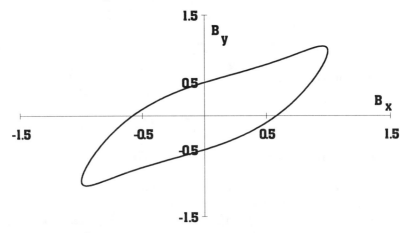

**Figure 3.9-1b Field Ellipse With One Non-Sinusoidal Field
(B$_y$ Has a 10 percent 3rd Harmonic Component)**

To take harmonics into account, the resultant field may be redefined in terms of the rms values of the fields along the orthogonal x, y, and z axes. Provided B$_x$, B$_y$, and B$_z$ are measured using a true rms detector, the rms value of the resultant is determined from these measurements by the familiar formula:

$$B_{\text{resultant}} = \sqrt{\left(B_{xrms}\right)^2 + \left(B_{yrms}\right)^2 + \left(B_{zrms}\right)^2} \quad \text{tesla, rms}$$

If electronic filtering is used ahead of the voltage detector to effectively eliminate the harmonic contents in the voltage v$_o$ (Figure 3.6-3), then only the power frequency field will be determined. In this case the voltage detector can be either a true rms detector or an rms calibrated average detector with equal accuracy.

The calculation of $B_{resultant}$ in the presence of harmonics is based on the mutual orthogonality of the harmonic frequency components and the coordinate axis components of the field. This calculation is shown in Example 3.9-1.

Example 3.9-1 Calculation of $B_{resultant}$ in the presence of harmonics.

The space components of the magnetic field contains harmonics as shown in the table below:

Component milligauss rms			Harmonic Number
x	y	z	
10	7	5	1
2	3	1	3
1	2	2	5

Find the value of $B_{resultant}$.

Solution:

$$B_{xrms} = \sqrt{10^2 + 2^2 + 1^2} = 10.2 \text{ milligauss rms}$$

$$B_{yrms} = \sqrt{7^2 + 3^2 + 2^2} = 7.9 \text{ milligauss rms}$$

$$B_{zrms} = \sqrt{5^2 + 1^2 + 2^2} = 5.5 \text{ milligauss rms}$$

$$B_{resultant} = \sqrt{(10.2)^2 + (7.9)^2 + (5.5)^2} = 14.0 \text{ milligauss rms}$$

It is apparent from this calculation that:

$B_{resultant}$ = square root of the sum of the squares of the rms values of all harmonics (including the first harmonic or fundamental) over the three space axes.

$$B_{resultant} = \sqrt{(10)^2 + (2)^2 + (1)^2 + (7)^2 + (3)^2 + (2)^2 + (5)^2 + (1)^2 + (2)^2}$$
$$= 14.0 \text{ milligauss rms}.$$

3.10 Commercially Available Field Meters

A great many field meters are commercially available. They range from the very complex and expensive to the simple and inexpensive. We will discuss these generically, indicating the capabilities of both the simple single axis meter and the complex three axis meter.

Single Axis Field Meters. The single axis field meter utilizes a single search coil; thus, it must be rotated in space, at the point of interest, in order to obtain the direction and magnitude of the maximum field. Its use is most appropriate if the field is known to be linearly polarized as from a single phase source. If this is not known, it can be determined by noting that the meter will have a null or zero reading along the two directions orthogonal to the line of polarization in a linearly polarized field.

Field range settings are typically from 10 milligauss to 100 gauss or more, with a resolution of about 1 percent of full range.

The meter, which is hand held, may incorporate the search coil internally or utilize an external search coil. In the latter case, the voltage detector (see Figure 3.6-3) may be simply a battery-powered multimeter, shielded internally against magnetic fields.

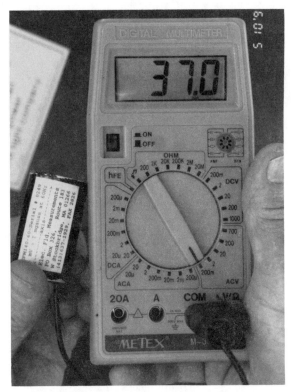

Figure 3.10-1 Single Axis Field Meter With External Search Coil Reading 37.0 Milligauss

The search coil is connected to the voltage measurement terminals of the meter. Without filtering, the average detecting meter will register $\dfrac{\pi}{2\sqrt{2}}$ times the average value of v_c. If no harmonics are present, the meter, with proper scale selection (per manufacturer's instructions), will read directly in rms milligauss. However, in the presence of harmonics, the scale factor of the meter is incorrect, and large errors may result, depending on the harmonic frequencies and magnitudes.

A typical single axis field meter with external search coil is shown in Figure 3.10-1 [4].

Three Axis Meters . A three axis meter allows us to make instantaneous measurement of the three orthogonal components of $\overline{B}$ at the point of measurement. These components are the projections of $\overline{B}$ along the meter axes, usually aligned by the user to earth related axes such as horizontal, vertical, and North. Using true rms instrumentation, the value of $B_{resultant}$ is calculated from:

$$B_{resultant} = \sqrt{\left(B_{xrms}\right)^2 + \left(B_{yrms}\right)^2 + \left(B_{zrms}\right)^2} \ .$$

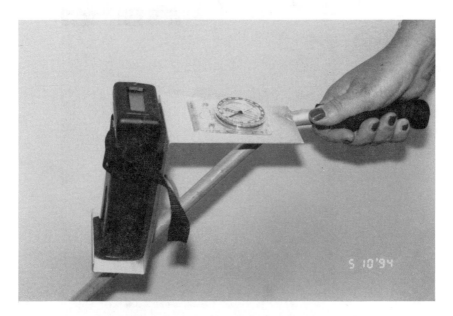

Figure 3.10-2 Three Axis Magnetic Field Meter With Internal Search Coils

A hand held meter is shown in Figure 3.10-2. A meter such as this may form the *heart* of a magnetic field measurement system. Such a system, utilizing distance measuring and direction sensing in conjunction with a microprocessor, allows the user to collect data needed for field surveys [5]. Figure 3.10-3 shows how field surveys can

50

be made in residences or offices, under transmission lines, and along property lines or rights of way.

The meter stores in memory the three axis magnetic field values, with a resolution of 0.05 milligauss in the 0-10 milligauss range, at intervals of 0.3 meters. The stored data can then be transferred to a computer for permanent storage and analysis. These data may be used to map magnetic field contours in a defined area, as shown in Figure 3.10-4a. Figure 3.10-4b shows a plan view of the magnetic field contours surrounding the transformer, at 100 percent load. The contours were plotted from a computer program that converted the measurements obtained by the meter into an equivalent contour map. Further data display capabilities are described in Chapter Six.

Figure 3.10-3 Using a Three Axis Meter to Perform a Field Survey

Figure 3.10-4a 25 kVA Single Phase Pad Mounted Transformer

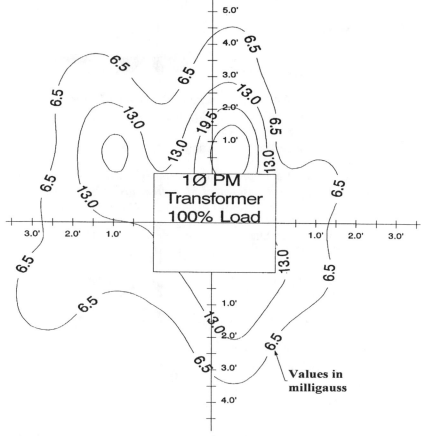

Figure 3.10-4b Measured Magnetic Field Contours for 25 kVA Single Phase Pad Mounted Transformer

3.11 Summary

The nature of a power frequency magnetic field depends on its source or sources (current carrying conductors). The magnetic field is linearly polarized when the source is a single phase alternating current or elliptically polarized when the sources are multi phase alternating currents, typically three phase, of a specific frequency.

The field measuring device or instrument is based on Faraday's Law: the voltage induced in a search coil is proportional to the rate of change of magnetic flux linkages. Two types of instruments are commercially available—a single phase instrument primarily used for measuring linearly polarized fields and a three axis instrument which is generally needed to measure elliptically polarized fields. It is possible to measure single frequency fields of the order of 10 milligauss with a resolution of about 0.1 milligauss or larger fields to an accuracy of about one percent of full scale.

If harmonics are present, the nature of the field is more complicated. Each of the harmonic components may be elliptically polarized and in various planes.

3.12 References

[1] Marshall, S. V. and Skitek, G. C., *Electromagnetic Concepts and Applications*, 3rd Edition, Prentice Hall, Englewood Cliffs, New Jersey, 1990.

[2] *A Protocol for Spot Measurements of Residential Power Frequency Magnetic Fields*, A Report of the IEEE Magnetic Fields Task Force of the AC Fields Working Group of the Corona and Field Effects Subcommittee of the Transmission and Distribution Committee, pp 1386-1393.

[3] Sicree, R. M., et al., *Comparison of Magnetic Flux Density Meter Responses Over a Database of Residential Measurements*, IEEE Transactions on Power Delivery, Vol. 8, No. 2, April 1993, pp 657-619, including discussions.

[4] Single Axis Meter Manufactured by Electric Field Measurements, P. O. Box 326, Route 183, West Stockbridge, MA 01266.

[5] Douglas, John, *Taking The Measure of Magnetic Fields*, EPRI Journal, April/May 1992, pp 16-17.

3.13 Exercises

[1] Single phase magnetic field.

Extend Example 3.3-1 to find the magnetic field vector $\overline{B}$ at a point 20 feet to the right of the center of the line, three feet above the ground. Let $\hat{x}$ be a unit vector in the horizontal (x) direction and $\hat{y}$ be a unit vector in the vertical (y) direction.

Answer: $(.4\ \hat{x} + 8.38\ \hat{y}\ )\sin 377t$ milligauss

[2] Multiphase magnetic field.
Extend Example 3.4-1 to find the horizontal (x) component of the magnetic field vector $\overline{B}$ at a point 20 feet to the right of the center of the line, three feet above the ground.

> Answer: 25.50 milligauss

[3] Measures and units.
What is the ratio of $\dfrac{B_{\text{resultant}}}{B_{\text{maximum}}}$ if $B_{s\,major} = k \times B_{s\,minor}$, where $k \geq 1$?

> Answer: $\sqrt{1+\left(\dfrac{1}{k}\right)^2}$

[4] Measurement.
Assume that harmonic currents are negligible. A single axis field meter calibrated in root mean square (rms) milligauss is oriented to maximum deflection—a field of magnitude 10 milligauss, rms is indicated.
The axis of the meter is then rotated into a plane orthogonal to the original axis.
(a)   No meter deflection is found in this plane.
(b)   The maximum deflection in this plane is 5 milligauss, rms.
Explain (a) and (b).

> Answer: (a) A single phase field, $R_{\text{resultant}} = 10$ milligauss, rms
> (b) A multiphase field, $R_{\text{resultant}} = 11.2$ milligauss, rms

[5] Harmonics.
A three axis meter employs switchable low pass filters in the signal processing circuits to effectively eliminate frequencies above 60 hertz.
A single phase magnetic field is found to have a resultant value of 10 milligauss rms, with the filters switched in. When the filters are switched out, the resultant field is 12 milligauss rms. Explain.

> Answer: Harmonics account for 6.6 milligauss rms.

4 Sources of Power Frequency Magnetic Fields in the Electric Energy Delivery System

4.0 Overview

The sources of power frequency magnetic fields are power frequency currents. Or, conversely, a 60 hertz current flow produces a corresponding 60 hertz magnetic field. Accordingly magnetic fields result from the generation, transmission, transformation, distribution and utilization of electric energy. In this chapter, we will examine a typical energy delivery system (or power system) and describe the fields associated with each stage of energy delivery. Generation will be omitted since generating stations are usually isolated from the general public.

A sketch of the electric energy delivery system is shown in Figure 4.0-1. For the purposes of discussion, we will divide it into the parts listed below and discuss magnetic fields as they emanate from these parts:

- Transmission System

    Transmission Line

    Distribution Substation

- Primary Distribution System

    Distribution Line

    Distribution Transformer

    Primary Switch

- Secondary and Service

    Secondary Line

    Service Line

Detailed characterizations of fields generated by the parts of the electric energy delivery system are found in Chapters Five and Six.

A sketch of a portion of a transmission system is shown in Figure 4.1-1. A single circuit three phase transmission line is shown, supplied from a generating station and tapped to feed a single distribution substation. Three phase lines are used to transmit large amounts of power, and a given line will typically serve a number of distribution substations.

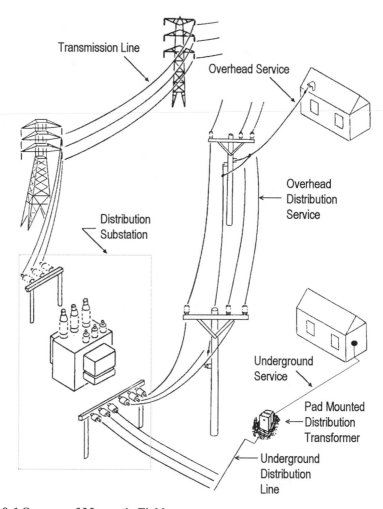

Figure 4.0-1 Sources of Magnetic Fields

Transmission line voltages range from 110 kV to 765 kV. A 345 kV line tower with typical dimensions is shown in Figure 4.1-2.

A magnetic field profile as calculated for the line exemplified by the tower shown in Figure 4.1-2 is shown in Figure 4.1-3. The calculation is based on the formulas developed in Appendix B for determining the magnetic field in the vicinity of a transmission line. The actual calculation is carried out by a computer program. Of the many programs available for such calculation, we used FIELDS 2.0 [3].

4.1 Transmission System

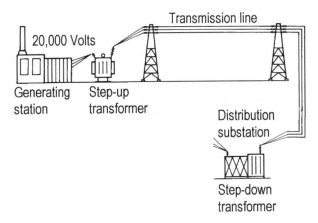

Figure 4.1-1 Portion of a Transmission System [1]

Transmission Line

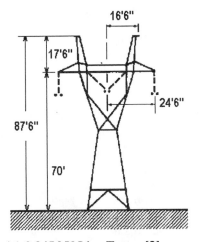

Figure 4.1-2 345 kV Line Tower [2]

The magnetic field is shown in terms of milligauss/100 amperes of line current. A 345 kV line might be designed to transmit approximately 600 megawatts of real power. At that power level the line current would be about 1000 amperes and the actual magnitude of the magnetic field would be ten times the value shown on the vertical scale.

An important consideration here is that the magnitude of the magnetic field decreases rapidly with lateral distance from the line. As is shown in Appendix C, the field strength can be expected to decrease according to an inverse square law with distance, if the three phase line currents are balanced. (Line currents are in fact almost

58

always well balanced in transmission systems.) Examples 4.1-1 and 4.1-2 illustrate the use of the magnetic field profile.

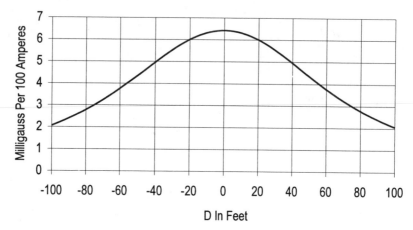

Figure 4.1-3 Magnetic Field Profile for 345 kV Line

Example 4.1-1   Scaling the magnetic field profile of Figure 4.1-3.

A 345 kV line, such as the one shown in Figure 4.1-2, carries balanced currents of 500 amperes rms per phase. What is the resultant field at a lateral distance of 100 feet?

Solution:

From the magnetic field profile of Figure 4.1-3, the resultant field is 2.1 milligauss rms, at a current of 100 amperes rms per phase. At 500 amperes the field would be 10.5 milligauss rms.

Example 4.1-2   Extending the magnetic field profile of Figure 4.1-3.

For the case of Example 4.2-1, estimate the field at a lateral distance of 200 feet.

Solution:

At twice the lateral distance, the field is approximately 1/4 x 10.5 milligauss rms = 2.6 milligauss rms.

Distribution Substation. Transformation from transmission voltages and currents to distribution voltages and currents occurs within the distribution substation. Since the magnitude of magnetic field intensity is proportional to current magnitude, fields will be greatest on the low voltage/high current side of the substation.

A 250 MVA, gas insulated substation is shown in plan view in Figure 4.1-4 [4].

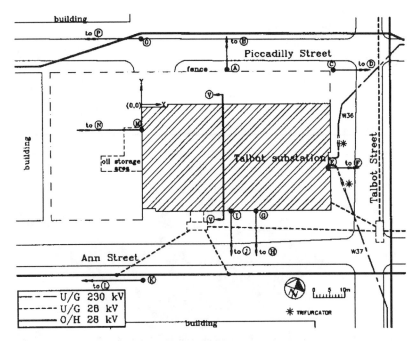

Figure 4.1-4 250 MVA Gas Insulated Substation

The purpose of this substation is to transform power from a transmission voltage of 230 kV to a distribution voltage of 28 kV, serving eleven outgoing feeder cables. It is one of eight large gas-insulated substations operated by Ontario Hydro.

A three-dimensional plot and a contour plot of the magnetic field both inside and outside the substation are shown in Figures 4.1-5 - 4.1-6. These plots were made from measurements using a three axis field meter over an extensive area around the station equipment. Only one of two 125 MVA station transformers was in operation when these measurements were made. Magnetic field magnitudes shown are resultant values, which reach some 400 milligauss in the immediate vicinity of the energized transformer.

4.2 Primary Distribution System

The primary distribution system distributes electrical power from the distribution substation to distribution transformers adjacent to customers. The primary distribution voltage is in the range from 4 to 35 kV; however, a range from 12 to 25 kV is the most common in United States utilities. Primary distribution constitutes one of the most important interfaces between utility generated magnetic fields and the public. The magnetic fields produced are the result of currents carried by distribution lines, both overhead and underground, and by associated equipment, such as distribution transformers, switches and fuses. We will first consider line current sources of magnetic fields, followed by equipment sources.

60

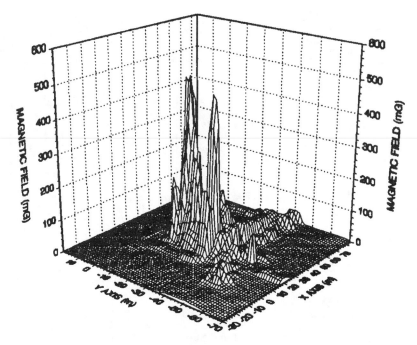

**Figure 4.1-5 Magnetic Field Inside and Outside Substation
Three Dimensional Plot**

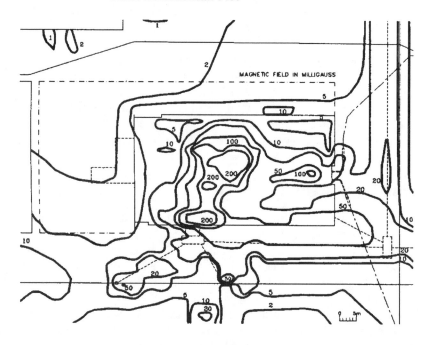

Figure 4.1-6 Magnetic Field Inside and Outside Substation Contour Plot

Primary Distribution Lines. Primary distribution lines emanate as feeders from the distribution substation. There may several such feeders, each with a capacity of the order of 10 MVA, radiating from the substation. Each feeder is radially connected, which is to say that there is normally no interconnection between feeders. Interconnections by normally open switches between adjacent feeders may be made if one feeder has suffered a fault. In this case the interconnections may provide service during repair of the faulted feeder. Typically, a distribution feeder will branch like a tree, rooted at the substations, into its service area. In its main trunk, the line will be three phase, carrying hundreds of amperes of current. Laterals branch off the trunk, carrying smaller currents. Principal laterals carry three phase currents, and lesser laterals or sublaterals carry single phase currents. The tree analogy is illustrated in Figure 4.2-1.

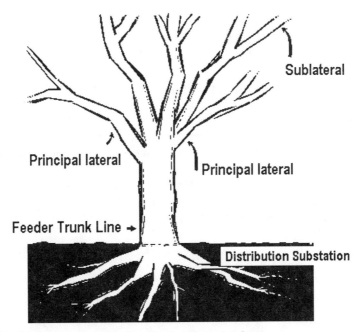

Figure 4.2-1 Primary Distribution Feeder Tree Analogy

The actual means of distribution vary widely between utilities and even within a single utility. Considerations such as conductor configuration and spacing, metallic or non metallic returns and grounding depend on practice, both present and past. A major division is between overhead and underground lines.

Overhead Primary Distribution Lines as Sources. A current-carrying, three phase overhead distribution line produces a magnetic field at ground level that is, typically, at maximum directly under the line. The magnitude of the field will depend on the type of construction, conductor height above ground and separation of conductors. In the case of a three phase line, the field will further depend on the balance of currents between the three phases.

The magnetic field profile of a three phase overhead line with cross arm construction is shown in Figure 4.2-2 [5].

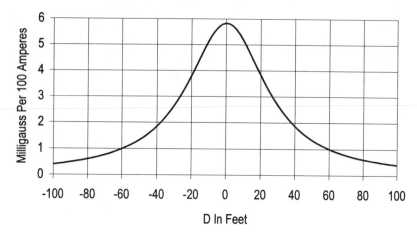

Figure 4.2-2 Magnetic Field Profile of a 12 kV Overhead Line With Balanced Three Phase Currents

This profile is a computer-generated plot of the resultant value of the magnetic field in milligauss rms versus lateral distance (D) from the center of the line. The magnetic field is computed at a height of three feet above the ground level. We note that the field profile is computed for balanced three phase sinusoidal currents of 100 amperes rms per phase. Since the magnetic field is proportional to current magnitude, the profile can be easily scaled to any desired set of balanced line currents.

Example 4.2-1. Scaling the magnetic field profile of Figure 4.2-2.

Using the field profile of Figure 4.2-2, find the resultant magnetic field at a lateral distance of 60 feet if the three phase current magnitudes are 300 amperes rms.

Solution:

Figure 4.2-2 shows a resultant field of 1 milligauss per 100 amperes at a lateral distance D of 60 feet. (Due to symmetry, D may be either to the left or right of the center of the line, ±60 feet, with the same result.) For a phase current magnitude of 300 amperes, the resultant magnetic field is three milligauss rms.

A detailed investigation of the effects of construction, conductor height, current balance, grounding and number of phase conductors on the magnetic fields due to overhead lines is given in Chapter Five.

Underground Primary Distribution Lines as Sources. Underground primary distribution lines consist, typically, of insulated cables in plastic duct, buried in a trench about three feet below ground level, as shown in Figure 4.2-3. Neither plastic duct nor earth affects the propagation of magnetic fields. The current carrying underground distribution cable produces a magnetic field at ground level which is at maximum directly over the line. As compared with an overhead distribution line, the field profile is higher and narrower, with the field strength dropping steeply as lateral distance increases. The shape of the profile, its height and steepness, is explained by the nearness of the cable to the observer (about six feet) and the close spacing between current carrying conductors (about two inches center to center).

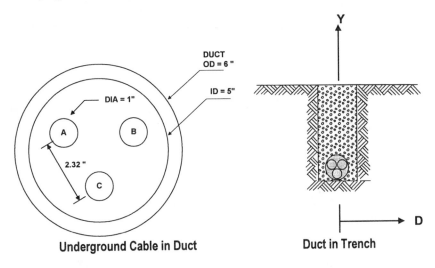

Figure 4.2-3 Underground Cable Configuration

The magnetic field profile of a three phase underground cable is shown in Figure 4.2-3 [5]. The magnetic field is computed at a height of three feet above ground level for balanced three phase sinusoidal currents of 100 amperes rms per phase.

The underground primary distribution line is typically buried in corridors along streets or sidewalks. The field strength in a corridor is examined in Example 4.2-2.

Example 4.2-2.   Maximum field in a corridor.

Using the field profile of Figure 4.2-4, find the maximum value of the resultant magnetic field in a cable corridor of ±5 feet, if the three phase current magnitudes are 500 amperes rms per phase.

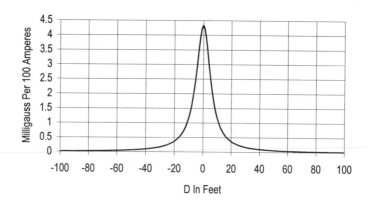

Figure 4.2-4 Magnetic Field Profile of a 12 kV Underground Cable With Balanced Three Phase Line Currents

Solution:

Figure 4.2-3 shows the maximum field (4.3 milligauss) to be at a lateral distance of 0. (D = 0). For a phase current magnitude of 500 amperes rms per phase, the resultant magnetic field is 5 x 4.25 = 21.25 milligauss rms. At the edges of the corridor the resultant field is approximately 15 milligauss.

The effects of trench depth, current balance, grounding and number of phase conductors on the magnetic fields due to underground lines are presented in Chapter Five.

Primary Distribution Equipment. Primary distribution lines terminate at distribution transformers, which reduce voltages from primary levels (for example, 4 kV to 35 kV) to user levels (for example, 120 V to 460 V). The primary distribution system employs switches for circuit reconfiguration, if needed, and fusing for circuit protection. These are placed in relatively large numbers throughout a feeder. Automatic switching devices such as reclosers are found relatively infrequently, as are capacitors and voltage regulators.

Distribution Transformers as Sources. Distribution transformers come in many forms and sizes. The most recognizable are the pole top transformers in the overhead distribution system. In a residential area the pole top distribution transformer may be single phase with a rating of 10 kVA mounted singly or in a cluster as shown in Figure 4.2-5. The underground distribution system equivalent is a pad mounted transformer rated 10 to 50 kVA, sited on the ground adjacent to a sidewalk or alley. Distribution transformers serving commercial or industrial customers will normally be three phase with ratings of 100 to 1,000 kVA.

The magnetic fields emanating from distribution transformers depend on the electrical load carried by the transformer, the larger the load the larger the field. The portion of the field due to stray core flux is relatively small as compared to the portion due to load currents. In any case the magnitude of the stray field drops off rapidly with distance.

Public interface with transformer fields is probably most important in the case of pad mounted transformers--since they can be approached more closely than pole top transformers.

The field contours surrounding a 25 kVA pad mounted transformer at full load are shown in Figure 4.2-6 [5]. Distances in Figure 4.2-6 are measured from the transformer sheet metal enclosure. The question of enclosing the transformer with a barrier, such as a fence, is considered in Example 4.2-3.

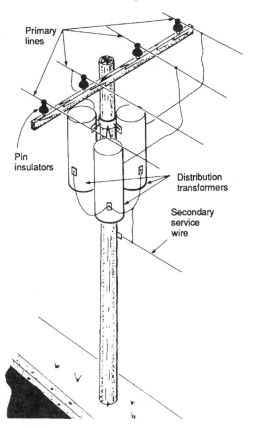

Figure 4.2-5 Modern Cluster Mounting of Distribution Transformers on a Pole [1]

Example 4.2-3.   Enclosing a transformer to limit exposure.

Estimate the size of a square enclosure, such as a fence, which would ensure that public exposure due to the 25 kVA transformer of Figure 4.2-6 would not exceed 10 milligauss.

Solution:

The fence should extend about three feet from the sides of the transformer case. The transformer dimensions are about three feet x three feet. A square fence, nine feet on a side, centered on the transformer would be adequate. A smaller enclosure could be used if it were centered some what to the left.

Switches/Fuses as Sources. Switches and/or fuses are not significant sources of magnetic fields in the overhead system. This is because the line geometry is generally not changed at the point of insertion of this equipment.

On the other hand, when placed in the underground system, switches/fuses require large changes in the line geometry. The three underground cables (which are in close proximity in the duct) must be separated at the switch/fuse, and this increased separation is responsible for an increase in the magnetic field in the vicinity of the equipment. This effect is examined in Example 4.2-4.

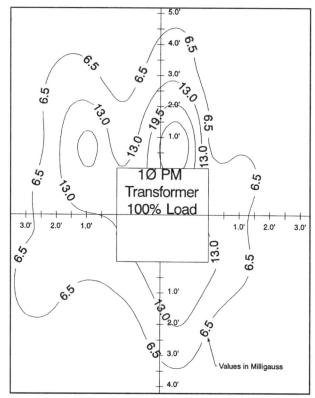

Figure 4.2-6 Measured Magnetic Field Contours for 25 kVA Single Phase Pad Mounted Transformer

67

Example 4.2-4. The effect of increasing conductor spacing.

A three phase underground cable, in a 3-foot-deep trench, has center to center distances between conductors of 2.3 inches and h equals -3 feet, as shown in Figure 4.2-3. If the center to center distances are increased to 12 inches, what is the effect on the magnetic field at a point three feet above the earth, directly above the cable?

Solution:

With 2.3 inches center to center distances, the resultant magnetic field is 4.3 milligauss /100 amperes rms of current. If the center to center distances are increased to 12 inches, the resultant field is 20.4 milligauss /100 amperes rms of current. This is an approximate solution based on a long line with 12 inches of separation between conductors, whereas the conductors are only separated for a short distance (on the order of several feet).

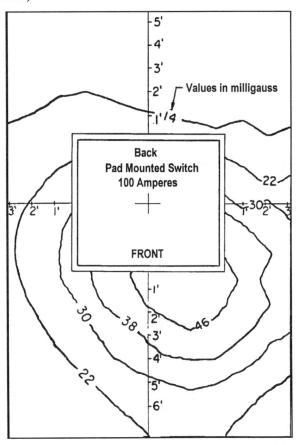

Figure 4.2-7 Measured Magnetic Field Contours in the Vicinity of a Pad Mounted Switch

Pad mounted switches in the underground system produce fields surrounding the switch cabinet. Figure 4.2-7 shows the magnetic field contours as measured in the vicinity of a pad mounted switch carrying 100 amperes [5]. The field is relatively large because the three phase current carrying cables must be separated at the switch.

The magnetic fields extending from primary distribution equipment will be characterized in greater detail in Chapter Six.

4.3 Secondary and Services

The secondary and services system distributes electric power from distribution transformers to customer service entrances. The secondary voltages for residential customers are 120/240 V single phase and 120/240/480 V single phase and three phase for commercial customers.

Secondary and Services as Sources. Since secondary voltages are relatively low, electrical insulation requirements are easily satisfied with modern insulating materials. As a result, modern secondary and services systems utilize triplex and/or quadruplex cable with an inherently close conductor spacing. If all currents are confined to these cables (that is, no ground return current), the net magnetic fields produced by the currents are insignificant.

Overhead secondary and services may be of the triplex or the open wire type. Open wire secondary produces much higher magnetic fields than does triplex cable. This is illustrated in Figure 4.3-1, which shows computed magnetic field profiles for a cross arm overhead secondary configuration and an overhead triplex configuration. While the open wire secondary produces a maximum field of the order of 10 milligauss /100 amperes, the triplex field is of the order of 0.1 milligauss /100 amperes--a 100:1 reduction.

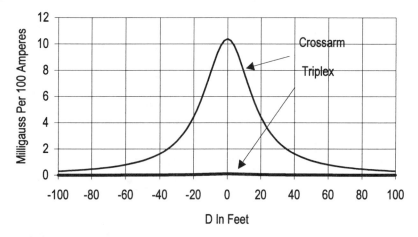

Figure 4.3-1 Magnetic Field Profile Comparing Cross Arm Secondary Configuration with Triplex Secondary Configuration

Meter Panels as Sources.. Electrical services terminate at the meter panel. Measurements have been made at the panels themselves to provide an indication of magnetic field magnitudes under peak demand conditions. These are shown in Table 4.3-1 [5].

Table 4.3-1 Measured Fields at Meter Panels

Panel	Parameters	Maximum Magnetic Field
Residential	@ 100A, one foot	96 milligauss
	@ 100A, three feet	12 milligauss
Commercial	@ 250A, one foot	128 milligauss
	@ 250A, three feet	15 milligauss

For both residential and commercial panels, the maximum magnetic fields at a distance of one foot from the front of the panel is of the order of 100 milligauss. The magnitudes at three feet are about 1/8 of the values at one foot.

4.4 Summary

We have seen that the electric utility energy delivery system contains a number of sources of magnetic fields. These fields interface with the public in a variety of ways. The most significant sources, in terms of field magnitudes and public exposure, are:

- primary distribution lines, both overhead and underground

- pad mounted distribution transformers and switch/fuses

- distribution substations

It should be noted that the sources discussed in this chapter were idealized. The idealizations were based on the assumption that three phase currents were balanced (no ground return current) and that all secondary currents flowed in the secondary system (again, no ground current).

In Chapter Five we will investigate sources with unbalanced three phase currents and secondary currents with return through the ground.

4.5 References

[1] Anthony J. Pansini, *Guide to Electrical Power Distribution Systems*, Prentice Hall, Englewood Cliffs, New Jersey, 1992.

[2] *Transmission Line Reference Book (Red Book), 345 kV and Above,* Second Edition, Electric Power Research Institute, 3412 Hillview Avenue, Palo Alto, California, 1982.

[3]   *The FIELDS 2.0 Program*, Southern California Edison Company, 6090 North Irwindale Avenue, Irwindale, California, 1992.

[4]   Wong, P. S., Rind, T. M., Harvey, S. M., and Scheer, R., *Power Frequency Electric and Magnetic Fields From a 230 kV Gas-Insulated Substation, IEEE 1993 Summer Power Meeting, Vancouver, B. C. Canada, July, 1993.*

[5]   Pacific Gas and Electric Company, *EMF Distribution Guidelines*, March 1993.

4.6 Exercises

[1]   **Transmission Lines.** Use the methods developed in Appendix B to calculate an exact value for the resultant field in Example 4.1-2.

Answer:	3.2 milligauss

[2]   **Distribution Lines.** Using the data of Figure 4.2-2 and 4.2-4, determine the current required in the underground cable to produce the same field as the overhead line carrying 100 amperes at a lateral distance of 10 feet.

Answer:	About 500 amperes

[3]   **Distribution Lines.** Repeat Exercise #1 only at the center of the line.

Answer:	About 130 amperes

[4]   **Conductor spacing.** Verify the solution of Example 4.2-4 using applicable formulas from Appendix B.

5 Characterization of Fields from Transmission and Distribution Lines

5.0 Overview

In Chapter Five we discuss the magnetic fields produced by transmission and distribution lines, focusing on three phase and single phase, overhead and underground, single and double circuit lines. Using the long line model (see Appendix B), the magnetic fields due to the currents in any of these lines are effectively described by the field profile—a plot of the resultant field magnitude versus lateral distance from the center of the line. All of the profiles presented here are computed for points of observation at a height of three feet above a flat earth, the standard height for such plots.

The field profiles for all of the lines considered in this chapter are similar in shape; they vary in magnitude depending on the factors listed below:

- conductor current
- conductor separation
- height (or depth) of conductor
- current balance (three phase)
- phasing (double circuit three phase)
- earth return portion

We offer a range of field profiles to illustrate the significance of these factors for both transmission and distribution lines.

5.1 Introduction to the Characterization (Description) of Fields from Transmission and Distribution Lines

Transmission line voltages range from 115 kV to 765 kV, whereas distribution line voltages range from 4 kV to 35 kV. Lines rated above 35 kV but less than 115 kV are termed sub-transmission lines. As sources of magnetic fields, transmission lines have a number of distinguishing characteristics [1]:

1. Transmission lines are invariably multiphase circuits. Typically the line is a single circuit three phase line, but double circuit three phase and multicircuit three phase lines are often used as well.

2. Transmission line currents are very nearly sinusoidal. The resulting magnetic fields have low harmonic content.
3. Transmission line currents are normally well balanced. A five percent difference between the magnitudes of the phase currents in a three phase circuit would be considered a large unbalance in transmission line currents.
4. Transmission lines are predominantly overhead lines. A relatively small number of underground transmission lines exist in the United States.
5. Phase conductors are normally located from 30 to 100 feet above ground with conductor separation from ten to 30 feet, depending on the line voltage.

Distribution lines, although similar in many respects to transmission lines, have important differences influencing their function as sources of magnetic fields:

1. Distribution lines may be three phase or single phase circuits.
2. Distribution line currents may contain appreciable harmonics of the 60 hertz power frequency. The predominate harmonics are the third, fifth and seventh. Total harmonic distortion may be significant but is normally less than five percent.
3. Distribution line currents may not be well balanced. A five percent zero-sequence component is not unusual for a three phase distribution circuit.
4. Distribution lines may be overhead or underground.
5. Conductors are typically 20 feet above ground, if overhead, with conductor separation on the order of three feet. If underground, the line is typically three feet below grade.

The currents that flow in transmission and distribution lines produce power frequency magnetic fields. Our characterization of these fields is based on the long line model as described in Appendix B. In the long line model, the field is oriented in the plane perpendicular to the line. Consequently the field is characterized by its profile—a plot of the resultant value of the field at a height of three feet versus lateral distance from the center of the line. A profile is readily computed from a computer software program such as FIELDS [2]. Using this method, we have described the fields associated with a number of line constructions (triangular, horizontal, vertical), circuits (single, double), and types [(overhead (OH), underground (UG))]. In addition we have determined the effects on magnetic fields of conductor separation, conductor height and phase current balance as well as earth versus metallic neutral return. The results of these parametric studies are shown graphically in the remainder of this chapter.

It should be remembered that the long line model is predicated on conductors whose height above a flat earth and separation are constant. Since this is never true we might ask the question: How well do the computed profiles, as shown in this chapter, represent actual (or measured) profiles? There is no single answer to this question. However, we estimate that if the conductor height and separation are reasonably constant (± five percent) for a specified section along the line, then the profiles at the center of the section will represent the actual profile to an accuracyof ± ten percent.

The specified section must be approximately a distance greater than or equal to 20 times the height of the line.

5.2 The Effect of Line Construction (Three Phase Overhead)

Transmission Lines. Line construction refers to the configuration of the line conductors. For example, Figures 5.2-1 and 5.2-2 show typical 500 kV lattice-type structures or towers supporting a triangular configuration of conductors (Figure 5.2-1), and a horizontal configuration (Figure 5.2-2) [1]. As these are transmission lines, we will compute field profiles for balanced phase currents only. The magnetic field profiles for the triangular construction are shown in Figure 5.2-1. These profiles correspond to three values of the bottom conductor height (h). The largest value of h, 72 feet, is its value at the tower. It is important to note that the line sags between towers. Profiles are also shown for values of h equal to 60 and 48 feet to indicate the effect of sag on the magnitude of the field. Similar profiles for the horizontal construction are shown in Figure 5.2-2. For both constructions the profiles peak directly below the center of the line. The horizontal configuration results in larger fields at all lateral distances when compared with the triangular configuration; however, the field magnitude differences become smaller as the lateral distance D increases.

Note that the field magnitudes are in terms of milligauss per 100 amperes of phase current. A major transmission line will typically carry phase currents of 1000 or more amperes rms. For phase currents of 1000 amperes rms, the magnetic fields are the values shown in the figure multiplied by a factor of 10. For example, in Figure 5.2-1, if the actual phase current flowing is 1000 amperes rms, then the magnetic field at D = 60 feet is approximately $3 \times 10 = 30$ milligauss under the line that is 72 feet high.

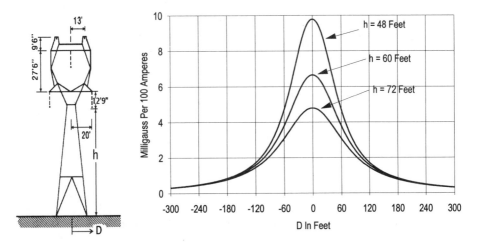

Figure 5.2-1 Magnetic Field Profiles for a 500 kV Transmission Line, Triangular Construction

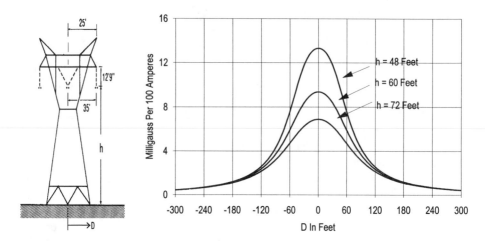

Figure 5.2-2 Magnetic Field Profiles for a 500 kV Transmission Line, Horizontal Construction

As transmission voltage is reduced, conductor separation and conductor height are likewise reduced. Figures 5.2-3 and 5.2-4 show typical 345 kV transmission line configurations [1]. The magnetic field profiles for the triangular configuration are shown in Figure 5.2-3, and those for the horizontal configuration are shown in Figure 5.2-4. The profiles have the same shapes as their 500 kV counterparts and about the same magnitudes at the point of observation.

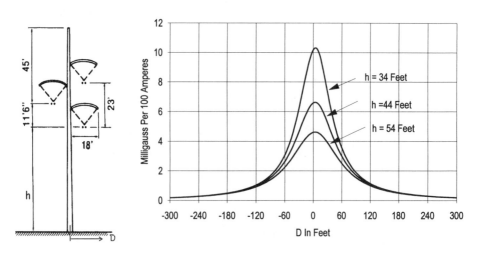

Figure 5.2-3 Magnetic Field Profiles for a 345 kV Transmission Line—Triangular Construction

This is explained as follows. Reducing conductor separation reduces the magnitude of the field while reducing conductor height increases the magnitude of the field. In going from 500 kV to 345 kV construction, these two effects are potentially canceling, resulting in a similar field magnitude at a given level of line current.

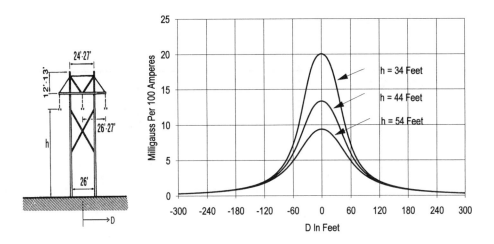

Figure 5.2-4 Magnetic Field Profiles for a 345 kV Transmission Line—Horizontal Construction

Figure 5.2-5 shows the magnetic field profile for a 230 kV line, horizontal construction.

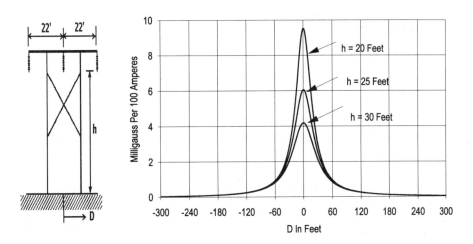

Figure 5.2-5 Magnetic Field Profiles for a 230 kV Transmission Line—Horizontal Construction

A 115 kV line [4] and corresponding magnetic field profiles are shown in Figure 5.2-6. The tower is of the so-called compact design. Accordingly, the profiles tend to be narrow when compared with horizontal construction.

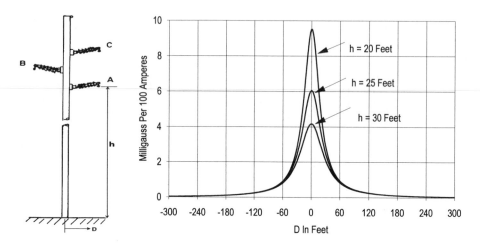

Figure 5.2-6 Magnetic Field Profiles for a 115 kV Transmission Line

Example 5.2-1   Calculation of current for a specified field level—triangular construction.

Suppose that the transmission corridors for 500 kV, 345 kV and 115 kV lines are 300 feet, 200 feet and 100 feet, respectively. If the allowable magnetic field magnitude at the edge of the corridor is 2 milligauss rms, how much phase current can be carried by each line? (Assume minimum conductor heights and triangular construction.)

Solution:

For the 500 kV line, at a value of D = 150 feet, phase currents of 200 amperes rms will produce about 2 milligauss rms.

For the 345 kV line, at a value of D = 100 feet, phase currents of 125 amperes rms will produce about 2 milligauss rms.

For the 115 kV line, at a value of D = 50 feet, phase currents of 200 amperes rms will produce about 2 milligauss rms. In Summary:

Line—kV	500	345	115
Phase Current—amperes rms	200	125	200

Example 5.2-2   Superposition of resultant magnetic fields.

A 500 kV line and a 115 kV line are constructed in parallel corridors. The centers of the corridors are 200 feet apart. Both lines are conducting phase currents of 300 amperes rms. What is the value of the (resultant) magnetic field midway between the two lines.

Solution:

Assume minimum conductor heights and triangular construction. At a value of D = 100 feet, and with a phase current of 300 amperes rms, the 500 kV line produces a resultant magnetic field of about 6 milligauss rms. Correspondingly, the 115 kV line produces a field of about 0.4 milligauss rms. Although we know both of these values, we cannot immediately calculate the field due to the two lines operating together. If we could use the concept of superposition in this case, then the field resulting from the two lines would be (6 + 0.4) or 6.4 milligauss rms. However, superposition does not apply to resultant values. The most we can assert that the resultant magnitude of the two lines will not be greater than 6.4 milligauss rms. The exact solution would require application of the theory developed in Appendix B and knowledge of the relative phase angles of all currents.

Example 5.2-3   Calculation of current for a specified field level—horizontal construction.

Suppose that the transmission corridors for 500 kV, 345 kV and 230 kV lines are 300 feet, 200 feet and 150 feet, respectively. If the allowable magnetic field magnitude at the edge of the corridor is 2 milligauss rms, how much phase current can be caused by each line? (Assume minimum conductor heights and horizontal construction.)

Solution:

For the 500 kV line, at a value of D = 150 feet, phase currents of 140 amperes rms will produce about 2 milligauss rms.

For the 345 kV line, at a value of D = 100 feet, phase currents of 70 amperes rms will produce about two milligauss rms.

For the 230 kV line, at a value of D = 75 feet, phase currents of 60 amperes rms will produce about two milligauss rms.

These currents, for horizontal construction, should be compared with those in Example 5.2-2 for triangular construction of the 500 kV and 345 kV lines.

Distribution Lines Primary distribution lines, while of lower voltage ratings, may carry as much or more current than transmission lines. Typically a distribution feeder will carry 600 amperes rms as it leaves a distribution substation. Lesser currents occur as the feeder fans out into its service area as discussed in Chapter Four.

The effect of line construction in distribution lines is shown by the line configurations and profiles of Figures 5.2-7 through 5.2-9. In these figures, with balanced currents, conductor height (h) is a parameter. The lines are representative of 21 kV construction. Lower voltage lines may have less separation between phase conductors while higher voltage lines will have greater separation.

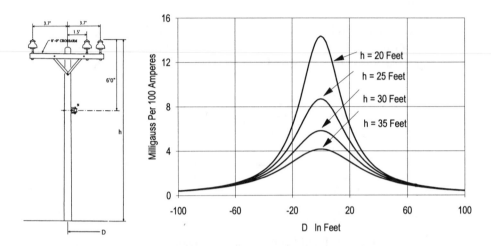

Figure 5.2-7 Magnetic Field Profiles For Overhead Primary Three Phase Horizontal Construction

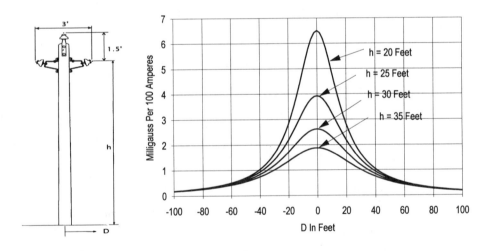

Figure 5.2-8 Magnetic Field Profile For Overhead Primary Three Phase Triangular Construction

A direct comparison of profiles from horizontal (or cross arm) and triangular construction is shown in Figure 5.2-10 at a common conductor height (h) of 30 feet. We see immediately that the triangular construction produces lesser fields. This effect is the direct result of the physically closer placement of the three phase conductors in triangular construction than in horizontal construction. Line construction that favors smaller separation between conductors and larger conductor height results in lesser fields at the point of observation (three feet above ground and at the lateral distance

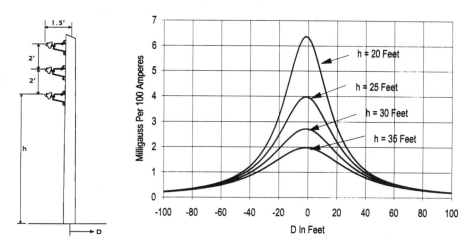

Figure 5.2-9 Magnetic Field Profiles For Overhead Primary Three Phase Vertical Construction

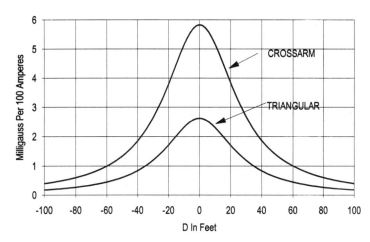

Figure 5.2-10 Magnetic Field Profiles as a Function of Construction Balanced Three Phase

Double Circuit Lines—Transmission In double circuit three phase lines it is possible to reduce the magnetic field at the point of observation by suitable arrangement of the phase conductors. This effect is shown in Figure 5.2-11 for a double circuit transmission line with horizontal construction [1]. Here the two circuits are separated vertically, each carrying balanced and equal phase currents. If the circuits are arranged horizontally, ABC/ABC, the field profile shows a maximum field of about 17 milligauss per 100 amperes. When arranged horizontally, ABC/CBA, the maximum field is about 8 milligauss per 100 amperes. This reduction is the result of a cancellation between fields in the ABC/CBA arrangement. The theory involved in this type of cancellation is discussed in Appendix C. The ABC/ABC arrangement is called the *super bundle* arrangement, whereas ABC/CBA is the so-called *low reactance* arrangement.

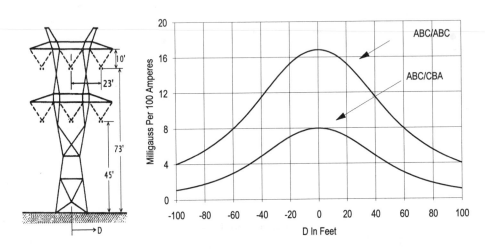

Figure 5.2-11 Magnetic Field Profiles For 345 kV Transmission Line Horizontal Construction Double Circuit

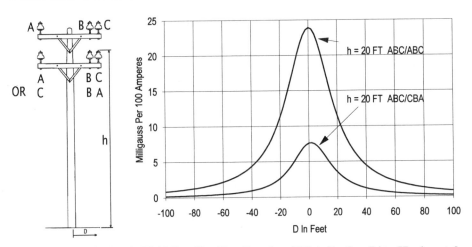

Figure 5.2-12 Magnetic Field Profiles For Overhead Distribution Line Horizontal Construction Double Circuit

Double Circuit Lines—Distribution Figure 5.2-12 shows the magnetic field profiles for a double circuit distribution line with horizontal construction. The two circuits are separated vertically, each carrying balanced and equal phase currents. If the circuits are arranged horizontally, ABC/ABC, the field profile shows a maximum field of about 24 milligauss per 100 amperes. When arranged horizontally, ABC/CBA, the maximum field is about 7 milligauss per 100 amperes.

It should be noted that the large decrease in the field seen here was predicated on perfectly balanced phase currents, with equal currents in each circuit. These conditions are more likely to occur in double circuit transmission lines than in distribution lines. Distribution lines have substantial unbalance between phases and unequal loading between circuits. As a result the relative phasing of the two circuits may be comparatively unimportant.

5.3 The Effect of Unbalanced Phase Currents (Three Phase Overhead)

A balanced set of three phase currents is described by the three equations below:

$$i_A = \sqrt{2}I_A \cos(377t + 0°) \quad \text{amperes}$$
$$i_B = \sqrt{2}I_B \cos(377t + 240°) \quad \text{amperes}$$
$$i_C = \sqrt{2}I_C \cos(377t + 120°) \quad \text{amperes}$$

Where t is time in seconds and $I_A = I_B = I_C$ is the rms value of the phase currents. In phasor notation:

$$\bar{I}_A = I_A \angle 0° \quad \text{amperes rms}$$
$$\bar{I}_B = I_B \angle 240° \quad \text{amperes rms}$$
$$\bar{I}_C = I_C \angle 120° \quad \text{amperes rms}$$

The phasor method is a technique used in analyzing Alternating Current circuits as a means of simplifying calculations. Referring to the phasor currents, if the phase angles between the currents are not 120 electrical degrees and/or the rms values of the phase currents are not equal, the set is said to be unbalanced. The most significant type of unbalance, with regard to magnetic fields, is that of unequal value of phase currents occurring in Y connected circuits. This type of unbalance can create a so-called zero-sequence[*] component as shown below.

$$\begin{aligned} I_A &= 110 \text{ amperes rms} \\ \text{Suppose:} \quad I_B &= 100 \text{ amperes rms} \\ I_C &= 100 \text{ amperes rms.} \end{aligned}$$

---

[*] The zero-sequence component is one of the three symmetrical components used to describe unbalanced three phase voltages or currents. A complete description of the symmetrical component decomposition of unbalanced phasors is given in Reference [5].

I_A is 10 percent higher than I_B or I_C. (If $I_A = 100$ amperes rms the current would be balanced.) The zero-sequence component, $\bar{I}_o$, is computed from the sum of the three current phasors:

$$\bar{I}_o = \frac{1}{3}\left(\bar{I}_A + \bar{I}_B + \bar{I}_C\right)$$

$$= \frac{1}{3}(110\angle 0° + 100\angle 240° + 100\angle 120°)$$

$$= \frac{10}{3} \angle 0° \text{ amperes rms.}$$

We see that the magnitude of $I_o \left(\frac{10}{3}\text{amperes rms}\right)$ is 3.3 percent of the magnitudes of the balanced set (100 amperes rms). In the computed profiles which follow, an x percent unbalance in phase currents is equivalent to an $\frac{x}{3}$ percent zero-sequence component.

The unbalanced current $\left(3\bar{I}_o\right)$ in a three phase circuit must be returned to the source. The return path may be a metallic neutral or a path through the earth or a combination of these. The earth path is assumed to be deep within the earth. Thus a current flowing in this path produces no field at the point of observation.

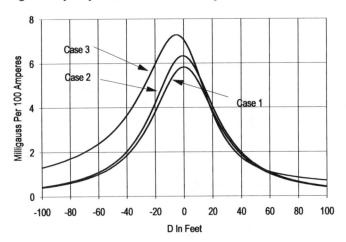

Figure 5.3-1 Magnetic Field Profiles for Overhead Primary Three Phase Horizontal Construction

The effect of unbalance in phase current magnitudes is shown in Figure 5.3-1. The configuration is horizontal with a conductor height of 30 feet. Figure 5.3-1 compares the balanced case (Case 1) with two 5 percent zero-sequence cases—Case 2, with current return through a metallic neutral as shown in Figure 5.2-7 and Case 3, with current return through the earth. We note the differences between the magnetic fields of the three cases for this Overhead configuration. The effect of zero-sequence current is to increase the magnitude of the field. The increase is moderate if a close-by metallic return is provided and relatively large if the return is through the earth. The fields are clearly reduced if a metallic return close to the phase conductors carries the neutral current. In an overhead line the minimum distance between the phase and neutral conductor is dictated by voltage considerations.

The increase in the magnetic field, with earth return, as a function of zero-sequence current is shown by the magnetic field profiles in Figure 5.3-2.

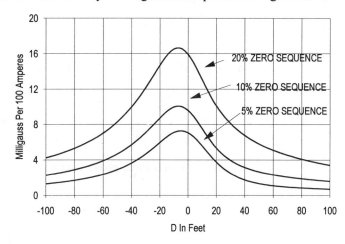

Figure 5.3-2 Magnetic Field Profiles for Overhead Primary Three Phase Horizontal Construction With 100 Percent Earth Return

As would be expected, the magnitudes of the profiles increase with increasing zero-sequence current. It should be noted that a zero-sequence component as large as 20 percent is very unlikely. Utilities normally maintain phase current balance in distribution circuits to produce less than 10 percent zero-sequence current.

Example 5.3-1   Unbalanced phase currents.

A three phase distribution line (see Figure 5.2-7) carries the following phase currents.

$$I_A = 300 \, \text{amperes rms}$$
$$I_B = 200 \, \text{amperes rms}$$
$$I_C = 200 \, \text{amperes rms}$$

Unbalanced currents are returned through the earth. Find the resultant magnetic field 20 feet to the right from the center of the line (D = 20 feet).

Solution:

The zero-sequence current is

$$I_0 = \frac{300\angle 0° + 200\angle 240° + 200\angle 120°}{3}$$

$$= 33 \text{ amperes rms}$$

This is $33/200 = 16.7$ percent of the magnitude of the balanced set (200 amperes rms). Using Figure 5.3-2, and interpolating between the profiles for 10 percent and 20 percent, the resultant magnetic field at 20 feet is about 18 milligauss.

5.4 The Effect of Line Construction (Single Phase Overhead)

Single phase overhead circuits may consist of two metallic conductors, providing a metallic path for return currents. This configuration is shown in Figure 5.4-1 with the corresponding field profiles for several conductor heights (h). Such construction is common for both primary and secondary Overhead circuits.

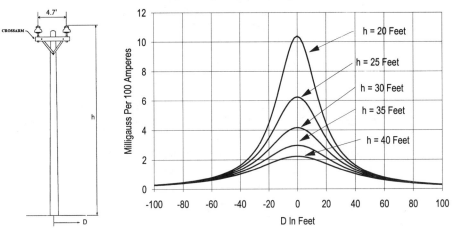

Figure 5.4-1 Magnetic Field Profiles for Overhead Single Phase Horizontal Construction With Conductor Separation of 4.7 Feet—Metallic Return Path

As in three phase lines, single phase conductor height has a significant effect in reducing the field at the point of observation. A second significant parameter is conductor separation. The conductor separation for Figure 5.4-1 is 4.7 feet. Figure 5.4-2 shows the same profiles as in Figure 5.4-1 with the exception that the conductor separation is reduced to 2.35 feet.

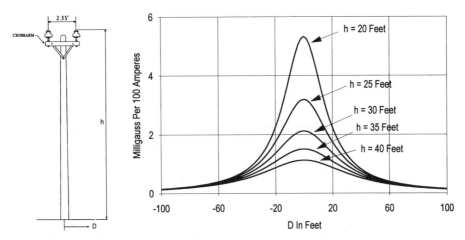

Figure 5.4-2 Magnetic Field Profiles for Overhead Single Phase Horizontal Construction With Conductor Separation of 2.35 Feet—Metallic Return Path

In comparing the profiles of Figure 5.4-2 and 5.4-1, we see that reducing the conductor separation by one half reduces the height of the field profiles by approximately one half. In other words, the magnitude of the magnetic field is approximately proportional to conductor separation. However, conductor separation is required for insulation, with the distance depending on the line voltage. In some cases, the distance of separation may be dictated by maintenance needs and personnel safety concerns rather than voltage rating.

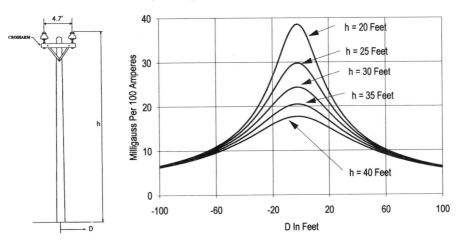

Figure 5.4-3 Magnetic Field Profiles for Overhead Single Phase Horizontal Construction Earth Return

Under some circumstances the return current may flow through the earth rather than through a metallic conductor. The field profiles for this case are shown in Figure 5.4-3. The magnitudes of the fields are seen to be much larger than in the case of a metallic return. This was to be expected, since the earth return path is assumed to be at a large depth—thus the return current in this case provides no field to cancel the field of the outgoing current.

5.5 The Effect of Line Construction (Three Phase Underground)

Three phase underground cable is quite compact in comparison to an overhead line with the same voltage rating. A 21 kV overhead distribution line has a conductor separation of approximately three feet; whereas in a 21 kV underground cable, the conductor separation is approximately two inches. Close proximity of the phase conductors promotes field cancellation. If the phase currents are balanced, the cancellation is almost complete except at short distances from the cable, as shown in the field profiles of Figure 5.5-1 for a 21 kV cable carrying balanced currents. In this figure the depth of the cable (h, which for underground cable is a negative number) is varied, and the observer is at three feet above the earth. As compared with the profile due to Overhead triangular construction, the Underground cable profile shown in Figure 5.5-2 is narrow but more intense at the center of the line. We note that the field due to the cable exceeds that due to the triangular overhead line for a distance D less than six feet to the left or to the right of the center of the line ($|D| \leq 6$ feet)[*]. This result, of course, is for a cable depth of three feet and a conductor height of 30 feet. The value of $|D|$ will be less if the cable depth is increased or the conductor height is decreased.

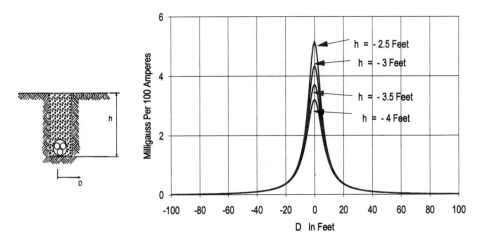

Figure 5.5-1 Magnetic Field Profiles for Underground Primary Three Phase 21 kV Construction

---

[*] This means the magnitude of D is less than or equal to 6 feet.

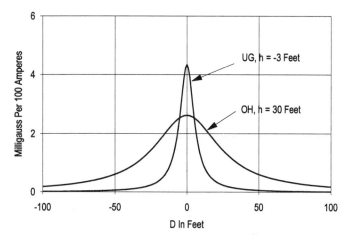

Figure 5.5-2 Magnetic Field Profiles of a Triangular Overhead Line and Underground Line with Balanced Three Phase Currents

Example 5.5-1   Comparing the fields of Underground cables with those of Overhead lines.

In Section 5.5 we noted that the field due to the cable exceeds that due to the triangular overhead line for $|D| \leq 6$ feet. For what values of D do the field, due to the cable, exceed:

(a)   the field due to a horizontal line?
(b)   the field due to a vertical line?

Assume a cable depth of three feet and a conductor height of 30 feet.

Solution:

(a) The horizontal line profiles are shown in Figure 5.2-7. Compare the profile for h = 30 feet with the cable profile in Figure 5.5-2. We see that the cable field never exceeds the field due to the horizontal line. In other words, the field due to the line is greater than that of the cable for all values of D.

(b) The vertical line profiles are shown in Figure 5.2-9. Compare the profile for h = 30 feet with the cable profile in Figure 5.5-2. We see that the cable field exceeds the field due to the horizontal line for $|D| \leq 2$ feet.

5.6 The Effect of Unbalanced Phase Currents (Three Phase Underground)

Unbalanced phase currents in the cable produce relatively small fields if the unbalanced current is returned through the metallic neutral of the cable. On the other hand, if the unbalanced current is returned through the earth, the field at the point of observation (three feet above the surface) may be relatively large (as shown in Figure 5.6-1). Figure 5.6-1 compares the field profile for balanced currents with the profile for unbalanced currents (refer to Section 5.3). The current unbalance (3.3 percent zero-sequence current) is returned through the earth.

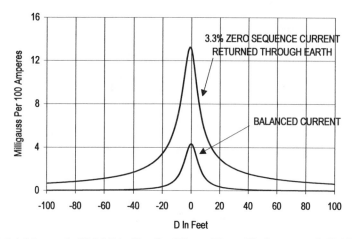

Figure 5.6-1 Magnetic Field Profiles for Three Phase Underground Cable Comparing Balanced with Unbalanced Case

As seen in Figure 5.6-1, a zero-sequence current of 3.3 percent returned through the earth increases the profile magnitude by approximately three times. We would expect that a 10 percent zero-sequence current returned through the earth would result in about a ten times increase in the profile magnitude.

5.7 The Effect of Earth Return (Single Phase Underground)

A single phase 12 kV underground cable is shown in Figure 5.7-1. This type of cable is typical of the cable found in underground distribution feeders, described in Chapter Four. Depending on the conductor size, the cable may carry currents in the range from 25 to 600 amperes. The concentric shield wire shown in the illustration, may carry a part or all of the return current.

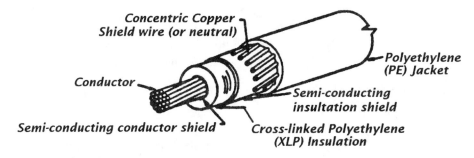

Figure 5.7-1 Single Phase Cable for Underground Distribution [6]

The part not carried by the shield wire (or neutral) returns in the earth. The magnetic field profile associated with a single phase underground cable depends very heavily on the portion of the return current which returns in the earth or conversely returns in the neutral. In Figure 5.7-2, this is shown graphically for a single phase feeder buried three feet below grade.

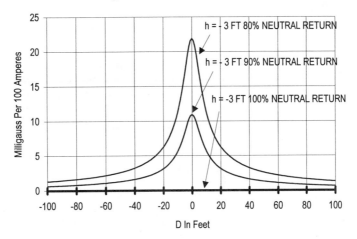

Figure 5.7-2 Magnetic Field Profiles For Underground Primary Single Phase 21 kV Construction

We note that the magnetic field is negligible if the entire return current is carried by the neutral. In fact, the magnitude of the profile is proportional to the percent of return current carried by the earth. For example, from Figure 5.7-2:

Percent Neutral Return	Percent Earth Return	Maximum Profile Magnitude
100	0	≈ 0 milligauss
90	10	11
80	20	22

Example 5.7-1   Magnetic field with earth return.

Estimate the magnitude of the magnetic field due to a single phase cable with 70 percent neutral return at a lateral distance (D) of 30 feet. The cable is buried three feet below the surface.

Solution:

If the neutral return is 70 percent, the earth return is 100 percent - 70 percent = 30 percent. Figure 5.7-2 shows a profile with 80 percent neutral return (20 percent earth return). The magnitude of this profile at D = 30 feet is about 4.5 milligauss. Using proportionality, the value for 30 percent earth return is approximately:

$$4.5 \times \frac{30}{20} = 6.8 \,\text{milligauss.}$$

5.8 Summary

The magnetic fields emanating from transmission and distribution lines, both three phase and single phase, are remarkably similar in profile. The significant variables that determine the magnitude and shape of the field profile are:

three phase
$\begin{cases} \\ \\ \\ \\ \\ \end{cases}$
current magnitude
separation between phase conductors
current balance and return path
distance from conductors to observer
phasing in double circuit construction

single phase
$\begin{cases} \\ \\ \\ \\ \end{cases}$
current magnitude
separation between metallic conductors (Overhead)
portion of current returning through earth
distance from conductors to observer

We have examined illustrative magnetic field profiles (computer generated using the FIELDS program) for several types of construction, amounts of unbalance, return paths and conductor heights (depths). As the FIELDS program utilizes the long line model as discussed in Appendix B, the profiles represent actual fields only where the conductor height and separation are reasonable constant over a distance of the order of 20 times the height of the conductors.

5.9 References

[1] *Transmission Line Reference Book 345 kV and Above,* Second Edition, Electric Power Research Institute, 1982.

[2] *IEEE Standard Procedures for Measurement of Power Frequency Electric and Magnetic Fields from Power Lines*, ANSI/IEEE Standard 644-1987, Institute of Electrical and Electronics Engineers, Inc., New York, New York, 1987.

[3] *The FIELDS 2.0 Program*, Southern California Edison Company, 6090 North Irwindale Avenue, Irwindale, California, 1992.

[4] *Transmission Line Reference Book, 115-138 kV Compact Line Design*, Electric Power Research Institute, 3412 Hillview Avenue, Palo Alto, California.

[5] *Elements of Power System Analysis*, Fifth Edition, by Grainger and Stevenson, McGraw Hill Book Company, 1993.

[6] *PG&E Standards Book: Cables for Underground Distribution*, Drawing Number 039955, 2/28/90.

5.10   Exercises

[1] Calculation of Field for Single Circuit Overhead Line. A three phase, 21 kV, OH circuit carries balanced currents of 500 amperes per phase. What is the resultant field at a lateral distance of 40 feet, if the construction is horizontal with a conductor height of 25 feet?

Answer:	10 milligauss

[2] **Calculation of Field for Double Circuit Line.** The line of Exercise 5.10-1 is split into two equal circuits on the same pole with the lower cross arm height of 20 feet. What is the resultant field at a lateral distance of 40 feet if the phasing is ABC/ABC? ABC/CBA?

Answer:	ABC/ABC: 11 milligauss	ABC/CBA: 4 milligauss

[3] **Calculation of Field for Underground Line.** The line of Exercise 5.9-1 is undergrounded at a depth of three feet. What is the resultant field at a lateral distance of 40 feet?

Answer: about 0.5 milligauss

[4] **Calculation of Field for Unbalanced Currents.** How is the answer of Exercise 5.3-1 changed if there is a 5.0 percent zero-sequence current returned through the earth?

Answer: 8 milligauss

[5] **Calculation of Field for Single Phase Line.** A single phase cable is buried three feet below grade. It carries 200 amperes. One hundred sixty (160) amperes are returned in the concentric neutral. Estimate the magnitude of the magnetic field directly above the cable, three feet above grade.

Answer: 4 milligauss

6 Characterization of Fields From Distribution Substations and Transformers

6.0 Overview

In Chapter Six we characterize magnetic fields produced by two major components of the electric distribution system:

- Distribution Substations
- Distribution Transformers

There are many distribution substation sizes, ratings and configurations. The data presented here serve to show the general characteristics of the magnetic fields associated with representative members of the substation family.

We provide measured data for the fields of a 50 MVA substation operating at about 60 percent of rated load. The data are displayed as field contours and a 3-D plot. In addition, we show a computer simulation of the fields of a 100 MVA substation operating at 40 percent of rated load. In both cases the principal sources of the magnetic field within the substation and along its boundaries are identified.

We describe measured magnetic fields of the following distribution transformers:

- 300 kVA, three phase, pad mounted
- 75 kVA, three phase, pad mounted
- 25 kVA, single phase, pad mounted
- 25 kVA, single phase, sub surface

Each of the transformers addressed produces fields which are characterized by a rapid decrease with distance from the transformer enclosures. Both contours and 3-D plots illustrate this variation of field strength with distance.

Pole mounted distribution transformers were not investigated since they are reasonably distanced from the public.

6.1 Introduction to Characterization of Fields of Distribution Substations and Transformers

As we saw in Chapter Four, the electric energy distribution system, which provides electric energy to industrial, commercial, agricultural and residential customers, is a ubiquitous source of magnetic fields. In Chapter Five we discussed the characterization of fields from distribution as well as transmission lines, recognizing that lines are an important, wide scale, source of power frequency fields. Equally important, but localized sources, in the distribution system are distribution substations and distribution transformers. Distribution substations are located in industrial areas to supply major energy customers, in commercial projects such as large malls and office buildings and adjacent to residential areas of single family homes and apartments. Distribution transformers are located close to points of utilization of electric energy: overhead on poles, at ground level on pads, below grade level in vaults and in the "electrical rooms" or power centers of commercial and industrial buildings. The configuration depends on the nature of the primary distribution line (overhead or underground) and the loads served by the transformer.

In the sections below describing the power frequency fields associated with some typical distribution substations and transformers, we will find that relatively high magnetic fields are associated with both. However, the high fields are largely localized to the immediate vicinity of the substation or transformer, unlike the fields of lines that cover wide areas.

6.2 Distribution Substations

Distribution substations have the role of receiving electric energy at a transmission or subtransmission voltage level (69 kV and above) and converting it to a primary distribution voltage level (35 kV and below) as needed for distribution feeders. (Refer to Section 4.2 of Chapter Four for a graphic description of the primary distribution system.)

Figure 6.2-1 Distribution Substation

Conversion of energy from a transmission voltage level to a primary distribution voltage level is carried out, within the substation, by one or more large step-down transformers. In addition, the substation may incorporate both high voltage and low voltage buses with provisions for switching. Typically, the substation will contain a control and metering room with protective relays for sensing and automatically sectionalizing portions of the distribution system in the event of a fault.

Distribution feeders are connected to the low voltage (primary distribution voltage) buses of the substation through station breakers-oil or air breakers that can be used to disconnect a faulted feeder if necessary. The feeders themselves, from four to ten in number, fan out from the substation toward points of utilization. While this configuration may describe a typical substation in a residential or commercial environment, as noted above, a number of other substation environments do exist.

We see in Figures 6.2-1 and 6.2-2, portions of a distribution substation located in a residential/agricultural area. This substation receives transmission energy at a voltage level of 115 kV and provides primary distribution voltage to seven feeders at 21 kV. Its load rating is 70 MVA. A three phase overhead feeder leaving the substation is shown on the right side of Figure 6.2-2.

Figure 6.2-2 Distribution Substation

Also shown in Figure 6.2-2 is a section of a chain link fence. The fence, in fact, surrounds the substation completely. All substations are isolated in this or some other way for public safety and equipment protection. Since the public may only advance to the substation perimeter or fence line, we are particularly concerned with magnetic fields along the perimeter. Of course electrical workers will be exposed to higher level fields within the substation perimeter as they carry out routine maintenance, repair,

96

metering and switching functions. Such higher internal fields are also illustrated in our discussion.

6.3 Characterization of Fields of Distribution Substations

The substation is a complex concentration of magnetic field sources. The fields due to sources such as the main buses of the station, vertical drops to switch gear, switch gear, distribution feeders, transformers, and high voltage transmission lines. Adding to the complexity of the problem is the fact that the feeder loads vary in time and may not be fully balanced. Calculation of the resulting magnetic field requires a large number of inputs to complex three dimensional field programs. A number of computer programs are available for the prediction of magnetic fields for proposed substations. For existing installations, measurements may be used to map the magnetic fields within and around the substation.

In this section, we characterize the power frequency magnetic fields of distribution substations both within and along the substation perimeter. The characterization of a 50 MVA substation is based on measurement, while that of a proposed 100 MVA substation is based on computation.

50 MVA Outdoor Substation. A 50 MVA, 115 kV / 21 kV outdoor substation, located in a rural area is shown in plan view in Figure 6.3-1 [1]. The substation incorporates two 25 MVA, step-down transformers, supplying three overhead and two underground feeders. It is supplied in turn by two 115 kV transmission lines.

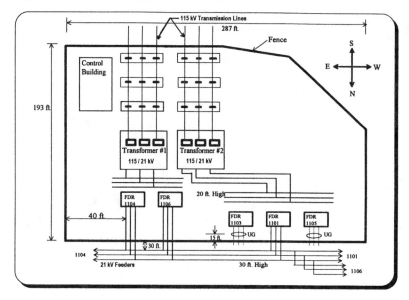

Figure 6.3-1 Plan View of 50 MVA Substation

The resultant magnetic field as measured at a height of three feet, within the substation fence is shown by field contours in Figure 6.3-2 and by a 3-D plot in Figure 6.3-3. The field measurements were made using an EMDEX II [2] instrument. The substation was providing a load of approximately 28 MVA when the measurements were made.

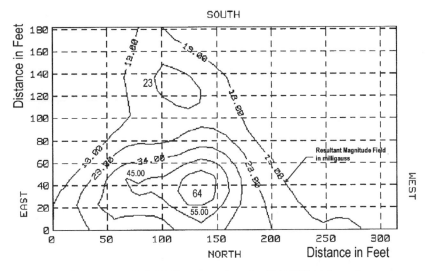

Figure 6.3-2 Measured Field Contours for the 50 MVA Substation

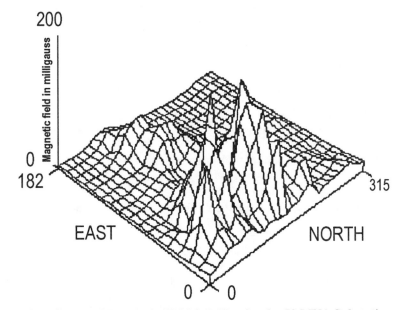

Figure 6.3-3 Measured Magnetic Field 3-D Plot for the 50 MVA Substation

The resultant magnetic field at the fence line is greatest along the north fence. The field here is seen to vary from about 10 milligauss to 40 milligauss. This is the low voltage/high current region of the substation and, as we would expect due to the high current, the fields here are the highest ones measured in the substation. Directly below the low voltage buses, within the substation, the measurements indicated a maximum field magnitude of 64 milligauss.

The relatively large fields along the north fence are due to both substation currents and to the concentration of feeders leaving the substation. We note that all five feeders leave the substation in this region. Along most of the east and west boundaries, the fields are small, since these boundaries are at a distance from the high current region of the substation and the feeders.

At other load levels the substation fields would change almost proportionately with load, providing that the loads on all feeders move together. In this case the contours would maintain their shapes and locations; only the field values associated with them would change. If the relative loads between feeders is changed, the contour shapes would also change.

Example 6.3-1   Calculation of 50 MVA substation field.

The field contours for the 50 MVA substation are shown in Figure 6.3-2, at a substation load of 28 MVA. We will assume that the loads on all five of the feeders increase together to an aggregate value of 50 MVA (the rated substation load). Determine the maximum magnetic fields at the north fence, the south fence, and within the substation.

Solution:

Under the assumption that the relative feeder loads are unchanged (only the sum changes), the fields increase approximately proportionately with load. Therefore:

$$\text{maximum field at north fence} = \frac{50}{28} \times 40 = 71 \text{ milligauss}$$

$$\text{maximum field at south fence} = \frac{50}{28} \times 15 = 27 \text{ milligauss}$$

$$\text{maximum field within substation} = \frac{50}{28} \times 64 = 114 \text{ milligauss}$$

100 MVA Outdoor Substation. A proposed 100 MVA, 115 kV / 21 kV outdoor substation, is shown in plan view in Figure 6.3-4a and in a 3-D View in Figure 6.3-4b. The substation incorporates three 35 MVA step-down transformers, each supplying three underground feeders. The substation is fed by two 115 kV transmission lines.

The resultant magnetic field as computed at a height of three feet above ground and within the substation fence is shown by the 3-D plot in Figure 6.3-5. Field computations were made using the program SUBCALC [3] authored by Enertech Consultants and Ohio State University under a contract issued by the Electric Power Research Institute. The substation load was 40 MVA, equally distributed between the step-down transformers and feeders. It should be noted that SUBCALC, as presently written, does not model the step-down transformers. The stray fields due to these transformers will attenuate rapidly with distance. Their effect along the substation boundary is considered to be small, justifying the fact that they are not modeled.

As shown in Figure 6.3-5, the computed field inside the substation is generally less than 25 milligauss except in the vicinity of the low voltage buses. Directly below the buses, sharp peaks of about 220 milligauss appear.

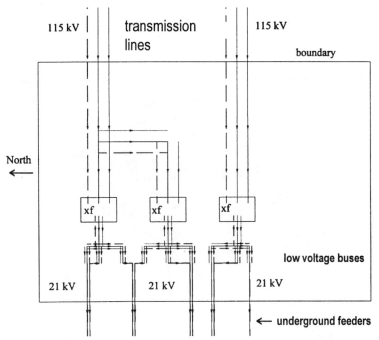

Figure 6.3-4a Plan View of 100 MVA Substation

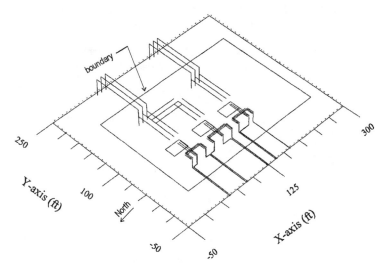

Figure 6.3-4b 3-D View of 100 MVA Substation

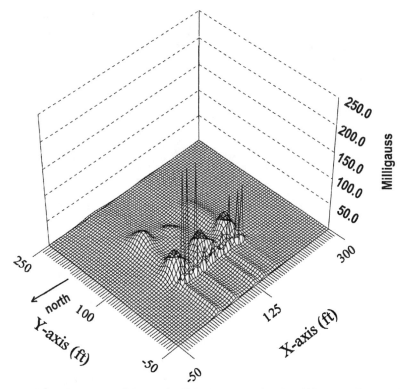

Figure 6.3-5 Computed Magnetic Field 3-D Plot for the 100 MVA Substation

Magnetic field profiles were computed along the east and west boundaries of the substation with the results shown in Figures 6.3-6 and 6.3-7.

Magnetic Field Along East Boundary
(40 MVA Load, Height of Measurement 3 Feet)

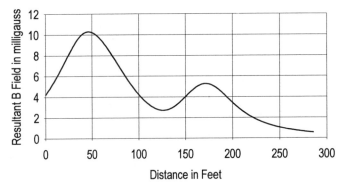

Figure 6.3-6 Computed Magnetic Field Profile for 100 MVA Substation (Clockwise Along the Boundary)

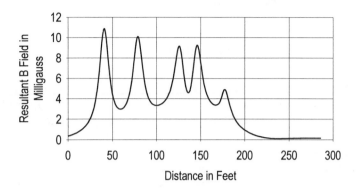

Figure 6.3-7 Computed Magnetic Field Profile for 100 MVA Substation (Counter Clockwise Along the Boundary)

The fields along the east boundary are largely determined by the transmission line currents while the fields along the west boundary are largely determined by the feeder currents. In fact the locations of the transmission lines and feeders, as they cross the boundaries, are seen as locations of maxima of the magnetic field profiles. Along the north and south boundaries the magnetic fields are vanishingly small as shown in Figure 6.3-5.

Example 6.3-2   Calculation of 100 MVA substation field.

If the loads on the feeders increase together to a peak at a substation load of 100 MVA, find the maximum resultant fields at the east and west boundaries and within the substation.

Solution:

As in Example 6.3-1, under the assumption that the relative feeder loads are unchanged (only the sum changes), the fields increase proportionately with load. Therefore:

$$\text{maximum field on east boundary} = \frac{100}{40} \times 10.3 = 26 \text{ milligauss}$$

$$\text{maximum field on west boundary} = \frac{100}{40} \times 11 = 28 \text{ milligauss}$$

$$\text{maximum field within the substation} = \frac{100}{40} \times 220 = 550 \text{ milligauss}$$

Distribution Substation Magnetic Fields. Our study of substation magnetic fields shows that the major impact on the public occurs at the boundary areas in which the distribution feeders exit the substation. If there is a concentration of feeders in this area or areas, we can expect a corresponding increase in the magnetic field strength, unless the utility has taken steps to mitigate this effect. We estimate that the fields emitted by the substation itself are effectively attenuated at a distance of some fifty feet from a substation boundary. However, the fields due to lines entering and leaving the substation may be significant at a much greater distance depending on the line configurations and currents. The calculation of fields due to transmission and distribution lines as discussed in Chapter Five applies.

6.4 Distribution Transformers

Distribution transformers convert electrical energy at primary distribution voltage levels (some typical values: 35 kV, 21 kV, 12.5 kV and 4.16 kV) to utilization or secondary levels (some typical values: 460 v, 240 v, 208 v and 120 v). Ratings range from 25 kVA to 500 kVA and sometimes even higher. The lower ratings, 25, 37.5 and 50 kVA, are normally single phase units, while those with higher power ratings are invariably three phase units. We find distribution transformers on pole tops, on concrete pads, in underground vaults and at the power centers of large buildings. They are always located close to the points of utilization because secondary voltage distribution is inherently inefficient. (Distribution of power at low voltage/high current, over long distance, results in unacceptable heating loss caused by high current flow.)

6.5 Characterization of Fields of Distribution Transformers

The purpose of distribution transformers is to convert electricity from a high voltage, low current state, to a low voltage, high current state. This is accomplished through the use of conducting coils wound on a magnetic core. A schematic diagram of a single phase, two winding (coil) transformer is shown in Figure 6.5-1.

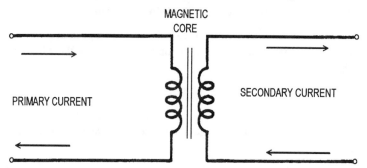

Figure 6.5-1 Schematic Diagram of a Two Winding Single Phase Transformer

An intense magnetic field exists within the transformer (thousands of gauss) but this field is effectively contained within the magnetic core of the transformer. The field external to the magnetic core is the result of the currents that flow in conductors into and out of the transformer and any stray flux that escapes from the core of the transformer. Of these three sources:

- primary current
- secondary current
- stray flux

the most important is the secondary current. This, of course, is true because secondary currents are much higher than primary currents (in the inverse ratio of the corresponding voltages) and stray flux is usually small. We see, then, that the highest external magnetic fields will be in the vicinity of the secondary conductors. In the sections below we address the fields associated with several distribution transformers, based on measurements. As stated earlier, such fields drop off rapidly with distance from the transformer. Accordingly, we have chosen to study only pad mounted and subsurface transformers since pole mounted units, some 20 feet above the earth, are effectively removed from the public...

Figure 6.5-2 300 kVA Pad Mounted Transformer

300 kVA, Three Phase, Pad Mounted Transformer. A 300 kVA, three phase, pad mounted transformer, installed on a university campus, is shown in Figure 6.5-2. External fields surrounding this transformer were measured [1] at a height of three feet above ground, with the transformer operating at about 50 percent of its full load rating. A three-dimensional plot of the resultant magnetic field is shown in Figure 6.5-3a followed by a contour plot in Figure 6.5-3b. We note that a rather high field is present immediately adjacent to the transformer, close to the secondary connections. The field drops off rapidly and is down to ambient levels at distances of ten feet or farther away from the transformer enclosure.

104

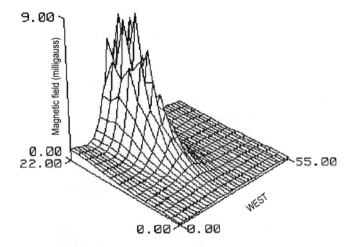

Figure 6.5-3a Measured Magnetic Field 3-D Plot for the 300 kVA, Three Phase, Pad Mounted Transformer

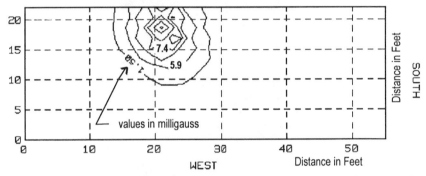

Figure 6.5-3b Measured Field Contours for the 300 kVA, Three Phase, Pad Mounted Transformer

It should be noted that the sheet metal transformer enclosure plays a role in containing the stray core flux, providing some shielding to the region surrounding the enclosure.

75 kVA, Three Phase, Pad Mounted Transformer. A 75 kVA, three phase, pad mounted, 22 kV, 208 Volt/120 Volt transformer is shown in Figure 6.5-4 [4]. This transformer might typically be found adjacent to a small shopping mall. It is part of an underground distribution system, being supplied at a primary distribution voltage of 22 kV, and providing three phase power for utilization at 208 volts / 120 volts.

The measured magnetic field contours in the vicinity of this transformer, when carrying about 20 kVA of load, are shown in Figure 6.5-5.

The contours of Figure 6.5-5 were measured at a height of three feet. The lower right region contains the secondary connections with the external field most intense in this region. Distances from the transformer enclosure are indicated in the figure. The transformer field will be almost entirely attenuated some ten feet from the enclosure.

Figure 6.5-4 75 kVA Three Phase Pad Mounted Transformer

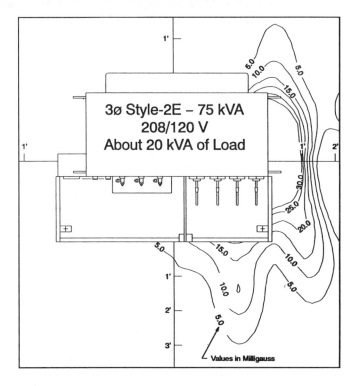

Figure 6.5-5 Measured Magnetic Field Contours For 75 kVA Three Phase Pad Mounted Transformer.

25 kVA, Single Phase, Pad Mounted Transformer. A 25 kVA, single phase, pad mounted, 12 kV / 240 V / 120 V transformer is shown in Figure 6.5-6 [4]. This type of transformer would be encountered in a residential area, adjacent to the sidewalk or parking strip. The measured magnetic field contours surrounding this transformer when it is carrying full load currents are shown in Figure 6.5-7. The fields were measured at a height of three feet; lateral distances from the enclosure are indicated in the figure.

Figure 6.5-6 25 kVA Single Phase Pad Mounted Transformer

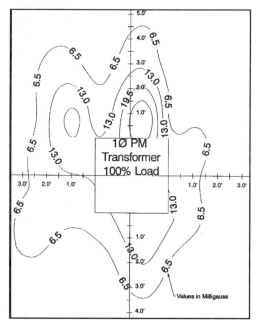

Figure 6.5-7 Measured Magnetic Field Contours for 25 kVA Single Phase Pad Mounted Transformer

25 kVA, Single Phase, Subsurface Transformer. . A subsurface transformer, as with the pad mounted transformer discussed above, may be located in a residential area and served by underground cable. The cover plate for a 25 kVA, single phase, subsurface transformer is shown in Figure 6.5-8 [4]. The cover plate is the only visible indication of the transformer location. The measured magnetic field contours, at a height of three feet above the plate, are shown in Figure 6.5-9. The contour values correspond to full load currents.

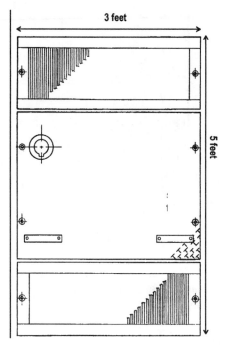

Figure 6.5-8 25 kVA Single Phase Subsurface Transformer

We note that the measured fields are quite small, even directly adjacent to the cover plate. This is because the secondary connections are some three feet below the surface, while the measurements were made three feet above the surface.

6.6 Summary

Distribution substations and distribution transformers are significant sources of magnetic fields for:

- personnel who work within substation boundaries

- the public who might live or work close to substation boundaries or fences

- the public who might live or work close to a distribution substation

As opposed to a line source, the fields of distribution substations and transformers are confined to their immediate vicinity. Our study shows that the public is largely removed from fields at distances of 50 feet from a distribution substation boundary and 10 feet from a distribution transformer enclosure. The exception to this statement occurs at substation boundary areas in which transmission lines entering the substation and/or distribution feeders leaving the substation are concentrated. Of these, the highest fields may be expected where feeders are concentrated.

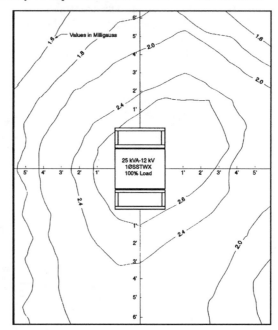

Figure 6.5-9 Measure Magnetic Field Contours for 25 kVA Single Phase Subsurface Transformer

6.7 References

[1] Ramirez, Steven and Lam, Hiep, *Measurement and Analysis of Power Frequency Magnetic Fields*, Senior Project Report, California Polytechnic State University, San Luis Obispo, CA., June, 1994.

[2] Enertech Consultants, Incorporated, 300 Orchard City Drive, Suite 132, Campbell, CA 95008.

[3] SUBCALC, Software to Calculate Magnetic Fields In and Near Electric Power Substations, Enertech Consultants of Santa Clara County Incorporated, 300 Orchard City Drive, Suite #132, Campbell, CA., 95008, October, 1993.

[4] *Electric and Magnetic Field Guidelines, A Handbook for Customers and Staff*, Pacific Gas and Electric Company, 77 Beale Street, San Francisco, CA 94156, March 11, 1993.

7 Power Frequency Magnetic Fields in the Home

7.0 Overview

Because people spend a large part of their lives in their homes there is a great deal of interest in characterizing power frequency magnetic fields in homes. This interest has led to studies to determine the chief sources of such fields and the nature of the fields produced. The principal sources are:

- power lines, including--transmission lines, primary distribution lines, secondary distribution lines and service connections

- ground currents

- appliances and lighting

As might be expected, magnetic field levels may vary widely between homes and even between rooms in a given home. The proximity of the home or room to transmission or distribution lines is a major factor in determining the magnitude of the ambient field. Ground currents, the result of multiple ground connections, may also be a significant source of the ambient field. Superimposed on the ambient field are the fields due to appliance and lighting sources, which are shown to drop off rapidly with distance from the device.

Finally, in Chapter Seven we present a protocol developed by a Task Force of the Institute of Electronic and Electrical Engineers (IEEE) for the measurement of residential power frequency magnetic fields. Designed to provide a degree of procedural uniformity in measurement, the protocol is particularly useful in obtaining statistically meaningful data on magnetic fields in homes.

7.1 Sources

The dominant sources of magnetic fields in homes, as determined by a survey of some 1000 homes [1], are:

- power lines

- ground currents

- appliances and lighting

Power lines include transmission lines, primary distribution lines, secondary distribution lines and service connections (the electrical connection between the secondary distribution line and the residential service entrance). Of these, the most important sources are transmission lines and primary distribution lines, which, although external to the home, do nevertheless produce fields within the living spaces of the home. The usual home, constructed of mostly wood, plaster, stucco, and glass, is transparent to magnetic fields because such materials are not effective in shielding

110

magnetic fields. Nonmetallic materials cannot shield the living spaces from external sources. Even reinforced concrete structures, with embedded steel, have poor shielding properties. Only certain metals such as ferromagnetic metals (the magnetic properties of many common materials are listed in Appendix A) and highly conductive materials such as copper and aluminum, can be used to provide effective magnetic shielding. The major factors affecting the magnitudes of the magnetic fields in the home, due to power lines, are simply the current in the line and the distance from the line to the home. This discussion is developed further in Section 7.2.

Ground currents are electrical return currents that flow through the earth or other structures, commonly water pipes. Ideally these currents would flow through metallic neutral conductors. However, the neutral to earth connection made in the secondary circuit and at the residential service entrance provide opportunities for the return currents to take alternate paths. If ground currents exist, the resulting magnetic fields may be significant. Again, ground currents are a source of fields external to the home itself. For further details, refer to Section 7.3 below.

Electric appliances and lighting produce the strongest magnetic fields found in the typical home, often in orders of magnitude greater than those of fields due to external sources. This condition is mitigated by the fact that the strengths of these fields drop off rapidly with distance from the appliance or lighting device. At distances of the order of three feet, appliance fields virtually disappear. That appliances and lighting are normally not *on* over a long portion of each day is a second mitigating factor. For example, an electric shaver, a high field producer at close range, is normally used no more than ten minutes each day. Appliances and lighting, sources that are internal to the home, are addressed in Section 7.4.

A graph [2] comparing the magnetic fields in the home, due to both external and internal sources, is shown in Figure 7.1-1. This figure is drawn from an Electric Power Research Institute survey [1] of 1000 homes. The median values of fields due to each source is shown as well as the range of values that were measured.

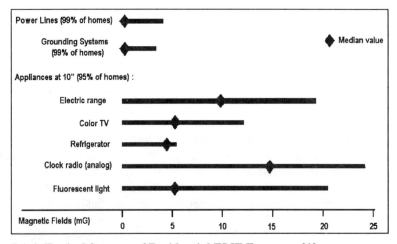

Figure 7.1-1 Typical Sources of Residential EMF Exposure [1]

It is of interest to see, in Figure 7.1-1, that power lines and ground currents are significant sources in less than 50 percent of American homes. At distances of 10 inches or less, appliances and lighting are the most intense sources—although often for short periods only.

We should also mention possible sources that are *not significant*. For example, internal house wiring [1] is not a significant source, with the exception of old-fashioned knob-and-tube wiring. This type of wiring was found in about seven percent of residences in the nationwide Electric Power Research Institute survey [1]. Appliances that do not draw appreciable current, such as digital clocks and transistor radios, are likewise insignificant sources.

7.2 Characterization of Power Line Fields in the Home

Power line is a generic term for any line in the energy delivery system. In the present context we are concerned with characterizing magnetic fields that occur in the home from power line sources. These power lines are:

- transmission lines

- primary distribution lines

- secondary distribution lines

- service connections

The role of each of these lines in the power delivery system is discussed in Chapter Four.

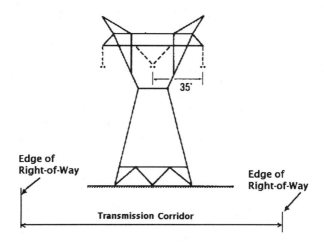

Figure 7.2-1 The Transmission Corridor

Transmission Lines. Homes may be built in the vicinity of transmission lines up to the so-called right-of-way of the line. The edges of the right-of-way establish a corridor directly under the lines as shown in Figure 7.2-1. Generally speaking, the higher the line voltage, the greater is the width of the corridor. Most high voltage transmission lines in the United States are rated at 115, 230, 345, 500, and 765 kV, respectively. At present, California regulations [3] require that the corridor shall be sufficiently wide such that the **electric** field due to the line shall be less than 1.6 kV/meter at the edges of the right-of-way.

To put the regulations in context, the minimum corridor widths, based on the above regulation, for two typical lines are shown in Table 7.2-1.

Table 7.2-1 Minimum Corridor Widths for Typical Lines Based on Electric Field Regulations

Configuration	Line Voltage	Minimum Width of Corridor
Figure 5.2-2	500 kV	180 Feet
Figure 5.2-4	345	120

Actual measurements recorded by the California Energy Commission have found average magnetic field values of less than 100 milligauss at the edges of the right-of-way. Florida and New York have placed regulatory limits ranging from 150 milligauss to 250 milligauss at the edge of the right-of-way, depending on line voltage [3]. The theory developed in Appendix C shows that the magnetic field, at a large distance from the line, decreases at a rate proportional to distance squared. This, coupled with the data of Table 7.2-1, enables us to determine approximately the field due to a transmission line source at a specified distance from the edge of the right-of-way.

Example 7.2-1   Field due to a 500 kV line.

A house is to be built at a distance of 100 feet from the edge of a minimum corridor of a 500 kV line. What average magnetic field could be expected in this house?

Solution:

We can only expect an approximate answer here. We suppose that the field at the edge of the corridor is 100 milligauss (per California Energy Commission findings). The minimum corridor width for a 500 kV line is 180 feet. Accordingly we are supposing a field of 100 milligauss at a horizontal distance of 90 feet from the center of the line. We wish to find the field at a horizontal distance of 190 feet from the center of the line as shown.

The distance of a point three feet above the center of the conductors is given by: $\sqrt{(h-3)^2 + D^2}$, where H is the height of the conductors and D is the horizontal distance of the point from the center of the line.

The field at 190 feet is approximately:

$$100\,\text{milligauss}\,x\left(\frac{90^2 + 45^2}{190^2 + 45^2}\right) = 27\,\text{milligauss}$$

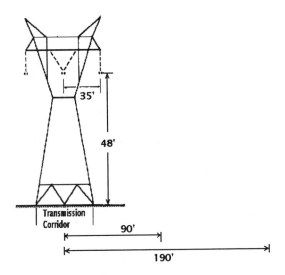

Figure for Example 7.2-1

We can obtain a more accurate, but not much different, answer by consulting the field profile of Figure 7.2-2.

Example 7.2-1 illustrates that homes built close to a transmission corridor may experience magnetic field magnitudes well above the median value shown in Figure 7.1-1. In fact, we would conclude that the Electric Power Research Institute survey did not contain homes located close to a transmission right-of-way.

Primary Distribution Lines. Primary and secondary distribution lines may follow streets, alleys or back lot lines in residential areas. Consequently they are much closer to living spaces in homes than are transmission lines. On the other hand, these lines usually carry substantially less current than transmission lines. Field profiles for a typical overhead primary distribution line and a typical underground distribution line are shown in Figure 7.2-2 and 7.2-3 respectively. We might suppose that either line (depending on whether this was an overhead or an underground area) was approximately 25 feet from the living area of a home. With phase currents of 300 amperes rms, the resultant magnetic field in the living area would be nine milligauss for the overhead line source and about one milligauss for the underground line source.

While the maximum values of the fields (D = 0) due to the overhead and underground lines are about the same, the underground line has a much narrower profile. Accordingly, at a reasonable distance from these lines, the underground line produces a much weaker magnetic field than the overhead line. We note that this result is calculated for balanced currents. If the three phase currents are unbalanced, our conclusion may no longer apply. (The effect of unbalanced currents is discussed in Sections 5.3 for overhead lines and 5.6 for underground lines.)

114

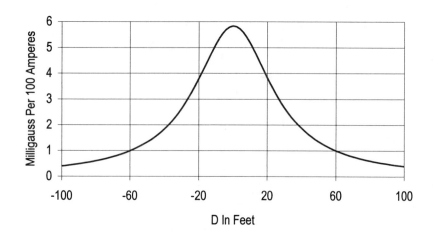

Figure 7.2-2 Magnetic Field Profile of a Primary Overhead Distribution Line With Balanced Three Phase Currents

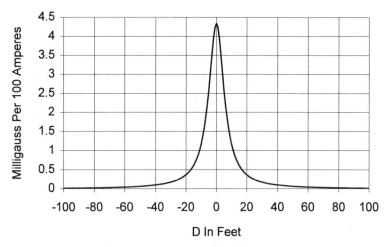

Figure 7.2-3 Magnetic Field Profile of a Primary Underground Distribution Line with Balanced Three Phase Currents

Example 7.2-1.   Overhead versus underground primary distribution.

Compare the magnetic fields in two homes due to primary distribution sources. Home A is at a lateral distance of 30 feet from an overhead primary distribution line carrying 200 amperes rms, balanced three phase currents. Home B is at a lateral distance of 15 feet from an underground primary distribution line carrying 400 amperes rms, balanced three phase currents. Both rooms are located on the ground floor of each home.

Solution:

Referring to Figure 7.2-2, the field in Home A is about five milligauss. Referring to Figure 7.2-3, the field in Home B is about two milligauss. Profiles will be shown in Chapter Five that would allow this field to be calculated for a room located upstairs.

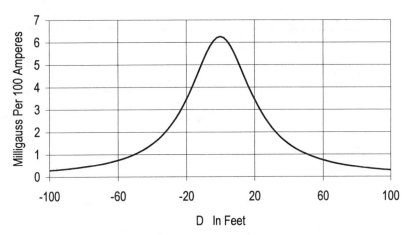

Figure 7.2-4 Magnetic Field Profile of an Overhead Secondary Single Phase Open Wire Distribution Line

Secondary Distribution Lines. Overhead secondary distribution lines in residential areas are usually single phase. The line itself often occupies a lower position on the same poles as the primary line. A field profile for a typical single phase overhead line is shown in Figure 7.2-4.

The secondary voltage is typically 240 volts between conductors and 120 volts to neutral (and earth ground). Secondary currents are relatively small, of the order of a few hundred amperes or less. At any current level, the field profile will be greatly reduced in magnitude if a triplex or quadraplex secondary construction is used. Universally used in underground secondary distribution lines, construction of this type may be used in overhead secondary lines.

Example 7.2-3.   Field due to a secondary overhead distribution line.

Find the magnitude of the magnetic field at a lateral distance of 25 feet from a secondary overhead distribution line carrying 150 amperes rms.

Solution:

Referring to Figure 7.2-4, the field is about 4.5 milligauss.

Service Connections. A service line (or service drop) connects the secondary distribution line to a service panel at the home. The service line is, typically, a triplex cable as shown in Figure 7.2-5. The line currents are less than 100 amperes rms, and the separation between conductors is quite small. Such a service line produces very little magnetic field if all of the current flows through the line. Substantial fields can result if return current flows through the ground as discussed in Section 7.3.

However, the service panel to which the service line is connected may be a larger source of magnetic field.

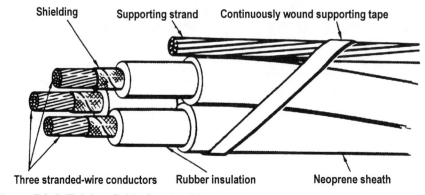

Figure 7.2-5 Triplex Cable Service Line

The following data [5] were measured at a residential service panel, such as the one shown in Figure 7.2-6, loaded to 100 amperes.

Distance From Panel	Maximum Field
1 foot	96 milligauss
3	12

Measured field contours in the region of a 200 ampere service panel, loaded at 100 amperes, are shown in Figure 7.2-7. This is a plan view of the region in the vicinity of a wall mounted panel.

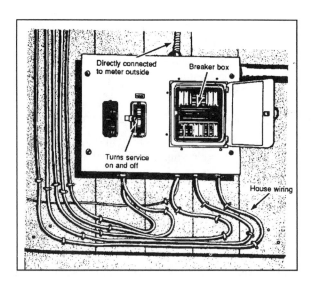

Figure 7.2-6 Residential Service Panel [6]

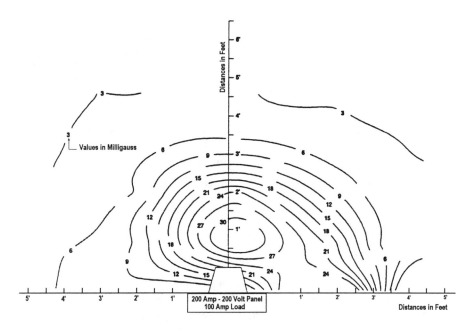

Figure 7.2-7 Magnetic Field Contours in the Region of a 200 Ampere Service Panel

In new home construction, utilities recommend ([4] and [5]) that the service panel be located remotely from bedrooms and other living spaces to minimize the field in those areas. For example, the service panel could be located in a garage wall or in a storage room.

Example 7.2-4.   Service panel fields.

At what distance from the service panel of Figure 7.2-6 is the field two milligauss or less?

Solution:

From Figure 7.2-6, the distance is about 4.2 feet. In other words, for distances of 4.2 feet or greater the field is two milligauss or less.

7.3 Fields Due to Ground Currents

The voltage levels for residential usage are typically 120 volt and 240 volt, single phase. The utility secondary system provides this voltage with three wires as shown in Figure 7.3-1.

We see in Figure 7.3-1 that the secondary system is connected to ground (*grounded*) by the utility at the distribution transformer. The grounded conductor is the neutral wire.

118

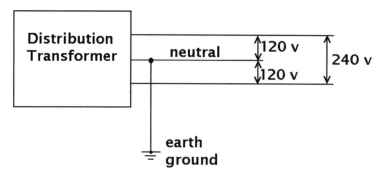

Figure 7.3-1 Typical Secondary for Residential Service

Neutral to ground connections are also made at the customer's residence. This is important because it protects the customer from shocks and/or fire in the event of a fault (accidental grounding of one of the other two conductors). Multiple grounding leads to ground currents in normal operation.

Grounding the neutral wire of an electrical distribution system protects customers against shock and fire by facilitating the fast operation of a fuse or a circuit breaker in the event of a fault. The National Electric Code (NEC) currently calls for grounding to a water pipe at the service entrance of a residence. This means that the return current can flow back to the distribution transformer through a parallel ground path instead of through the neutral conductor. When such currents are conducted by pipes inside a home, they can be a substantial source of residential magnetic fields.

Figure 7.3-2 shows a schematic of a typical residence and illustrates how ground currents may flow in this type of configuration. The house is connected to the electric distribution system with three wires—two *hot* wires and a neutral wire. Usually, the neutral wire is connected to the cold water pipe or a rod that is driven into the ground, or both. Figure 7.3-2 shows the case where the neutral is connected to the cold water pipe only. In this case there is both a neutral current I_N and a current flowing in the water pipe I_G. Notice that I_G may flow into neighboring water pipes and into the ground itself at points where the water pipe is grounded.

Finding ways to reduce the fields from ground currents in homes presents a special challenge to electric utilities, since the mitigation efforts will usually involve changes on the customer's side of the meter and will possibly require modification of the National Electrical Code. In addition, where grounding to water pipes is concerned, other utilities (for example, telephone, water, etc.) may need to be consulted.

Such multiple connections allow ground currents to flow on numerous paths, including water pipes, telephone cables and cable TV lines. Magnetic fields caused by closely spaced conductors (such as most house wiring) tend to cancel each other while magnetic fields caused by ground currents do not. Even a relatively small current flowing on a pipe can cause a significantly large magnetic field in a residence. Field values based on an Electric Power Research Institute survey of 1000 homes are shown in Figure 7.3-3. This figure compares the fields due to ground currents with those due to power lines, including the maximum and minimum values, as well as the 95 percent to five percent frequency distributions. The author of the Electric Power Research Institute survey found that in many residences a substantial amount of the ground return current flows on water pipes rather than on the intended neutral wire connected to the electric distribution transformer. He also found that it is not unusual for the ground current in one residence to flow through interconnected water pipes to another residence, creating magnetic fields there.

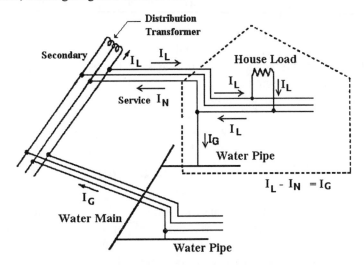

Figure 7.3-2 Grounding System That May Have Ground Currents [5]

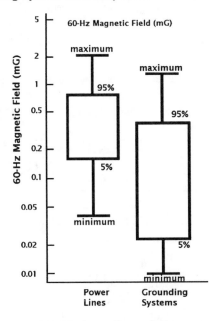

Figure 7.3-3 Comparison of Fields from Ground Currents and Power Lines [1]

It may be possible in some circumstances to reduce the pipe current and still follow the National Electric Code grounding guidelines. Customers, working with a licensed electrician, may be able to eliminate improper re-grounding inside a residence. Also, in some cases, a licensed plumber may be able to insert insulated joints in residential water lines. Such joints would electrically isolate each home and prevent the intrusion of currents from neighboring premises.

Changes in the grounding practice prescribed by the National Electric Code are also being suggested. One suggestion is based on the system common in some European countries: ground the neutral conductor only at the distribution transformer and run a separate ground wire (in addition to the neutral) to each residence. This would mean that the service connection to most homes would involve four wires rather than the three commonly used today in the United States. Such an arrangement would eliminate the connection to the water pipe at the home and thus reduce pipe currents. Appliances could also be grounded directly to the fourth wire, rather than to a water pipe. Such a scheme would probably require the use of additional protection equipment, which would respond to a fault current flowing over the ground wire.

7.4 Characterization of Fields Due to Appliances and Lighting

Residential utilization of electric power involves household appliances and lighting. During operation these become sources of magnetic fields. Research has determined that the fields produced by appliances and lighting vary considerably depending on size, type and manufacturer. Some indications of the measured fields due to common appliances as given in reference [6] are shown in Table 7.4-1. Table 7.4-2, a more extensive listing, is from reference [7]. In comparing the entries of Tables 7.4-1 and 7.4-2, we find wide variations in measured values. Actual fields will depend on the specific make and model of the appliance or lighting fixture being tested.

As shown by these tables, the magnitudes of magnetic fields drop off very rapidly as distances from the appliances increase. The variations of the field magnitudes with distance are shown graphically for nine specific appliances in Figures 7.4-1 through 7.4-9 [8].

Appliance	RANGE OF MAGNETIC FIELDS (milligauss) DISTANCE FROM SOURCE		
	1.2 inches	12 inches	39 inches
Electric Blanket	2 - 80	-	-
Clothes Washer	8 - 400	2 - 30	0.1 - 2
Television	25 - 500	0.4 - 20	0.1 - 2
Electric Range	60 - 2,000	4 - 40	0.1 - 1
Microwave Oven	750 - 2,000	40 - 80	3 - 8
Electric Shaver	150 - 15,000	1 - 90	0.4 - 3
Fluorescent Lamp	400 - 4000	5 - 20	0.1 - 3
Hair Dryer	60 - 20,000	1 - 70	0.1 - 3

Table 7.4-1 Magnetic Fields from Electrical Appliances and Lighting [6]

Appliance	RANGE OF MAGNETIC FIELDS (milligauss)		
	DISTANCE FROM SOURCE		
	4 inches	12 inches	36 inches
Clothes Dryers	4.8 - 110	1.5 - 29	0.1 - 1
Clothes Washers	2.3 - 3	0.8 - 3.0	0.2 - 0.48
Coffee Makers	6 - 29	0.9 - 1.2	<0.1
Toasters	10 - 60	0.6 - 7.0	<0.1 - 0.11
Crock Pots	8 - 23	0.8 - 1.3	<0.1
Irons	12 - 45	1.2 - 3.1	0.1 - 0.2
Can Openers	1300 - 4000	31 - 280	0.5 - 7.0
Mixers	58 - 1400	5 - 100	0.15 - 2.0
Blenders	50 - 220	5.2 - 17	0.3 - 1.1
Vacuum Cleaners	230 - 1300	20 - 180	1.2 - 18
Portable Heaters	11 - 280	1.5 - 40	0.1 - 2.5
Hair Dryers	3 - 1400	<0.1 - 70	<0.1 - 2.8
Electric Shavers	14 - 1600	0.8 - 90	<0.1 - 3.3
Televisions	4.8 - 100	0.4 - 20	<0.1 - 1.5
Fluorescent Fixtures	40 - 123	2 - 32	<0.1 - 2.8
Fluorescent Desk Lamp	100 - 200	6 - 20	0.2 - 2.1
Saber & Circular Saws	200 - 2100	9 - 210	0.2 - 10
Drills	350 - 500	22 - 31	0.8 - 2.0

Table 7.4-2 Magnetic Fields From Electrical Appliances and Lighting [7]

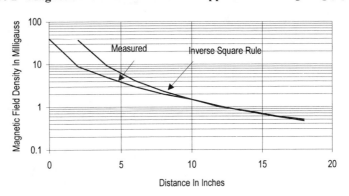

Figure 7.4-1 Magnetic Field for 50 Watt Electric Heating Pad

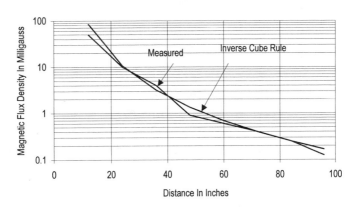

Figure 7.4-2 Magnetic Field for 26 Inch Color Television

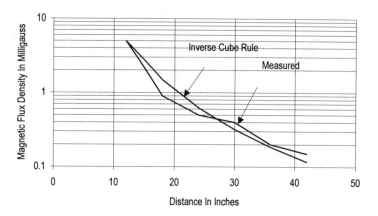

Figure 7.4-3 Magnetic Field for 400 Watt Amplifier

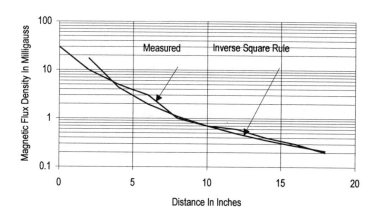

Figure 7.4-4 Magnetic Field for Hair Dryer

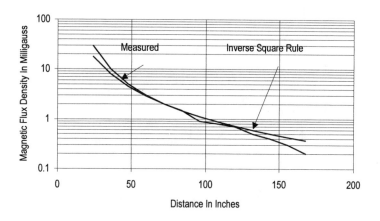

Figure 7.4-5 Magnetic Field for Microwave Oven

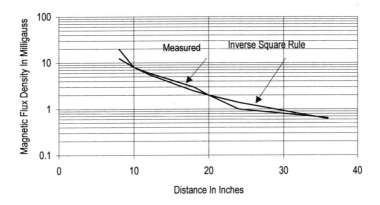

Figure 7.4-6 Magnetic Field for 12 Inch Computer Monitor

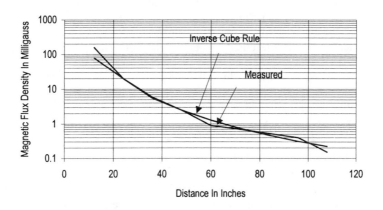

Figure 7.4-7 Magnetic Field for Blender

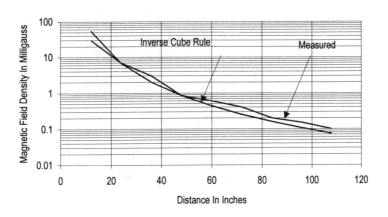

Figure 7.4-8 Electric Range Oven

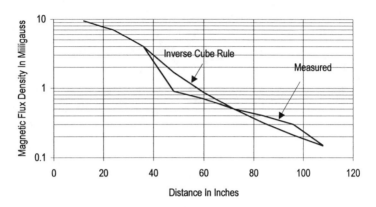

Figure 7.4-9 Magnetic Field for 16 Cubic Foot Refrigerator

We note that for four cases (the heating pad, the hair dryer, the microwave oven and the computer monitor), the variation with distance is best described by an inverse square rule: $\left(\dfrac{1}{D}\right)^2$. In the other five cases, (the television, the stereo amplifier, the blender, the electric range and the refrigerator), the variation is best described by an inverse cube rule: $\left(\dfrac{1}{D}\right)^3$. In any case field strength drops rapidly with increased distance from these appliances.

The magnetic fields in a home are the result of all of the sources acting together. At any given time, measurements can provide us with field values throughout the home. These may be used to generate a field contour map, such as the one shown in Figure 7.4-10, for a typical residential room [1]. In the living room shown here, field levels were usually below 0.5 milligauss. Exceptions were the peak fields produced by a clock, the utility service drop, an entertainment center and a fish tank motor.

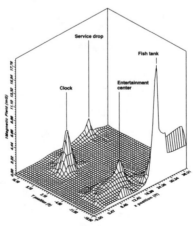

Figure 7.4-10 Field Contour Map for Residential Living Room [1]

A systematic and uniform method for obtaining measurements in a residential setting is discussed in Section 7.6.

7.5 Variations Among Types of Residences

Power line fields have been found to be correlated with residential areas [2].

Highest median values are found in urban areas and the lowest in rural areas. Among types of residences, apartment buildings and duplexes had the highest median fields; single family dwellings had the lowest. Median fields also tended to increase with the age of the residence: fields in homes less than 10 years old were about half those in homes more than 50 years old.

In epidemiological studies, homes have been assigned a *wire code* based on examination of the overhead lines near the home. Wire codes were treated as surrogates for the actual fields measured in homes. Electric Power Research Institute's 1000 home survey [2] has shown a correspondence, albeit overlapping, between wire codes and the median fields found in these homes. (Refer to Chapter Two for a discussion of wire codes and their definitions.)

7.6 Measurement of Residential Power Frequency Magnetic Fields

The measurement of power frequency magnetic fields in residences is no different, in theory, from measurement in any other setting. (The general principles of measurement of magnetic fields are described in Chapter Three.) The usual measurements in a residential setting, so-called *spot measurements,* are carried out for the purpose of informing residents about the magnetic field levels in their living spaces, for example, bedrooms, family rooms and kitchens. Residents may request that the electric utility make the measurements or alternatively hire an independent firm to perform this service. In fact, informed residents may choose to borrow a meter and take measurements themselves.

Since spot measurements are being requested and made in increasing numbers, an IEEE committee has designed a protocol to provide a degree of procedural uniformity [9]. The protocol is intended for use with hand-held meters of either the single axis or three axis type. The goal is usually to record the maximum value and/or resultant value of the magnetic field at certain discrete points in the home. This is greatly expedited if a three axis meter is used, as discussed in Chapter Three. The goals of the protocol do not provide for characterizing the polarization, spatial and temporal variations, or the harmonic content of the magnetic field. Note that if fluctuations of the field reading occur during the measurements, it is suggested that an approximate average value over a five second interval be recorded. Spot measurements represent a *snap shot* in time and space and do not of themselves provide information regarding temporal and spatial variations. However, a sequence of spot measurements at a given point, over a period of time (say 24 hours), would provide information on temporal variation—at that point. And a sequence of spot measurements at a given time over a region (say a room), would provide information on spatial variation—at that time.

To expedite spot measurements in the home the protocol specifies a data sheet with plan view as shown in Figure 7.6-1. The data sheet allows for entering measured field data for points in several rooms, all at a height of approximately three feet above the floor. Each point is identified in the plan view. If field levels along the coordinate axes are measured, a coordinate system, such as the one in Figure 7.6-1, is suggested.

RESIDENT

Name __________
Address __________
____
City, State, Zip __________
Phone __________ Date ______
Time: Start ______ Stop ______
Outdoor Temperature (°F) ______

MAGNETIC FIELD METER DATA

Meter Model ______ S/N ______
Frequency Response __________
Detector (true rms, average rms,..) ______
Accuracy __________
Date of last calibration __________
Probe (Dia., No. axes) __________

Spot Measurement Data   Recorded by __________

No.	B (μT_{rms} or mG_{rms}) B_{max}	x	y	z	Resultant	Location / Description
						Family Room
						Kitchen
						Bedroom
						Bedroom
						Bedroom

PLAN VIEW

Include outline of residence with rooms at each level and indicate measurement location with • and number.

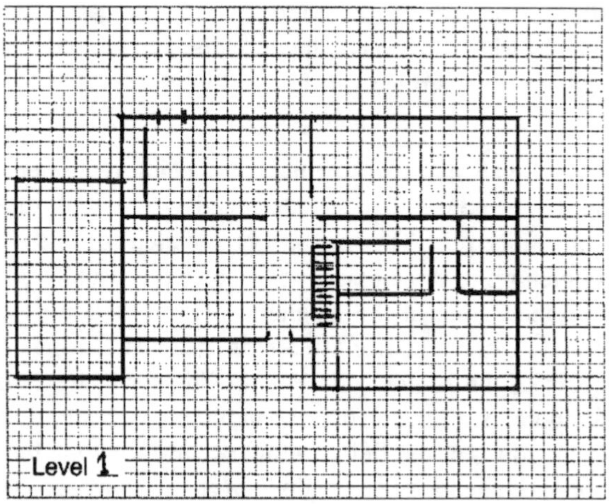

coordinate system

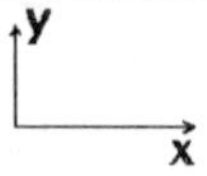

z-axis is vertical

Figure 7.6-1 Protocol Data Sheet With Plan View [9]

An important aspect of the IEEE protocol involves identifying the magnetic field meter and its characteristics. For example, the frequency response of the meter should be indicated because field meters with different bandwidths can yield different results if harmonics are present in the magnetic field. Television sets produce fields rich in harmonics. A wide bandwidth meter may read 20 percent higher than a meter sensitive only to the 60 Hertz field of a television set. Further, if there are harmonics, a true rms detector is required to determine the resultant value of the field. The accuracy of a field meter refers to the uncertainty in a measurement. Meter accuracy of the order of ±0.1 milligauss is desirable for residential measurements.

Example 7.6-1   Estimating the field in a living space.

Consider the plan view of a Kitchen/Breakfast Room shown in Figure 7.6-2. We will be interested in the fields from the three appliances listed below:

Appliance	Magnitude of Appliance Field 12 Inches From Front Panel
Refrigerator	10 milligauss
Electric Range	30
Dishwasher	40

The ambient field at point P is one milligauss.

Estimate the total field at point P.

Solution:

We will assume that the fields due to the three appliance sources drop with distance according to the inverse cube rule (see Figure 7.4-8). The point P is about three feet from the front of the electric range, three feet from the front of the dishwasher and four feet from the front of the refrigerator. Using the inverse cube rule we compute the following:

Appliance	Magnitude of Appliance Field at Point P
Refrigerator	$10 \times \left(\dfrac{1}{4}\right)^3 = 0.16 \text{ milligauss}$
Electric Range	$30 \times \left(\dfrac{1}{3}\right)^3 = 1.11$
Dishwasher	$40 \times \left(\dfrac{1}{3}\right)^3 = 1.48$

Since the magnetic field is a vector field and we only know its magnitude (but not its direction) from each source, we cannot combine the fields. Even further, we have no information on the time phase of each field. The best we can do is to establish the maximum possible field at point P. This would be the sum of all of the source fields at that point plus the ambient field. Doing that calculation we have:

maximum value of field at point P = 0.16 + 1.11 + 1.48 + 1.0 = 3.75 milligauss.

128

We would expect that a spot measurement of the field at point P would give a value no greater than 3.75 milligauss.

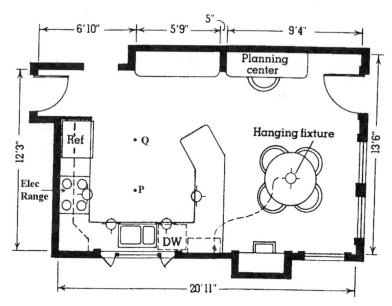

Figure 7.6-2 Plan View of Kitchen/Breakfast Room

7.7 Summary

The principal sources of magnetic fields in the home are:

- power lines
- ground currents
- appliances and lighting

A survey of some 1000 American homes, nationwide, reveals that the median ambient field is 1 milligauss or less. However, if the home is situated close by a transmission corridor or an overhead distribution line, the ambient field may be an order of magnitude larger. (The ambient field is measured at the center of a room to avoid significant contributions to the field from appliance sources.) Significant ground current may result in an ambient field of the order of 10 milligauss. Appliances and lighting produce much higher fields in close proximity. However, these fields decrease very rapidly with distance $\left(\text{as } \frac{1}{D^2} \text{ or } \frac{1}{D^3} \right)$.

To promote uniformity of measurement, useful in generating field data for absolute and comparative purposes, an IEEE Task Force has proposed a protocol for making spot measurements of magnetic fields in residences.

7.8 References

[1]   Douglas, John, *EMF in American Homes*, Electric Power Research Institute Journal, pp 18-25, April/May, 1993.

[2]   Douglas, John, *Managing Magnetic Fields*, Electric Power Research Institute Journal, pp 6-13, July/August, 1993.

[3]   High Voltage Transmission Lines Summary of Health Effects Studies, California Energy Commission, July 1992.

[4]   *Electric and Magnetic Field Guidelines, A Handbook for Customers and Staff*, Pacific Gas and Electric Company, 77 Beale Street, San Francisco, CA 94156, March 11, 1993.

[5]   *Distribution System EMF Guide*, Southern California Edison Company, Customer Service Department, 2244 Walnut Avenue, Rosemead, CA 91770, March 19, 1992.

[6]   Pansini, J. Anthony, *Guide to Electrical Power Distribution Systems*, Prentice Hall, Englewood Cliffs, NJ, 07632, 1992.

[7]   J. R. Gauger, *Household Appliance Magnetic Field Survey*, IEEE Transactions Power Apparatus and Systems, Vol. PAS-104, pp. 2436-2445, 1985.

[8]   Goldberg, S., Horton, W. F., and King, B. P., *Characterizing the Magnetic Field of Household Appliances*, pp 403-412, Proceedings of the North American Power Symposium, Reno, NV, Oct 1992.

[9]   *A Protocol for Spot Measurements of Residential Power Frequency Magnetic Fields*, A Report of the IEEE Magnetic Fields Task Force of the AC Fields Working Group of the Corona and Field Effects Subcommittee of the Transmission and Distribution Committee, IEEE Transactions on Power Delivery, Vol. 8, No. 3, July 1993.

7.9 Exercises

7.9-1   Field due to a 500 kV line.

If the field at the edge of the right-of-way for a 500 kV line is 100 milligauss, what is the line current?

Answer:	about 2500 amperes (using Figure 5.2-2)

7.9-2   Field due to a secondary overhead single phase distribution line.

A secondary overhead line produces a field of 10 milligauss at a lateral distance of 20 feet. What is the field at a distance of 50 feet?

Answer:	about three milligauss

130

7.9-3   Field due to a 100 watt electric heating pad.

The field due to a 100 watt electric heating pad is about 50 milligauss at a distance of 1 inch. Estimate the field at a distance of 10 inches.

Answer:	about 0.5 milligauss

7.9-4   Field in a living space.

Along the same lines as Example 7.6-1, find the maximum possible field at point Q of Figure 7.6-2?

Answer:	about two milligauss

8 Management of
Magnetic Fields by Shielding

8.0 Overview

In Chapter Eight we concern ourselves with the management of magnetic fields generated by a fixed current configuration. We assume that this source exists and consider the problem of managing the resulting field either globally or in a specific, restricted region. The discussion is carried out by means of examples:

- managing the field due to a three phase underground cable globally

- managing the field due to an overhead line in a specified region.

The means of management for these examples is magnetic shielding, which can be quite effective. Electromagnetic shielding is likewise effective in managing the field in a specific region. Finally we consider active shielding or cancellation, another technique applicable to restricted regions.

8.1 Introduction to Shielding

Magnetic fields are subject to *management* by several techniques. First, to clarify the term, management, in the sense of magnetic fields, means mitigation or reduction of the magnitude of the field in a specified region. This region may be large, for example, everywhere (globally). On the other hand, the region may be very restricted, for example, a school room or a bedroom (locally). The first example requires that the management technique be applied at the source of the field. The second example implies that the technique need only be applied in the restricted region that is affected by the field.

We have seen the effects of management at the source in Chapter Six. There, the sources were lines, both for the transmission and distribution of electric power. The fields in the vicinity of such lines are reduced by the techniques of:

- decreasing spacing between conductors as in triangular construction versus horizontal construction for overhead lines.

- increasing conductor height, for overhead lines.

- reducing unbalance of phase currents in three phase lines.

- forcing return currents to flow in the metallic neutral.

- using double circuit construction with the low reactance arrangement, for overhead lines.

(The uses of these techniques in management at the source are documented in detail in Chapter Six.)

Shielding techniques are applicable both to management at the source (globally) and management in a restricted region (locally). These techniques may be divided further into *passive shielding* techniques and *active shielding* techniques. Examples of passive and active shielding are described more fully as they are presented in this chapter.

8.2 Principles of Shielding

For the purposes of our discussion, we will consider two types of shielding. These are:

- passive shielding

- active shielding

Passive shielding involves the use of materials that interact with power frequency magnetic fields. Such materials have either large relative permeabilities or high electrical conductivities, or both. A material with a large relative permeability, such as a ferromagnetic material (for example, silicon steel) achieves shielding by diverting magnetic flux to a desirable path--as opposed to the path without shielding. A material with high electrical conductivity (for example, copper) achieves shielding by providing a medium for eddy currents, whose fields oppose the applied magnetic field. Eddy current shielding is especially effective if the shielding material has both a large relative permeability and a high electric conductivity.

Active shielding involves the use of current-carrying conductors whose magnetic field effectively cancels the applied field. The current through the canceling coils may be automatically controlled to maintain a canceling field even as the applied field varies. This property gives rise to the term active shielding. Generally, the active shielding technique is a local technique used to reduce or cancel the field in a restricted region.

In summary we will deal with two techniques of shielding, **passive** and **active**, and two types of regions, **local** and **global**. A matrix of these subjects showing the Sections in which they are discussed is presented in Table 8.2-1.

Technique	Region	
	Local	Global
Passive	8.4 and 8.5	8.3
Active	8.6	not applicable

Table 8.2-1 Matrix of Shielding Techniques and Regions with Applicable Sections

8.3 Shielding Applied to a Three Phase Underground Cable [1]

This section describes an example of passive shielding at the source (global). It concerns the magnetic field due to currents in a three phase underground distribution cable as commonly used in distribution systems. The use of a ferromagnetic duct greatly attenuates the stray field produced by these currents at the point of observation.

Distribution Feeder. The primary distribution system typically consists of many three phase radial feeders. Each feeder, as shown in Figure 8.3-1, is composed of a central stem or main (backbone), with branching laterals and sub-laterals. The main, as well as major laterals, provide paths for three phase power to flow. Some of the smaller laterals and sub-laterals may be single phase loads.

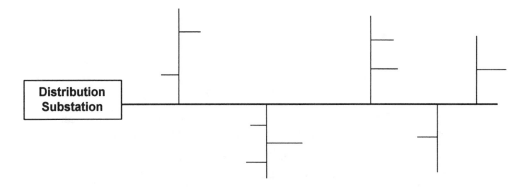

Figure 8.3-1 Distribution Feeder One Line Diagram

The three phase feeder may be grounded at the substation and at many points throughout the feeder, grounded only at the substation, or not grounded at all. The grounding strategy used depends on each utility's standard practices. For example, Pacific Gas and Electric Company, operating in Central and Northern California, operates both Y connected and multiply grounded feeders at 22 kV as well as Δ connected and ungrounded feeders at 12 kV. Either type of feeder may be an overhead feeder, an underground feeder or a combination of the two.

Magnetic Field Profiles. The calculation of magnetic fields due to overhead or underground current carrying conductors is usually based on the assumption that:

- the line is long and straight

- the relative permeability of all surrounding material is unity

- the conductivities of the surrounding materials are zero.

Using the *long line* approximation, the magnetic field is always orthogonal to the direction of the line. Thus the field is completely described by its values in any plane that intersects the line orthogonally. Magnetic field profiles for an overhead (OH) and underground (UG) feeder are shown in Figure 8.3-2.

The magnetic field is determined at a height of three feet above ground. The parameter h is the height of the conductor relative to ground (the earth's surface). A negative value of h indicates that the conductor is below the earth's surface. The parameter D is the lateral distance from the center of the line. (The dimensional configurations of both lines are shown in Figures 8.3-11 through 8.3-13, at the end of this Section. Note that the underground line consists of a three phase insulated cable in a six inch duct.

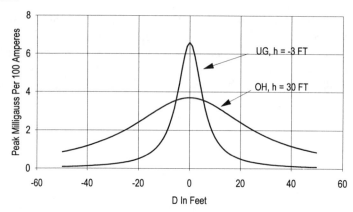

Figure 8.3-2 Magnetic Field Profiles of a Triangular Overhead Line and Underground Line With Balanced Three Phase Currents

The vertical scale of Figure 8.3-2 is in units of peak milligauss per 100 amperes, rms, of line current. Peak values are used to facilitate comparison with magnetic field plots derived from finite element programs. Finite element programs are used to compute magnetic fields in regions where magnetic characteristics (such as permeability) are not constant. The effective value of the sinusoidally varying magnetic flux density is its peak value divided by the square root of two. The profiles shown were calculated for balanced three phase currents (no zero-sequence current).[*] Because of symmetry, the fields are greatest directly under the overhead line or over the underground cable. As compared with the overhead line, the underground cable results in a field profile that is higher and narrower, with field strength dropping off rapidly as lateral distance (D) increases. As long as there is no zero-sequence current, the field profiles will be substantially the same (as in Figure 8.3-2), even if the phase currents become unbalanced. However, if zero-sequence current flows and is returned through the ground, the magnetic field becomes larger as is shown in Figure 8.3-3.

In this case, the field due to the underground cable is greater than that due to the overhead line for almost all values of lateral distance (D). In addition, the magnitudes of both fields are greater than in the case of balanced currents.

Magnetic Field Plots. The magnetic field profiles derived from such deterministic programs as FIELDS [2] and TL-Workstation [3] are based on the assumption that the conductors are surrounded by non-magnetic materials. In order to examine the field more closely and to evaluate shielding, we will utilize magnetic field plots based on finite element methods.

---

[*] For a detailed discussion of zero sequence current refer to Chapter 5, Section 5.3.

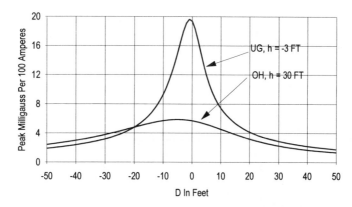

Figure 8.3-3 Magnetic Field Profiles of a Triangular Overhead Line and an Underground Cable with 3.3 Percent Zero-sequence Current Returned Through Ground

A magnetic field plot shows the magnitude and direction of the magnetic field intensity vector in the plane orthogonal to the line. The direction is indicated by flow lines that *encircle* the current carrying conductors.

Balanced Currents. Suppose the underground cable carries the following balanced three phase, 60 hertz currents:

$$i_A = 100\sqrt{2}\ \cos\omega t \qquad\qquad \text{amperes}$$

$$i_B = 100\sqrt{2}\ \cos(\omega t + 240°) \qquad \text{amperes} \qquad\qquad (8.3\text{-}1)$$

$$i_C = 100\sqrt{2}\ \cos(\omega t + 120°) \qquad \text{amperes}$$

where $\omega t = 2\pi \cdot 60 = 377$ radians/second and these currents enter the conductors labeled A, B, and C, as shown in Figure 8.3-12, at the end of this section.

The magnetic field plot is a function of t (seconds). A plot for $\omega t = 2\pi n$ (n = 0,1,2,...) is shown in Figure 8.3-4.

Figure 8.3-4a shows the field plot at the cable conductors while Figure 8.3-4b shows the plot over a wider area including the entire trench. (Cable dimensions are given in Figure 8.3-12.) These plots were computed using the finite element analysis program Maxwell 2D [4]. Particular values of the x and y components of the magnetic flux density vector at a height of three feet above the ground are shown in Figure 8.3-4b, with the x component listed first and followed by the y component. Values shown are peak values in units of tesla.

These vectors, in general, change their magnitude and direction with time, tracing out an ellipse every cycle of the 60 hertz frequency. The resultant value of the magnetic flux density vector is equal to the square root of the sum of the squares of the semi-major and semi-minor axes of the ellipse. (This is equivalent to the previous definition of the resultant detailed in Chapter Three, Section 3.2.) Resultant values as

136

computed from the finite element program are used to plot the field profile shown in Figure 8.3-5. This may be compared with Figure 8.3-2 to verify the close correlation between the results of the deterministic and finite element programs.

(magnetic flux densities shown in tesla)

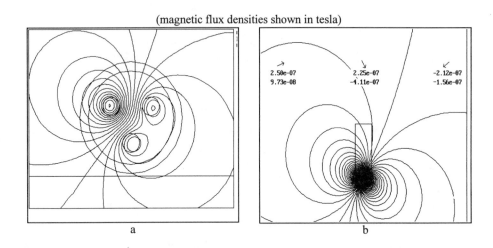

a b

Figure 8.3-4 Magnetic Field Plot for Cable/Plastic Duct at $\omega t = 2\pi n$ **, Balanced Currents**

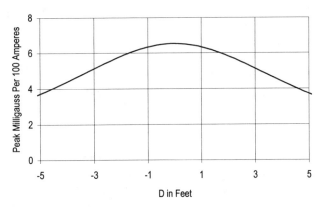

Figure 8.3-5 Magnetic Field Profile for Cable/Plastic Duct, Balanced Currents

Unbalanced Currents. Suppose the cable carries the following unbalanced three phase, 60 hertz currents:

$$i_A = 110\sqrt{2}\ \cos\omega t \qquad\qquad \text{amperes}$$

$$i_B = 100\sqrt{2}\ \cos(\omega t + 240°) \qquad \text{amperes} \qquad\qquad (8.3\text{-}2)$$

$$i_C = 100\sqrt{2}\ \cos(\omega t + 120°) \qquad \text{amperes}$$

There is a 10 percent unbalance in the magnitude of i_A, and consequently the zero-sequence current has a magnitude of 3.3 amperes rms.

The magnetic field plot at $\omega t = 2\pi n + 60°$ is shown in Figure 8.3-6.

Shielding the Magnetic Field of the Underground Cable. The purpose of shielding in this instance is to reduce the value of the resultant field above ground. A particular goal is to reduce its value at a height of three feet above the ground. The study of shielding requires the use of a computer program such as the finite element program with eddy current calculation. In this case the shielding material is assumed to have a relative permeability greater than one and a relatively large conductivity.

(magnetic flux densities in tesla)

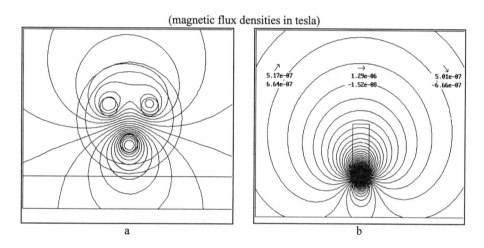

a   b

Figure 8.3-6 Magnetic Field Plot for Cable/Plastic Duct at $\omega\tau = 2\pi n + 60°$, 3.3 Percent Zero-sequence Current Returned Through Ground

The underground duct, shown in Figure 8.3-12, is typically formed from a non-magnetic, non-conductive plastic (i.e. a material whose relative permeability is unity and whose conductivity is essentially zero). The plastic duct can be replaced with a metallic duct for the purpose of shielding. The metallic duct is formed from a ferro-magnetic material, for example, iron or steel. (Alternately, ferro-magnetic particles might be added to the plastic when the duct is manufactured).

We will assume that the duct in Figure 8.3-12 has the following material characteristics:

relative permeability $\mu = 100$, conductivity $\sigma = 10^7$ Siemens/meter

This is a modest value of relative permeability [5] and an average value of conductivity for steel or iron. The corresponding skin depth at 60 hertz is:

$$\delta = \sqrt{\frac{2}{\omega\mu\sigma}} = \sqrt{\frac{2}{377 \times 100 \times 4\pi \times 10^{-7} \times 10^{7}}} = 2.05 \text{ millimeters} = 0.08 \text{ inches}$$

The metallic duct has a wall thickness of 0.5 inches, about six skin depths. However, we will see that this thickness does **not** result in attenuation of the zero-sequence magnetic field in the x-y plane, although it is effective in reducing fields produced by balanced currents.

Balanced Currents. A magnetic field plot for this cable/steel duct configuration and the balanced currents of Equation (8.3-1) is shown in Figure 8.3-7.

Figure 8.3-7a shows that the magnetic flux is concentrated within the steel duct. The maximum flux density within the duct is approximately 270 gauss.

A field profile is shown in Figure 8.3-8.

(magnetic flux densities in tesla)

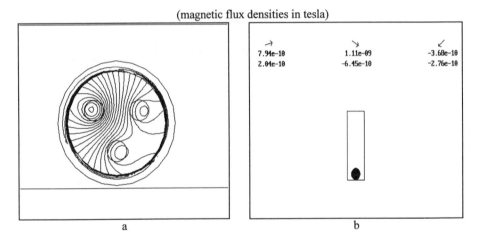

| a | b |

Figure 8.3-7 Magnetic Field Plot for Cable/Steel Duct at $\omega\tau = 2\pi n$, Balanced Currents

As a result of the shielding provided by the steel duct, the resultant magnetic flux density, at a height of three feet, is reduced by a factor of 200 as compared with the plastic duct (Figure 8.3-5). The steel duct is quite effective in shielding the magnetic flux produced by the balanced three phase currents.

We note that the profile in Figure 8.3-8 is not a smooth curve. We attribute the absence of expected symmetry to the very small field magnitudes in this case.

Unbalanced Currents. A magnetic field plot for the cable/steel duct configuration and the unbalanced currents of Equation (8.3-2) is shown in Figure 8.3-9.

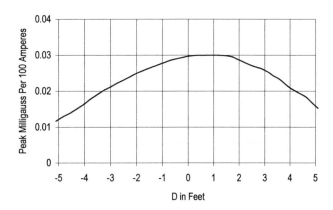

Figure 8.3-8 Magnetic Field Profile for Cable/Steel Duct, Balanced Currents

Figure 8.3-9 shows a concentration of flux within the duct with a maximum flux density of approximately 300 gauss. The resultant magnetic field profile at a height of three feet is shown in Figure 8.3-10. In comparing this profile with the underground profile of Figure 8.3-3, we see that the steel duct has very little effect on the resultant field densities at a height of three feet.

The steel duct is quite ineffective in shielding the magnetic flux provided by the unbalanced three phase currents. Of course, this was to be expected. Application of Ampere's Circuital Law to this case shows that the field due to the zero-sequence current will be unaffected by the steel duct.

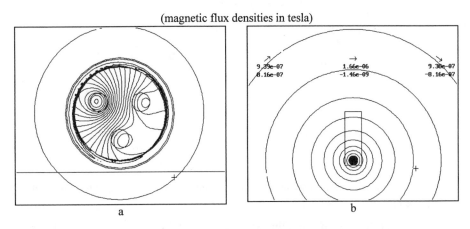

Figure 8.3-9 Magnetic Field Plot for Cable/Steel Duct at $\omega\tau = 2\pi n$ **, 3.3 Percent Zero-sequence Current Returned Through Ground**

140

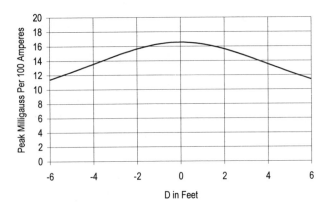

D in Feet

Figure 8.3-10 Magnetic Field Profile for Cable/Steel Duct, 3.3 Percent Zero-Sequence Current Returned Through Ground

Example 8.3-1   Calculation of field due to zero-sequence current in three phase cable.

In Figure 8.3-9, a flux density vector $B = 1.66 \times 10^{-6}\,\hat{x} - 1.46 \times 10^{-9}\,\hat{y}$ tesla, is computed at a point three feet above the earth's surface, directly over the cable at $\omega t = 2\pi n$. This point is six feet above the cable itself. What would the flux density vector be at $\omega t = 2\pi n$ if the only current in the cable was $10\sqrt{2}\cos\omega t$ amperes? fs(Assuming a single phase current corresponding to 3.3 percent zero-sequence current).

Solution:

Using Ampere's Circuital Law (Appendix A),

$$\overline{H} = \frac{I}{2\pi r}\,\hat{x}$$

$$= \frac{10\sqrt{2}}{2\pi \times \frac{6}{3.28}}\,\hat{x} \text{ ampere / meter}$$

and

$$\overline{B} = 4\pi \times 10^{-7}\overline{H}$$

$$= 1.55 \times 10^{-6}\,\hat{x} \text{ tesla}$$

This value is about the same as that due to the zero-sequence component of the three phase currents.

Application of Shielding to Distribution Feeders. Shielding of the type discussed above, utilizing a steel duct or the equivalent, is effective when cable currents are balanced but ineffective when currents are unbalanced. The zero-sequence components of the unbalanced currents produces a magnetic flux unattenuated by the shield. This effectively limits the application of steel duct type shielding to feeders that cannot produce zero-sequence currents, which are feeders grounded only at the substation or not grounded at all. This is not an insignificant result since the resultant field three feet above a three phase cable with balanced line currents is on the order of seven peak milligauss per 100 amperes of rms line current.

Eddy current losses will occur as a result of the time varying flux in the steel duct. For the steel duct of Figure 8.3-12, we estimate these losses to be 0.37 watts/meter [6] if the cable currents are 100 amperes rms and balanced. (Losses are assumed to increase with the square of the current.) It should be possible to reduce losses by reducing the conductivity of the shield material or laminating it. A thin layer of magnetic material on the plastic duct may be adequate to provide shielding and to minimize eddy current losses.

As an alternative to a steel duct, simple steel plates may be installed above and along the cable route. Durkin et al., in an IEEE paper *Five Years of Magnetic Field Management* [7], report that 4 feet x 2 feet x 3/8 inch ASTM 1010 steel plates installed in this way resulted in a two- to fourfold reduction in the magnetic fields above cable ducts. Considering the additional physical protection afforded the cable, this method has been adopted by Consolidated Edison Company of New York for all new projects.

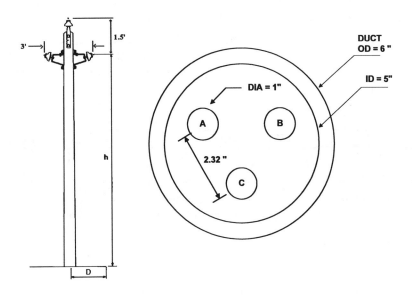

8.4 Shielding Applied to a Specific Region-Magnetic Shielding

In this section we examine passive shielding as applied to a specified region--a case of local magnetic shielding. A ferromagnetic shield can be very effective in reducing the magnitude of the field in a specified region. For example, the magnitude of the magnetic flux density in this case is reduced by a factor of about 11 after shielding. We introduce the theory of magnetic circuits to provide a basis for analysis and design of magnetic shields. In the study discussed below, conductivity of the shield is assumed to be small, thus eddy currents are not a factor. The important case of eddy current shielding is considered in Section 8.5.

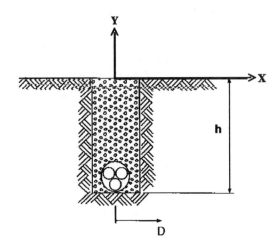

Figure 8.3-13 Underground Cable and Duct in Trench

Magnetic Circuit Principles. The principles of magnetic shielding as applied to a specified region are related to the theory of magnetic circuits as described in Appendix A. The magnetic circuit is analogous to an electric circuit in the sense of the following analogous quantities:

Magnetic Circuit Quantities		Electric Circuit Quantities
Magnetomotive force (Ampere-turns)	~	Electromotive force (Volts)
Magnetic flux (Webers)	~	Electric current (Amperes)
Magnetic reluctance (Ampere-turns/Weber)	~	Resistance (Ohms)

The magnetic circuit of Figure 8.4-1 is analogous to the electric circuit of Figure 8.4-2.

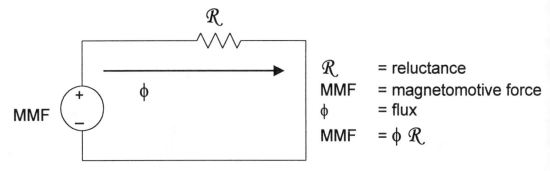

$\mathcal{R}$ = reluctance
MMF = magnetomotive force
ϕ = flux
MMF = $\phi \, \mathcal{R}$

Figure 8.4-1 Magnetic Circuit

EMF = electromotive force
R = resistance
I = electric current
EMF = I R

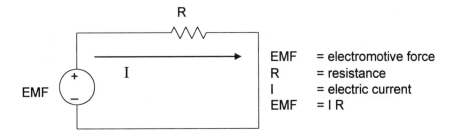

Figure 8.4-2 Electric Circuit

This analogy allows us to use the familiar concepts of electric (DC) circuits in analyzing and designing magnetic circuits. For example, magnetic flux divides between two paths according to their reluctances as shown in Figure 8.4-3.

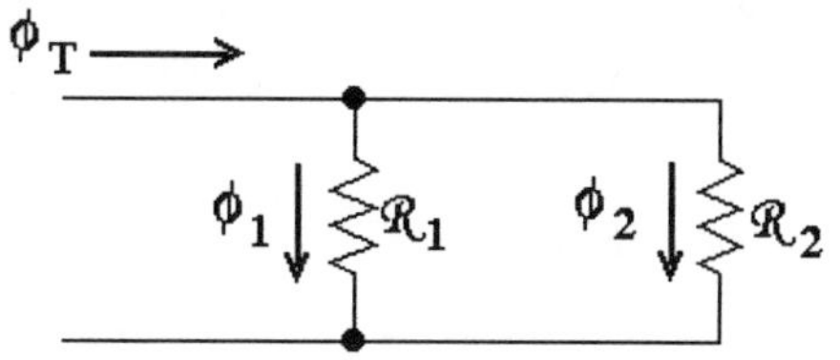

$$\phi_1 + \phi_2 = \phi_T$$

$$\phi_1 = \frac{\phi_T \mathcal{R}_2}{\mathcal{R}_1 + \mathcal{R}_2} \text{ and } \phi_2 = \frac{\phi_T \mathcal{R}_1}{\mathcal{R}_1 + \mathcal{R}_2}$$

Figure 8.4-3 Magnetic Flux Division

And magnetomotive force divides in a flux path according to reluctance, as shown in Figure 8.4-4.

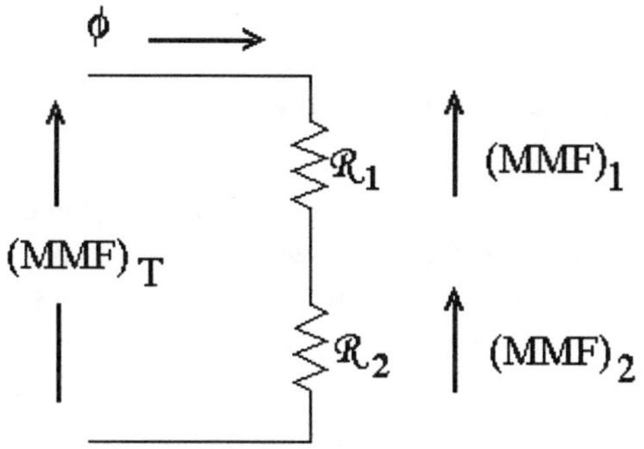

$$MMF_1 = \frac{\mathcal{R}_1}{\mathcal{R}_1 + \mathcal{R}_2}(MMF)_T$$

$$MMF_2 = \frac{\mathcal{R}_2}{\mathcal{R}_1 + \mathcal{R}_2}(MMF)_T$$

$$(MMF)_T = (MMF)_1 + (MMF)_2$$

Figure 8.4-4 Magnetomotive Force Division

The principles of flux division and magnetomotive force (MMF) division, derived from circuit concepts, are used either implicitly or explicitly in shield analysis and design. We will use them explicitly in an application that follows.

Application of Magnetic Circuit Concepts in Shield Design--An Example Study. To further examine magnetic shielding, we consider a situation in which shielding might be required as illustrated in Figure 8.4-5.

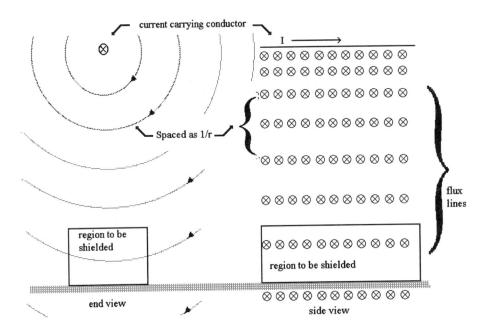

Figure 8.4-5 Shielding Example Showing Current Carrying Conductor and Region To Be Shielded

Referring to Figure 8.4-5, we see two views, one, an end view showing the cross section of a conductor carrying the current I and the cross section of a region to be shielded, directly below the conductor. The side view shows a lengthwise cross section of the conductor and region. Magnetic flux lines are indicated in both views, represented by concentric circles in the first and ⊗s in the second view. The region to be shielded might be a living or working space located directly below a current-carrying line. Note that the lines of magnetic flux as shown in Figure 8.4-5 would not be affected if the walls of this space or structure were constructed of such common building materials as wood and plaster. Nor would they be affected by the earth on which this structure rests.

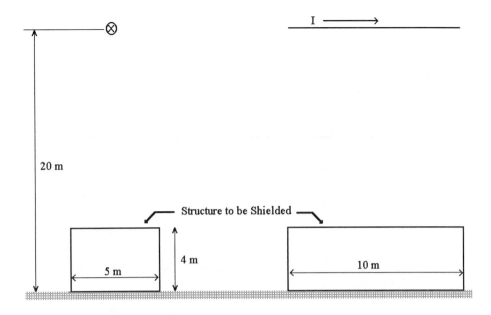

Figure 8.4-6 Example Dimensioned

To be more specific we have dimensioned our example, as shown in Figure 8.4-6.

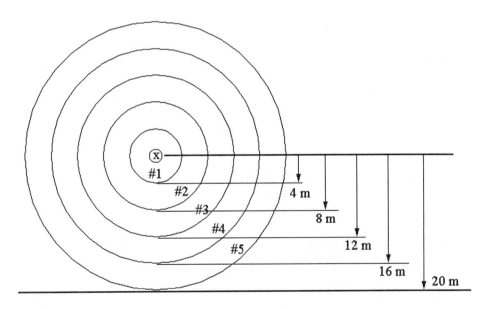

Figure 8.4-7 Concentric Shells

In Figure 8.4-7 we apply the magnetic circuit concepts by dividing the region surrounding the conductor into concentric shells of four meter widths.

The number of such shells is infinite, of course. The mean radii (from the conductor to the center of the shells) and the average reluctance of a unit length of each shell are given in Table 8.4-1. Length is measured along the structure, parallel to the current-carrying conductor.

Shell #	Mean Radius (r)	Reluctance of 1 meter length
1	2 meter	0.25×10^7 weber/ampere
2	6	0.75×10^7
3	10	1.25×10^7
4	14	1.75×10^7
5	18	2.25×10^7
6	22	2.75×10^7

Table 8.4-1 Shell Reluctances

$\mathcal{R} = \dfrac{2\pi}{\mu_0 \times 4 \times L}$ is the reluctance of a shell of mean radius r, permeability μ_0,

width four meters and length L meters, the reluctance of a shell of unit length is:

$\mathcal{R} = \dfrac{\pi r}{2\mu_0}$ webers/ampere, where $\mu_0 = 4\pi \times 10^{-7}$ Henry/meter (the permeability

of air).

For the purpose of our magnetic shielding example, we take I = 500 amperes, rms, at 60 hertz. Accordingly, the MMF = 500 amperes, and a magnetic circuit representing this MMF and the reluctances of the cylindrical shells is shown in Figure 8.4-8.

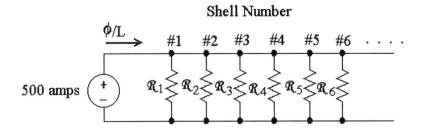

Figure 8.4-8 Magnetic Circuit Representing Conductor and Cylindrical Shells

We see that a complete representation would require an infinite number of parallel reluctances of increasing magnitudes. Clearly this representation is an approximation in that we have divided a continuous field into a number of discrete fields corresponding to the cylindrical shells. The smaller the shell widths, the more accurate the representation becomes. We have chosen four-meter widths to illustrate the method.

Example 8.4-1   Direct calculation of flux density compared with magnetic circuit calculation.

Find the magnetic field flux density at the center of the structure by two methods.

Solution:

- By direct calculation: $B = \dfrac{\mu_0 I}{2\pi r} = \dfrac{2 \times 500 \times 10^{-7}}{18} = 55.6$ milligauss.

$$\frac{\phi_5}{L} = \frac{electromotive\ force}{\mathcal{R}_5} = \frac{500}{2.25 \times 10^7} = 2.22 \times 10^{-5}\ webers/meter$$

- Magnetic circuit:

$$B = \frac{\phi_5}{L \times 4} = \frac{2.22 \times 10^{-5}}{4} = 55.6\ milligauss$$

The comparison is exact at the center of the structure.

Shielding the structure shown in Figure 8.4-6 will involve the reluctance of shell # 5. To a first approximation it will not involve the reluctances of any of the other shells. Shielding will consist of applying sheets of ferromagnetic material to two walls and the roof and floor of the structure, thus creating a low reluctance path around the shielded space. From a circuit viewpoint, shielding is represented as shown in Figure 8.4-9 for shell #5.

Figure 8.4-9 shows the reluctance of shell #5 is 2.25×10^7 webers/ampere, with no shielding. The corresponding magnetic flux in shell #5 is:

$$\frac{\phi_5}{L} = \frac{500}{2.25 \times 10^7} = 2.22 \times 10^{-5}\ webers/meter.$$

The reluctance of the shield will be in parallel with only 5 meters of the 113 meter circumference of shell #5. Accordingly, the flux in shell #5 will increase by about 5 percent after the structure is shielded.

The reluctance of 0.10×10^7 webers/ampere represents the reluctance of the interior of the structure. If the shield is to be effective the reluctance of the shield, $\mathcal{R}_{shield}$, should be small compared to 0.10×10^7 webers/ampere.

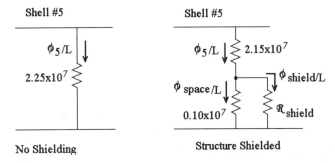

Figure 8.4-9 Representation of Shell #5 No Shielding and With Structure Shielded

For the flux directions shown in Figure 8.4-5 we will base the flux paths through the shield on Figure 8.4-10.

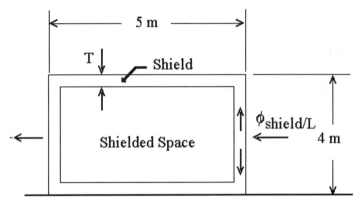

Figure 8.4-10 Flux Paths Through the Shield

Here the average flux path is about 7 meters. With a shield thickness of T meters and relative permeability μ_r, the reluctance of the shield is:

$$R_{shield} = \frac{7}{\mu_0 \mu_R (2T)} = \frac{0.28 \times 10^7}{\mu_R T} \text{ webers/ampere.}$$

The flux through the shielded space of Figure 8.4-10 is:

$$\frac{\phi_{space}}{L} = \frac{1.05 \times 2.22 \times 10^{-5} \times \dfrac{0.28 \times 10^7}{\mu_R T}}{0.10 \times 10^7 + \dfrac{0.28 \times 10^7}{\mu_R T}} \text{ webers/meter (using the principle of flux}$$

division).

And the flux density in the shielded space is:

$$B_{space} = \frac{\phi_{space} \times 10^7}{L \times 4} = \frac{58 \times \dfrac{0.28}{\mu_R T}}{0.1 + \dfrac{0.28}{\mu_R T}} \text{ milligauss.}$$

We see how the flux density through the shielded space depends on the parameters of the shield: its thickness T and relative permeability μ_R. A plot of B_{space} versus the value of ($\mu_R T$) is shown in Figure 8.4-11.

As an example of the use of Figure 8.4-11, if a value of B_{space} of 10 milligauss is selected, the required value of ($\mu_R T$) is 14 meters. Using a ferromagnetic shield with a relative permeability (μ_R) of 10^4, the required thickness (T) of the shield is 1.4 millimeter.

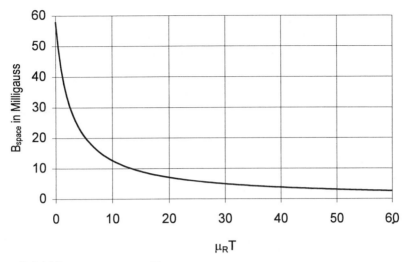

Figure 8.4-11 B$_{\text{space}}$ versus (μ_R T)

The above application of magnetic circuit concepts in shielding design is been based on the simplifying assumptions listed below:

characterization of the shield by a constant permeability--a linear model
discrete modeling of a continuous phenomenon
distortion of flux lines in the presence of the shield neglected

These assumptions are not easily changed for hand calculation. Accordingly a study of the type described, based on magnetic circuit concepts, may be considered as a first approximation or a guide to a more accurate computer based study

Example 8.4-2   Flux density in the shielded space.

For the shielded space described in Figures 8.4-6 and 8.4-10, with I = 500 amperes, T = 5 millimeter, and $\mu_R = 10^4$, calculate the flux density in the shielded space.

Solution:

Here $\mu_R T = .0050 \times 10^4 = 50$ milligauss.

The flux density in the shielded space is given by:

$$B_{\text{space}} = \frac{58 \times \dfrac{0.28}{\mu_R T}}{0.1 + \dfrac{0.28}{\mu_R T}} = \frac{58 \times \dfrac{0.28}{50}}{0.1 + \dfrac{0.28}{50}} = 3.1 \text{ milligauss.}$$

Alternatively we could use Figure 8.4-11 with $\mu_R T = 50$.

Example 8.4-3 Flux density in the shield.

It is important to calculate the flux density in the shield as well as in the shielded space. This is essential because ferromagnetic materials are actually non-linear, that is, μ_R depends on the value of B in the material.

From Figure 8.4-11, a value of T = 5.0 millimeters and $\mu_R = 10^4$ results in a value of

B_{space} = 3.1 milligauss. What is the flux density in the shield under these conditions?

Solution:

Using the principle of flux division:

$$\frac{\phi_{shield}}{L} = \frac{\phi_s / L \times \mathcal{R}_{space}}{\mathcal{R}_{space} + \mathcal{R}_{shield}}$$

$$= (1.05)\frac{2.22 \times 10^{-5} \times 0.1 \times 10^7}{0.1 \times 10^7 + \frac{0.28 \times 10^7}{\mu_R T}}$$

$$B_{shield} = \frac{\phi_{shield} \times 10^7}{L \times 2T}$$

$$= \frac{233 \times 0.1}{\left[0.1 + \frac{0.28}{\mu_R T}\right] \times \left[.5 \times 10^{-3}\right]}$$

with $\mu_R T = 10^4 \times 5.0 \times 10^{-3} = 50$

$$B_{shield} = \frac{46.6 \times 10^3}{0.1 + \frac{.28}{50}} = 4.41 \times 10^5 \text{ milligauss} = 441 \text{ gauss}$$

Application of Finite Element Methods in Shield Design--An Example Study. The finite element method entails the use of a computer to solve the magnetic field equations by dividing the study space into discrete lengths or triangular areas. A number of computer programs are available for this purpose including Maxwell 2D [4] which was used in the study presented below.

The example is that defined above in Figures 8.4-5 and 8.4-6. Again we are studying shielding of the space within a structure located directly below a current-carrying conductor as shown in Figure 8.4-12. This is the same problem that was solved (approximately) using magnetic circuit concepts in the example immediately above.

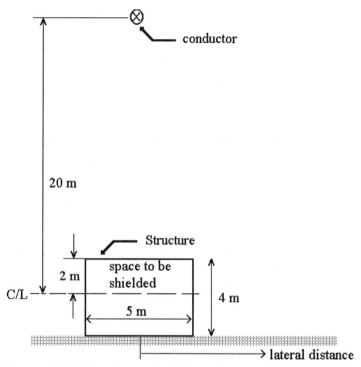

Figure 8.4-12 Coordinates and Dimensions of Example

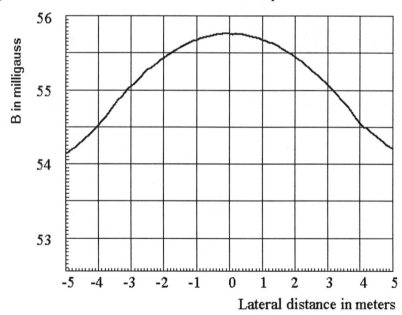

Figure 8.4-13 $\overline{\text{B}}$ Field Along Center Line versus Lateral Distance With No Shield

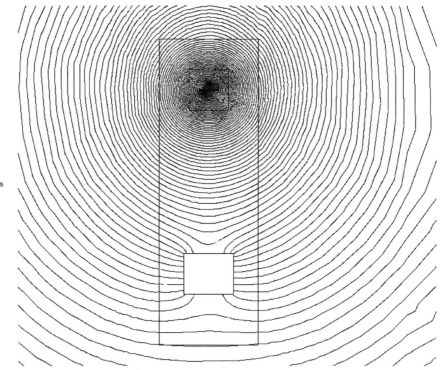

Figure 8.4-14 Plot of Flux Lines with Shield

Using the finite element method we compute the magnetic field density profile along the center line of the space to be shielded (2 meters above the ground). The profile is shown in

Figure 8.4-13. We see that the unshielded flux density is about 55.5 milligauss at the center of the structure and drops to about 54 milligauss at either side.

Using the previous study as a guide, we select a shield with a relative permeability of 10^4 and thickness of five millimeters. This yields a $(\mu_R T)$ product of 50 meters. Surrounding the structure with this shield, we utilize the computer program to obtain a plot of flux lines. This plot is shown in Figure 8.4-14. We can see from the plot that the flux lines are distorted by the presence of the shielding material. Lines from both above and below the structure are attracted to the low reluctance path offered by the shield. This effect was not considered in the hand analysis of the study based on magnetic circuit concepts. Consequently we would expect the value of B_{space} in the magnetic circuit solution to be appreciably less than that for the computer solution.

A field profile showing the magnetic flux density along the center line on both sides and through the shield is shown in Figure 8.4-15. Here we see the attenuation of flux density within the shielded space. The minimum flux density which occurs at the center of the space, and is about five milligauss. This is a reduction of about 11/1 as compared to the space unshielded. The hand analysis based on magnetic circuits predicted a value of 3.1 milligauss.

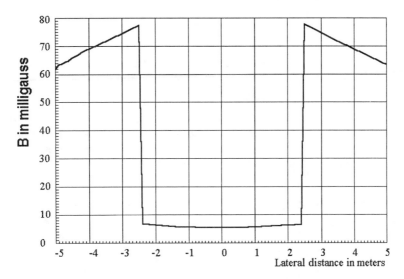

Figure 8.4-15 $\overline{B}$ Field Along Center Line versus Lateral Distance With Shield

Figure 8.4-16 shows a detail of the flux lines in the region of a corner of the structure. We can see the concentration of lines within the ferromagnetic shield.

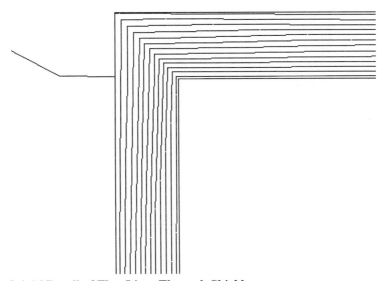

Figure 8.4-16 Detail of Flux Lines Through Shield

Example 8.4-4   Flux density in the shield revisited.

The flux density in the shield was computed in Example 8.4-3 based on the magnetic circuit technique. We now see in Figure 8.4-14 that there is some flux concentration in the shield. Re-estimate the flux density in the shield.

Solution

It appears from Figure 8.4-14 that a 1.6/1 concentration of lines occurs in the shield. Accordingly, the flux density in the shield is approximately 706 gauss.

Discussion. The study based on magnetic circuit concepts provided a guide for the shield design. A more accurate analysis technique, utilizing a finite element computer program, can be used to verify and/or refine the design. Clearly both methods are mutually supporting in the design of a shield for a specified region. While the magnetic circuit method identifies the critical variables and how they are related to B_{space}, the computer program provides a more accurate solution.

The example study was defined in terms of simple geometry, involving a single current-carrying conductor and a structure directly below the conductor. Can the methods used here be applied in more general cases? The answer depends on the complexity of the geometry. As the complexity increases, the magnetic circuit concepts become more difficult to apply. The finite element methods also become more cumbersome, requiring, for example, longer solution times.

An important variation of the above example study is based on the geometry of Figure 8.4-17.

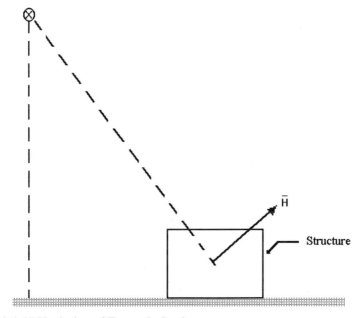

Figure 8.4-17 Variation of Example Study

Here the structure is not directly below the conductor. Further generalizing, we may think of the current in the conductor as the net or zero-sequence current in a three phase transmission or distribution line. To design a shield for this case, we must consider the direction of the magnetic field intensity vector, $\overline{H}$, relative to the walls of the structure. Shielding attached to the vertical walls would provide a path for flux resulting from the vertical component of $\overline{H}$ while shielding attached to the horizontal walls (roof and floor) would provide for flux resulting from the horizontal component of $\overline{H}$.

In an application of magnetic shielding, Durkin et al., discuss shielding an office within a generating station [7]. Before shielding, fields as high as 0.15 tesla (1.5×10^6 milligauss) were measured inside the office. A sandwich type shield consisting of a 0.25 inch thick A1010 steel plate as the first layer, an air gap of two inches and a 1/16 inch thick layer of mu-metal were added to the office outer wall adjacent to the source of the field. The field attenuation produced within the office as a result of this shield was reported to be 1/1500.

8.5 Shielding Applied to a Specific Region—Electromagnetic Shielding

When a good electric conductor, which may or may not be magnetically permeable, is immersed in a time-varying magnetic field, electric currents are induced in the conductor by magnetic induction. These currents are commonly called *eddy currents*. The magnetic field produced by the eddy currents tends to oppose the inducing magnetic field with the result that the magnetic field in a space surrounded by a conductor is less than the inducing field. This type of shielding is called electromagnetic shielding [8].

The problem considered in Section 8.4, shielding the space inside a structure, as shown in Figure 8.4-6, is amenable to electromagnetic shielding. That is, we might surround the structure with a good conductor in the same way we surrounded it with a highly permeable magnetic material in Section 8.4. We may recall that surrounding the structure with 5 mm of magnetic material with a relative permeability of 10,000 reduced the flux density of the space within the structure by a factor of 11. What thickness of aluminum or copper would be required to reduce the flux density by a similar ratio? To answer this we introduce the concept of *skin depth*.

Generally, the most important parameter to be considered in electromagnetic shielding is *skin depth* [5]. The skin depth (δ) of a material, characterized by permeability (μ, Henry/meter) and conductivity (σ, Siemens/meter), is given by:

$$\delta = \frac{1}{\sqrt{\pi\mu\sigma f}} \text{ meters}$$

where f is the frequency of the inducing field in hertz. Typical values for δ at 20°C are :

Material	δ
aluminum	0.120 meter
copper	0.007

An electromagnetic wave, incident on a conducting material, is attenuated as it propagates through the material. For example, a plane electromagnetic wave normally incident on this material is attenuated by the factor $1/e$ for each distance δ that it traverses in the material.

Returning to the question, an attenuation of $1/11$ would require a conductor thickness of 2.4δ, that is, 2.9 centimeter of aluminum or 1.7 centimeter of copper.

A detailed discussion of electromagnetic shielding for various structural shapes is given by Rikitake [8]. In particular, he shows that a conducting square shell becomes highly effective as a shield for wall thickness of $\sqrt{2}\delta$ or greater. Cylindrical shells are likewise effective for similar wall thickness.

Example 8.5-1   Skin depth.

Calculate the number of skin depths provided by five millimeters of copper at 60 hertz.

Solution:

$$k\delta = 0.5 \, \text{centimeter}$$

$$k = \frac{0.5 \, \text{centimeter}}{1.7 \, \text{centimeter}} = 0.3 \, \text{skin depths}.$$

The attenuation of the magnetic field for this thickness of copper is given by: $e^{-.3} = 0.74$.

8.6 Active Shielding or Cancellation

The magnetic field intensity at a specific point is equal to the vector sum of the $\overline{H}$ vectors due to all currents. (This assumes that the media of propagation is linear so that the mathematical principle of superposition can be applied.) Consequently the possibility exists for reducing the field existing in a specific region by adding a canceling field. This is known as active shielding as opposed to the use of shielding materials as employed in passive shielding techniques. The process is illustrated in Figure 8-6-1.

Referring to Figure 8.6-1 we suppose that a current I_1 flows in conductor ① as shown. This produces a circumferential magnetic field in the specified region.

$$H_1 = \frac{I_1}{2\pi R_1}$$

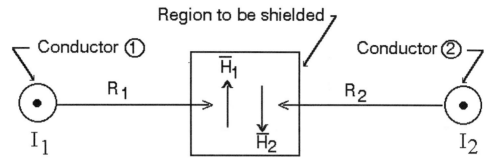

Figure 8.6-1 Illustration of Active Shielding

We wish to reduce the magnetic field in the region by active shielding. This involves the use of conductor ②. A current I_2 is caused to flow in conductor ② to produce H_2.

$$H_2 = \frac{I_2}{2\pi R_2}$$

If $I_1 = I_2$, $R_1 = R_2$ and the dimensions of the region are small compared to R_1, the resulting field in the region will be much less in magnitude than H_1. While conductor ① may be a transmission line conductor transmitting large amounts of power, conductor ② is simply part of a low voltage system carrying very little power. Further if R_2 is made less than R_1, I_2 can be reduced proportionately.

A practical use of active shielding would involve several coils enclosing the region to be shielded. A field detecting device would continuously measure the field in the region and adjust the coil currents to minimize the magnitude of the field at one or more points within the region.

Commercial versions of this scheme offer to reduce fields to one milligauss or less in 90 percent of the volume of a home.

Walling et al., propose short circuited coils underlying sections of a transmission line and located above the area to be protected as a magnetic field mitigation method [9]. The shield circuit configuration is shown in Figure 8.6-2. A capacitor in series with the shield circuit is used to reduce the reactance of the coil circuit. The voltage induced into the shorted coils by time varying magnetic flux produces a current and resulting flux that can be adjusted to reduce the magnetic flux due to the transmission line currents. This is not unlike a large eddy current shield.

The shield configuration shown has the greatest relative magnetic field reduction near the right-of-way edges but also reduces the field substantially at distance. The penalty for this type of shielding is in capital investment and continuous power losses in the shorted coil.

158

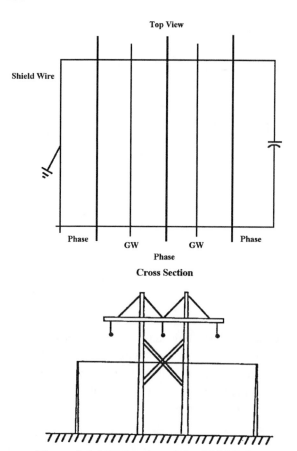

Figure 8.6-2 Mitigation of the Field Below a Transmission Line [9]

Another approach to reducing the magnetic field in the vicinity of a transmission line was proposed by Jonsson et al., [10]. In an IEEE paper, *Optimized Reduction of the Magnetic Field near Swedish 400 kV Lines by Advanced Control of Shield Wire Currents. Test Results and Economic Evaluation*, the authors report on field reduction possibilities by introducing controlled shield wire currents. The concept is applied to the relatively new delta configuration of Swedish 400 kV transmission lines. Three shield conductors are placed in a triangular configuration surrounding the three phase conductors as shown in Figure 8.6-3. With three external and adjustable voltage sources connected to the shield conductors, the shield conductor currents can be controlled to minimize the magnetic field in the vicinity of the line. The minimized field magnitude is far below the magnitude normally found near such lines. The paper further reports that this result can be achieved at an additional cost of about twenty percent of the line costs per kilometer. However, it must be kept in mind that the proposed measures are intended to be used only on a limited section of line up to a few kilometers in length.

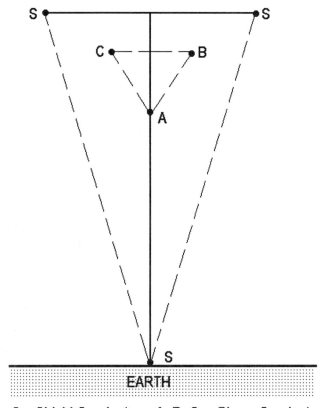

S = Shield Conductors A, B, C, = Phase Conductors

Figure 8.6-3 A Delta Phase Configuration with Surrounding Shield Conductors [10]

8.7 Summary

Magnetic fields may be managed by shielding under appropriate conditions. For example, the field due to a zero-sequence current or net current **cannot** be shielded globally. However, the field due to balanced three phase currents, as in a duct, is readily shielded. Where zero-sequence fields occur, the best we can do is to shield specific restricted regions. In this case, either magnetic and/or electromagnetic shielding is effective. Various forms of active shielding are also useful for restricted regions.

Magnetic shielding is studied using both magnetic circuit concepts and the finite element analysis method. The magnetic circuit technique is approximate but is valuable for developing relationships between variables to guide the shield design. The finite element method provides better accuracy; in addition it can be used to solve shielding problems that are just not amenable to pencil and paper methods.

8.8 References

[1] Horton, W. F. and Goldberg, S., *Shielding the Power Frequency Magnetic Fields Produced by Underground Distribution Cables*, Proceedings of the North American Power Symposium, Kansas State University, 1994.

[2] FIELDS Program 2.0, Southern California Edison Research Center, 6090 Irwindale Avenue, Irwindale, CA 91702.

[3] TL WORKSTATION Code, ENVIRO Manual Version 2.2, EPRI Project 2472, July 1989.

[4] MAXWELL 2D Field Simulator, Ansoft Corporation, Four Station Square, Suite 660, Pittsburgh, PA 15219, June 1991.

[5] S. V. Marshall and G. G. Skitek, *Electromagnetic Concepts and Applications*, Third Edition, Prentice Hall, Englewood Cliffs, NJ 07632, 1990.

[6] K. Kawasaki, M. Inami and T. Ishikawa, *Theoretical Considerations on Eddy Current Losses in Non-Magnetic and Magnetic Pipes for Power Transmission Systems*, IEEE Transactions on Power Apparatus and Systems, Vol. PAS-100, No. 2, pp 474-484, Feb. 1981.

[7] Durkin, C. J. et al, *Five Years of Magnetic Field Management*, IEEE Power Engineering Society 1994 Summer Meeting, San Francisco, CA, 1994. Paper Number 94 SM 392-1 PWRD.

[8] T. Rikitake, *Magnetic and Electromagnetic Shielding*, D. Reidel Publishing Company, A Member of the Kleuwer Academic Publishers Group, Boston, MA, 1987.

[9] R. A. Walling, et al, Series-Capacitor Compensated Shield Scheme for Enhanced Mitigation of Transmission Line Magnetic Fields, IEEE Transactions on Power Systems and Apparatus,

[10] U. Jonsson, A. Larsson and J-O. Sjödin, Optimized Reduction of the Magnetic Field near Swedish 400 kV Lines by Advanced Control of Shield Wire Currents. Test Results and Economic Evaluation, IEEE Transactions on Power Delivery, Vol. 9, No. 2, April 1994.

8.9 Exercises

[1] Underground cable.

An underground coaxial cable, buried 1 meter below the surface, carries a single-phase current of 100 amperes rms. Find the magnitude of the magnetic flux density vector 1 meter above the surface, directly over the cable if:

        (A)   the return current is carried by the shield

        (B)   the return current is in the earth.

Answer:	(A)	essentially zero
	(B)	10 milligauss rms

[2] Magnetic circuit.

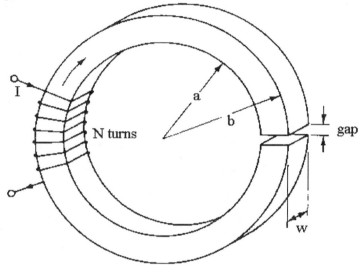

Figure 8.10-1 Magnetic Circuit

Show an equivalent magnetic circuit for the iron core inductor with air gap.

Answer:

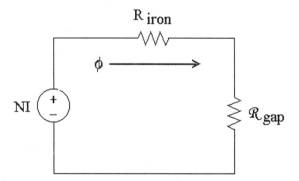

Figure 8.10-2 Equivalent Magnetic Circuit

[3] In Exercise [2], a = 1 meter, b = 1.2 meter, w = 0.2 meter, gap = 0.002 meter, I = 10 amperes, N = 50 turns, $\mu \cdot \mathcal{R}_{iron} = 10^4$.

Find $\mathcal{R}_{iron}$, $\mathcal{R}_{gap}$, and the magnitude of the magnetic flux density in the air gap.

Answer: $\mathcal{R}_{iron} = 1.38 \times 10^4$ A/weber

$\mathcal{R}_{gap} = 3.98 \times 10^4$ A/weber

B = 0.234 tesla (neglecting fringing at the gap)

[4] Magnetic shield.

Referring to Figure 8.4-7, divide the region surrounding the conductor into twice as many shells, that is, each shell is to have a width of 2 meters. Draw the magnetic circuit representing the conductor and cylindrical shells as in Figure 8.4-8.

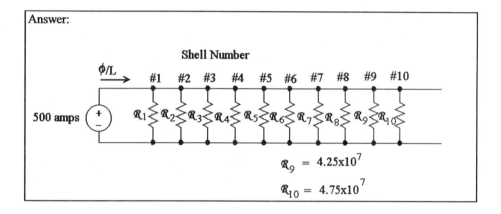

Figure 8.10-3 Equivalent Magnetic Circuit Showing Shells

[5] Magnetic shield continued.

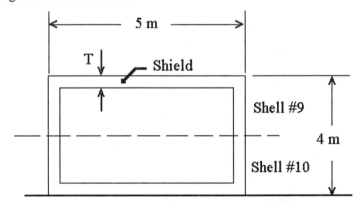

Figure 8.10-4 Flux Paths Through the Shield

Shells 9 and 10 intersect the shield as shown above. The average flux path through the shield is 7 meters. Find the $\mathcal{R}_{shield}$ for these paths.

> Answer:
>
> $$\mathcal{R}_{shield} = \frac{0.56 \times 10^7}{\mu_R T} \text{ weber/ampere}$$

[6]   Shell reluctances.
Represent shells #9 and #10 with the magnetic shield included.

> Answer:
>
>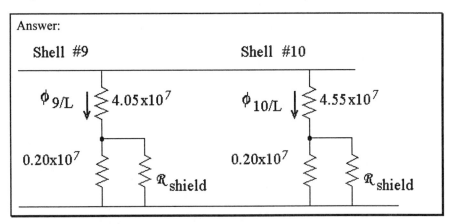

Figure 8.10-5 Equivalent Magnetic Circuit Showing Shells

[7]   Flux density in the shielded region.
Find B_{space} if the shielding material has a relative permeability of 5000 and a thickness of 10 mm (see Exercise [6])

> Answer:
>
> 3.0 milligauss. This is essentially the same answer as was found previously in the text. It can be improved by noting that the flux in the shield is actually about 60 percent larger than calculated as shown in the finite element analysis.

9 Regulatory Issues and Prudent Avoidance

9.0 Overview

In Chapter Nine we examine the magnetic field-adverse health effects issue from a regulatory perspective. We consider the difficulty of regulating magnetic field exposure in the face of uncertainty. The principal uncertainty, in spite of many wide-scale epidemiological studies, it develops, is of dose-response assessment.

In the United States, only two states, New York and Florida, have set magnetic field exposure limits, and these are related only to transmission lines in New York and to newly constructed transmission lines and substations in Florida. A number of state commissions have established guidelines of "prudent avoidance". The meaning of this policy and its implications are discussed below. An accompanying appendix (Appendix F) provides a listing of prudent avoidance actions that can be taken on a personal level.

We look at the actions of other countries and find that Sweden has wrestled with some of the same magnetic field exposure concerns facing United States agencies. Australia and the United Kingdom, as well, have established their own broad guidelines and standards.

Finally we consider international guidelines and find that they have been instituted on the basis of the well established effects of induced currents as opposed to relevant epidemiological studies.

9.1 Introduction to Regulatory Issues

The concerns of the public regarding the health effects of power frequency magnetic fields have been heard throughout the United States and much of the industrialized world. These concerns are based on the results of several epidemiological studies but largely on episodic reports (for example, cancer cluster accounts appearing in newspapers, magazines and books). Probably, the majority of the public that is aware of the health effects issue and many in the scientific community believe that a link between cancer and magnetic field exposure will be found. The issue is viewed, rightly or wrongly, as similar to the tobacco issue: a cause and effect relationship between public health and magnetic field exposure will be discovered sooner or later--the smoking gun analogy.

All manner of institutions, including courts, legislatures and administrative agencies, have addressed the magnetic field health issues, all wrestling with a situation that involves a high degree of scientific uncertainty. Unfortunately, science has not yet been able to provide sufficient information on which to base a regulatory policy--despite a number of large-scale epidemiological studies.

Regulatory agencies normally address scientific uncertainty, such as the magnetic field-adverse health effects issue, through procedural mechanisms. Studies are undertaken, evidence is assembled, arguments are heard and decisions are reached much as in a judicial proceeding. The process varies among states and countries and with the charter of the agency. Political pressures may modify or hasten the regulatory procedures, fueled by the public demand to "**do something**".

In the United States, legislative or administrative agencies in at least 17 states have formally considered the electric and magnetic field (EMF)-adverse health effects issue [1]. Responses range from dismissal of the question due to lack of evidence (Wyoming) to codification of formal EMF limits for transmission lines (Florida). In most cases the agencies have initiated studies and/or hearings on the issue. At least one other country (Sweden) has prescribed EMF limits. Other countries have issued guidelines and standards.

In this chapter we will look at the regulatory process as it applies to magnetic fields and the actions of regulatory agencies in establishing standards, limits and guidelines.

9.2 Standards, Limits and Guidelines

Regulatory agencies act by establishing standards, limits and guidelines, all of which have been enacted in some form with respect to magnetic fields by various agencies.

In regulatory language, a standard may be either a qualitative or quantitative measure for comparison. A qualitative standard might specify a characteristic such as "transmission line current shall be well balanced", while a quantitative standard could specify a tower height in feet. When adopted by a governmental body, standards become mandatory.

"Limit" is a narrower term than standard, referring to a restrictive level. For example, an electric field limit may be the maximum allowable field in kV/meter, a measurable quantity.

Guidelines are optional standards or limits, exerting some pressure for conformance without requiring compliance.

In the context of magnetic fields, standards can be established for siting of transmission lines, conductor configurations, or double circuit arrangements. Limits can be established on the measured or calculated values of magnetic fields at the edges and within transmission rights of way. Further guidelines may be set down to guide design of transmission lines where no standard or limits have been established. The principle of "prudent avoidance" discussed later in this chapter is a widely used guideline.

In the magnetic field-adverse health effects arena, regulatory agencies find themselves in a quandary. As shown in Chapter Two, there is not yet a definitive indication that magnetic field exposure can affect health negatively, and there are no data to show that it cannot. From this perspective, we will examine the actions that have been taken by regulatory agencies.

9.3 Regulation in the Face of Uncertainty

It is a well established principle that there is no such thing as 100 percent environmental protection. As there is a health risk associated with the emission of many types of contaminants, a role of public health agencies is to limit the emission to "acceptable" levels. The scientific community may determine an acceptable level by means of risk assessment and a risk estimate. However, risk assessment involves assessing dose-response--that is, the determination of the relation between the magnitude of exposure and the probability of occurrence of the health effects in question. In spite of numerous epidemiological studies, a conclusive dose-response assessment for magnetic fields has not yet been determined. Regulating agencies must deal with the uncertainties associated with this.

Regulation in the face of such uncertainty involves weighing public health against public expense. Table 9.3-1 is a simplified decision matrix that illustrates the possible relevant regulatory decisions and the effects of these decisions.

Ultimate Finding	Regulatory Action	
	None	Set Limits
Magnetic fields are benign.	Correct. Public health is unaffected. No expenditures.	Incorrect. Public expenditure for nothing.
Magnetic fields are harmful.	Incorrect. Public health endangered.	Correct. Public health is protected.

Table 9.3-1 Regulatory Decision Matrix

In the case of magnetic fields, setting limits may involve very large public expenditures. Without a clear risk estimate, regulatory agencies have generally chosen to take no action.

9.4 Regulatory Actions in the United States

Long-term exposure of the public to power frequency magnetic fields is generally seen to be directly linked to the electric energy delivery system. This system, as a source of magnetic fields, is described in Chapter Four.

The electric energy delivery system in the United States is the business of numerous electric utilities, both investor owned and publicly owned. Electric utilities are regulated at the federal level by the Federal Energy Administration and at the state level by agencies called public utility commissions or public service commissions. For example, the California Public Utility Commission (CPUC) is responsible for regulation of investor owned utilities within the state of California. The public utility commissions have purview of the electromagnetic fields associated with the utility owned electric energy delivery systems. As such they may impose regulatory limits on such quantities as the magnitude of the resultant field at the edge of a transmission line right of way.

Since health effects are involved, agencies responsible for public health have also taken an active interest in the issue. At the federal level the National Institute of Environmental Health Standards (NIEHS) shares leadership with the Department of Energy (DOE) in a federally coordinated program of EMF research. State departments of public health are responsible for evaluating the results of EMF research programs and reporting to the public and the state regulatory agencies. This in turn allows the regulatory agencies to take more fully informed regulatory actions.

State	Electric Field Limit
California	1.6 kV/m at edge of ROW[*]
Florida	10 kV/m for existing 500 kV transmission lines 8 kV/m for existing 230 kV transmission lines 2 kV/m for new transmission lines at edge of ROW
Minnesota	8 kV/m for existing 230 kV transmission lines
Montana	1 kV/m at edge of ROW in residential area 7 kV/m at edge of ROW at road crossing 2.5 to 3.5 kV/m in areas such as parking lots
New Jersey	3 kV/m at edge of ROW
New York	1.6 kV/m in ROW
North Dakota	9 kV/m at edge of ROW
Oregon	9 kV/m in ROW 7 kV/m at edge of ROW at road crossing

[*] Right of Way

4-1 Electric Field Limits by State [1-2]

Table 9. Table 9.4-1 provides a listing of states whose commissions impose limits on the electric field and those limits.

Federal Level. Section 2118 of the Energy Policy Act, signed by former President George Bush in October 1992, established a federally coordinated program of EMF research with provisions for communication of results and information to the public. The United States Department of Energy (DOE) and the National Institute for Environmental Health Sciences (NIEHS) oversee this research. DOE is responsible for engineering research, while NIEHS oversees health effects. To date, no national standards exist for the regulation of magnetic fields based on long-term health effects. Nor does a federal agency have a clear mandate or specific authority to regulate. In the absence of federal direction, the states have reacted to the issues.

Item	Utility	Case/Project Description	Status
1	Arizona Public Service Company	SDG&D 500-kV Interconnection	No direct ruling on EMF issue but committee granted Certificate of Environmental Compatibility.
2	Atlantic Electric Company	Application to Construct and Operate a 230-kV Transmission Line and Related Substation Facilities (1984)	Landowners raised issue of EMF health effects; AEC submitted reports on EMF health effects; landowners dropped EMF claims and 230-kV line was approved when a new route was agreed upon.
3	Atlantic Electric Company	Mickleton 230-kV Line	Line approved; Board found that fields from this line would be similar to fields from existing lines of same voltage class and that field levels from this line fall within field standards in New Jersey and other states.
4	Cheyenne Light, Fuel and Power Company	Coriette 115-kV Line	Commission found no evidence of adverse health effects. Landowners appealed decision to Wyoming Supreme Court. Supreme Court found sufficient record of no health risks.
5	Commonwealth Electric Company	Application for 345-kV Substation: No. DPU 86-257	Department of Public Utilities found no evidence that EMF from the substation would increase health risk.

Table 9.4-2 Results of EMF Survey [1] (Continued on next page)

State Level. Public utility commissions in several states have, for many years, regulated one of the two aspects of electromagnetic fields--the electric field. Maximum levels of the electric fields at the edges of transmission line right of ways are regulated as a safety concern. The second aspect, the magnetic field, was ignored historically since it was not suspected of causing or enabling disease.

These limits apply at ground level. The electric field is much greater as we approach a transmission line conductor. (The electric field strength in volts per meter is proportional to line voltage and varies inversely with distance.) We note that high electric fields at ground level are an issue with transmission lines (voltages of 115 kV and higher) but not for distribution lines (voltages of 35 kV and lower). Thus the electric field limits generally refer to the transmission line right of ways.

It would seem that regulations limiting the magnitudes of magnetic fields would be much more far reaching than those limiting the magnitudes of electric fields. Unlike electric fields, of which the highest fields are associated with transmission lines, the highest magnetic field exposures that the public experiences are associated with the distribution system. As we saw in Chapter Four, the distribution system brings electric power and thus magnetic fields into almost every aspect of life. Meaningful regulation would have to be applied at the substation level, the feeder lever, distribution transformer level and secondary/services level. However, public utility commissions generally have cognizance only over transmission lines.

Item	Utility	Case/Project Description	Status
6	Hawaii Electric Light Company, Incorporated	Puna-Pohaiki Certification for two 69-kV Lines	Commission found insufficient evidence to conclude that a health risk exists and approved line.
7	Minnesota Power Company	Application for Exemption From Siting Requirements for Duluth Area	Application approved; Board found that proposed upgrade would result in lower magnetic field levels and there is "no definitive evidence" that EMF causes health problems.
8	New England electric System	Certificate to Install 450-kV Line. No. DSF81-349	Commission found that effects on public health, if any, fall within acceptable limits.
9	Northern States Power Company	Applications to Modify and Upgrade Transmission Lines	Application approved in May 1991; Board found that there is "no evidence that the increased magnetic fields associated with the increased power flow in 500-kV line will present a risk to human health".

Table 9.4-2 (Continued) Results of EMF Survey [1] (Continued on next page)

Item	Utility	Case/Project Description	Status
10	Pacific Power and Light	Eugene-Medford 500-kV Line	Application granted; "COPE" group subsequently raised EMF issues prior to construction; site certificate amendments sought for 4 route changes; EFSC approved August 29, 1990. PP&L will use "Delta" Line configuration for EMF mitigation.
11	Potomac Edison Company	Brighton-High Ridge 500-kV Line	Commission found no basis to concluded that power line fields cause adverse health effects. Rejected proposals for field standards that have no scientific basis. Staff will monitor ongoing EMF research and report on semi-annual basis.
12	Public Service Company of Colorado	Daniels Park Transmission Line	Board of Commissioners held that undergrounding did not constitute "Prudent Avoidance" under the circumstances. Overhead line approved with conditions offered by Utility. State Court overturned decision on ground unrelated to the merits of the EMF claim.
13	San Diego Gas and Electric Company	Application for Certification: Decision #93785	Commission found that available information did not indicate that EMF causes adverse health effects. Ordered utility to continue to fund EPRI studies and inform Commission of results.
14	Southern California Edison Company	Kramer-Victor 230-kV Line	CPUC granted approval to construct transmission line; required utility to "minimize" magnetic fields associated with line.
15	Wisconsin Public Service Corporation (WPSC)	Bayport-Mason Street upgrade (69-kV double circuit 138-kV Line and Substation)	Approved by PSC-magnetic fields not likely to adversely affect human health; condition: magnetic field measurements before and after project.

Table 9.4-2 (Continued) Results of EMF Survey [1]

In the transmission area, commissions have dealt with the magnetic field-health effect issue on a case by case basis. Concise descriptions of several transmission line and substation siting cases and relevant commission decisions are given in Table 9.4-2 [1]. These were selected from among a large number of such cases as a representative sample.

We note that state utility commissions are generally willing to approve applications for projects that do not result in greater fields than those typically encountered by the public.

In contrast to the electric field limits established by several states (as shown in Table 9.4-1), magnetic field limits have been established by only two states. These limits are shown in Table 9.4-3.

State	Magnetic Field Limit
New York	200 milligauss at edge of ROW* for lines of over 125 kV and more than 1 mile in length.
Florida	200 milligauss at edge of ROW for single circuit 500 kV lines. 250 milligauss at edge of ROW for double circuit 500 kV lines. 150 milligauss at edge of ROW for lines of 230 kV or less.

*Right of Way

Table 9.4-3 Magnetic Field Limits by State [1]

An excerpt of the standards adopted by the Florida Public Service Commission specifically for new transmission lines and substations is shown below.

In view of the difficulty in establishing hard (concrete) limits, regulatory agencies have moved toward establishing guidelines. A prominent and widely used guideline is known as "prudent avoidance".

Prudent Avoidance. Prudence suggests that people be kept out of magnetic fields whenever it is inexpensive to do so. In fact "prudence" can be conceived as a proportionate response--a response proportionate to the risk.

As an alternative to direct regulatory actions with "**hard limits**", some regulatory agencies have adopted a policy vis-à-vis magnetic fields, termed "**prudent avoidance**".

This policy, as implemented by a public utilities commission, might require a utility to select, for example, from two alternative designs of a substation project, the design that exposes the public to the lesser magnetic fields. This design would be approved providing its cost is not substantially higher than the competing design. As an example, the Colorado Public Utilities Commission in a November 1992 Statement of Adoption [4] included the following statements:

Florida Statute 17-814.450 Paragraph (3) Electric and Magnetic Field Standards [3]	
(a)	The maximum electric field at the edge of the substation shall not exceed 0 kV/m
(b)	The maximum electric field on the ROW of a 230 kV or smaller transmission line shall not exceed 8 kV/m.
(c)	The maximum electric field on the ROW of a 500 kV transmission line shall not exceed 10 kV/m.
(d)	The maximum magnetic field at the edge of a 230 kV or smaller transmission line ROW or at the property boundary of a new substation serving such lines shall not exceed 150 milligauss.
(e)	The maximum magnetic field at the edge of the transmission line ROW for a 500 kV line or at the property boundary of a new substation serving a 500 kV line shall not exceed 200 milligauss, except for double circuit 500 kV lines to be constructed on ROWs existing on March 21, 1989, as identified below where the limit will be 250 milligauss.
(f)	For existing ROWs extending from the Andytown substation to Orange River substation, Andytown substation to the Martin Generating Plant, and the Martin Generating Plant to the Midway substation, where the facility owner has acquired, prior to March 21, 1989, a ROW sufficiently wide for two more 500 kV transmission lines and has constructed one or more 500 kV transmission lines on this ROW prior to March 21, 1989, the maximum magnetic field at the edge of the ROW or property boundary of a new or modified substation shall not exceed 250 milligauss.

The Colorado Public Utilities Commission shall include the concept of prudent avoidance with respect to planning, siting, construction, and operation to achieve a reasonable balance between the potential health effects of exposure to magnetic fields and the cost and impacts of mitigation of such exposure, by taking steps to reduce the exposure at reasonable or modest cost. Such steps might include, but are not limited to: (1) Design alternatives considering the spatial arrangement of phasing of conductors; (2) Routing lines to limit exposures to areas of concentrated population and group facilities such as schools and hospitals; (3) Installing higher structures; (4) Widening right of way corridors; and (5) Burial of lines.

The California Public Utilities Commission has mandated prudent avoidance measures that apply to the investor owned utilities within California. Among these measures are that utilities take all no-cost or low-cost steps that might reduce EMF exposure in new and upgraded projects, such as wide easements or underground wiring. Low-cost is defined as a maximum of 4 percent of the project cost. Utilities must establish guidelines reflecting the Public Utility Commission policy. The order also specifies that the utilities have uniform residential and workplace measurement programs. (Many utilities already offer free measurements for customers.)

The term "prudent avoidance" was first developed by Dr. M. Granger Morgan and his colleagues, Drs. Indira Nairand and H. Keith Floring at Carnegie Mellon University [5]. Dr. Morgan used the term when presenting policy options for risk management of public health effects from magnetic field exposure to the United States Congressional Office of Technology Assessment. "Prudent avoidance" was originally defined as "the avoidance of any field that can be avoided without significant cost to the quality of life". In practice, the term means that when a prudent investment (say five percent of the cost of a project) will result in a significant avoidance of public exposure to magnetic fields, that investment will be made. Of course the words prudent and significant are open to subjective interpretation.

At present a number of state public utility commissions employ a policy of prudent avoidance of magnetic fields. These states include, in addition to Colorado and California: Hawaii, North Carolina, Illinois, Michigan, Vermont and Wisconsin [6]. For public utility commissions that traditionally have employed the setting of rules that call for avoidance of population centers and existing facilities, prudent avoidance is in place.

Prudent avoidance of magnetic fields can be achieved by personal action as opposed to legal or regulatory fiat. For example, an action of prudent avoidance on the personal level would be to choose not sleep under an energized electric blanket. A listing of possible prudent avoidance actions which can be taken on a personal level is given in Appendix F.

Consumer Products. As we saw in Chapter Seven, consumer products such as appliances generate some of the highest magnetic fields the general public encounters. Only electrical workers might be exposed to higher fields in the workplace than in their homes. However, the fields, which are only present when the appliance is in use, decrease rapidly with distance from the appliance. We know of no concerted effort to limit, by regulation, the maximum field that an appliance can produce. Further, appliance manufacturers have not used low magnetic field emission as a selling point for their products. This may come some time in the future if and when low magnetic field emission for an appliance becomes a significant factor to the consumer.

On the other hand, manufacturers of computer terminals have been the subject of regulation. For example, in the state of New York, the maximum magnetic field emitted by computer monitors is limited to 2 milligauss at a distance of 30 centimeters (about one foot) or more.

9.5 Regulatory Actions in Other Countries

Australia. Australia has established "Interim Guidelines on Limits of Exposure to 50/60 hertz Electric and Magnetic Fields (1989)". These are summarized in Table 9.5-1.

Note that these guidelines are consistent with the international guidelines discussed in detail in Section 9.6.

Circumstance	Electric Field	Magnetic Field
Public-all day	5 kV/m	1 gauss
Public-limited day	10	10
Occupational-all day	10	5
Occupational-limited day	30	50 (2 hours per day)

Table 9.5-1 Australian Standards [7]

The United Kingdom. The United Kingdom approved a guidance standard in 1988 for a 50/60 hertz magnetic field [8]. The standard is 20 gauss for both public and occupational exposure, very significantly higher than the limits considered in Sweden and the United States.

We see that these standards are much higher than those in place or contemplated in the United States. They reflect the conclusion of the National Radiological Protection Board that "there is at present insufficient biological and epidemiological data to make a health risk assessment or even to determine whether there is a potential hazard to health with regard to a thermal effect of electromagnetic fields."

Sweden. In July 1991, the Swedish government implemented a new standard (called MPR2) requiring that extremely low frequency (ELF) magnetic emissions from computer monitors not exceed 2.5 milligauss at 50 centimeters from the screen. The Swedish national labor union and others protested that the MPR2 rule is not an adequate protection. The Swedish union adopted its own standard of 2.0 milligauss at 30 centimeters from the screen. Most older monitors do not meet that standard; however, monitor manufacturers are currently producing monitors that do meet the standard.

In January 1993 the Swedish National Electrical Safety Board was formed and given the responsibility of setting standards and new measures related to magnetic fields. In response to both recent and previous epidemiologic studies, the National Electrical Safety Board has revised assessment of the health hazards associated with magnetic fields. The board has declared that "considerations around new standards for electrical installations will be based on the assumption that there is a connection between the exposure to power frequency magnetic fields from transmission lines and childhood cancer." [9] The National Electrical Safety Board has emphasized that it does not conclude that an absolute correlation between exposure to magnetic fields and childhood cancer exists. The board does clearly state that such a correlation based known facts cannot be excluded and that further studies must be carried out. The National Electrical Safety Board in the future will analyze and compare the cost of reducing magnetic fields with the possible health benefits that can be accomplished.

New guidelines under consideration are:

- Directives on new installations
- Directives on new houses, schools, day care centers in the proximity of existing installations.

Another government agency, the Swedish National Institute of Radiation Protection (SSI), has also reached preliminary conclusions on the basis of recently presented research results. SSI has concluded that some results support the hypothesis of a relationship between cancer and magnetic fields although many factors also demonstrate the opposite of such relationships. SSI, however, have also concluded that magnetic field mitigation efforts have lower priority than, for instance, measures against radon, UV radiation, and radiation from medical X-ray investigations. SSI has stated that mitigation efforts that can be achieved at reasonable expense and that reduce magnetic field exposure to the general population may be justified.

To summarize, Swedish authorities are not convinced that a correlation between magnetic fields and adverse human health has been established, but they will take a cautious standpoint and will support further studies on the matter.

9.6 International Guidelines

The International Radiation Protection Association (IRPA) in 1977 formed the International Non-Ionizing Radiation Committee (IRPA/INIRC). IRPA/INIRC, in cooperation with the Environmental Health Division of the World Health Organization (WHO), has undertaken responsibility for the development of health criteria documents on non-ionizing radiation. In line with this responsibility, IRPA/INIRC published *Interim Guidelines on Limits of Exposure to 50/60 Hz Electric and Magnetic Fields*, a document that was approved by IRPA in 1989 [10]. The electric and magnetic field exposure limits of the Interim Guideline are summarized as shown in Table 9.6-1. As stated in the document, the basic criterion used to determine the exposure limits is to limit current densities induced in the head and trunk by continuous exposure to 50/60 hertz electric and magnetic fields to no more than about ten milliamps per meter-squared.

In words "members of the general public should not be exposed on a continuous basis to rms magnetic flux densities exceeding 0.1 milli-tesla (1 gauss). This applies to areas in which members of the general public might reasonably be expected to spend a substantial part of the day. Exposures to magnetic flux densities between 0.1 milli-tesla (1 gauss) and ten milli-tesla (100 gauss) rms should be limited to a few hours per day."

The criterion for these exposure limits is based on health effects produced by currents induced in the body by external electric and magnetic fields. Induced current densities between one and ten milliamps per meter-squared (induced by 0.5 to five milli-tesla magnetic fields at 50/60 hertz) are reported to produce minor biological effects, while for current densities between ten and 100 milliamps per meter-squared (induced by five to 50 milli-tesla magnetic fields at 50/60 hertz) well established effects, including visual and nervous system effects occur. Clearly the exposure limits based on this criterion are some three orders of magnitude greater than those contemplated in the United States and Sweden, based on epidemiological studies. The IRPA/INIRC concludes that the association between cancer incidence and 60 hertz field exposure is still not established and remains a hypothesis.

Exposure Characteristics	Electric Field Strength in kV/meter (rms)	Magnetic Flux Density in milli-tesla (rms)
Occupational		
Whole working day	10	0.5
Short term	30[a]	5[b]
For limbs	—	25
General Public		
Up to 24 hours per day[c]	5	0.1
Few hours per day[d]	10	1

Table 9.6-1 IRDA/INIRC Interim Guidelines: Exposure Limits

a   The duration of exposure to fields between 10 and 30 kV per meter may be calculated from the formula, $t < 80/E$, where t is the duration in hours per work day and E is the electric field strength in kV per meter.

b   Maximum exposure duration is 2 hours per work day.

c   This restriction applies to open spaces in which members of the general public might reasonably be expected to spend a substantial part of the day, such a recreational areas, meeting grounds, and the like.

d   These values can be exceeded to a few minutes per day provided precautions are taken to prevent direct coupling effects.

9.7 Summary

Ever since the magnetic field-public health issue first stirred public consciousness in the United States, pressure has been directed toward the regulators of the power industry—state public utility commissions—to regulate the emission of magnetic fields. Informed by public health agencies, these commissions have dealt with the issue in a number of ways--from complete dismissal to fixing limits on exposure. These actions, taken in the face of uncertainty, show no clear-cut pattern of regulatory action. The most common response has been to require that new power system design and construction satisfy a policy of prudent avoidance. Prudent avoidance is clearly a policy an individual may also choose for himself/herself in everyday life.

Countries other than the United States--notably Sweden, Australia and the United Kingdom, have also acted vis-à-vis the magnetic field-public health issue. Sweden has established restrictive limits on the emission of power frequency fields from computer monitors. Australia and the United Kingdom have established field limits that are a factor of 1000 times less restrictive than those that have been considered by regulatory agencies in the United States. These limits are based on heating of tissue as opposed to more subtle cellular aspects and are not helpful in the current area of regulatory issues.

An international agency has likewise based exposure limits on induced currents rather than the possible health effects associated with low level fields.

9.8 References

[1]   *Health Effects of Exposure to Power-Line Frequency Electric and Magnetic Fields*, Report of Electro-magnetic Health Effects Committee, Public Utility Commission of Texas, Austin, Texas, March 1992.

[2]   *High Voltage Transmission Lines, Summary of Health Effects Studies,* California Energy Commission, Staff Report, July 1992.

[3]   Florida Public Service Commission, *Chapter 17-814, Electric and Magnetic Fields*, 1993.

[4]   Colorado Public Utilities Commission, *Statement of Adoption: Decision No. C92-1381, Docket No. 92R-259E*, November 1992.

[5]   Morgan M. G., H. K. Florig, I. Nair, and G. L. Hester, *Controlling Exposure to Transmission Line Electromagnetic Fields: A Regulatory Approach That Is Compatible with the Available Science*, Public Utilities Fortnightly, March, 1988, pp. 49-58.

[6]   Keeney, Timothy R. and Susan S. Addiss, *Connecticut 1993 Report on Task Force Activities to Evaluate Health Effects from Electric and Magnetic Fields*, March, 1993.

[7]   Commonwealth of Australia, *Interim Guidelines on Limits of Exposure to 50/60 Hz Electric and Magnetic Fields (1989)*, National Health and Medical Research Council, Canberra, 1989.

[8]   National Radiological Protection Board (NRPB), *Guidance as to Restrictions on Exposures to Time Varying Electromagnetic Fields and the 1988 Recommendations of the International Non-Ionizing Radiation Committee*, NRPB-GS 11, Oxfordshire, UK, 1989.

[9]   IEEE Power Engineering Review, Vol. 13, No. 10, October 1993, Stig Goethe, *EMFs and Health Risks: Research and Reactions in Sweden*.

[10]   *Interim Guidelines on Limits of Exposure to 50/60 Hz Electric and Magnetic Fields*, Health Physics, January 1990, Vol. 58, No. 1, Pgs. 113-122.

Appendices

Appendix A
The Physics of Magnetic Fields

A.1 Introduction to Fields

The concept of a field is actually a mathematical abstraction--a convenience for describing physical phenomena. The magnetic field construct is utilized to make it easier to talk about the physical things that happen in conjunction with the flow of electric currents. The field construct is so useful that it is employed very broadly in engineering.

We may be concerned with two types of fields--one type is a scalar field, and the other is a vector field. In the case of a scalar field, each point in a three dimensional space is assigned a number that represents the magnitude of the field at that point. In other words, for every coordinate point there is a scalar number that describes the magnitude of the field at that point. On the other hand, a vector field is one in which each coordinate point is assigned a vector describing the magnitude and direction of the field at that point.

The concepts of both the scalar and vector field have been widely used in many engineering and scientific applications. Examples of scalar fields are those of the temperature throughout a room or the electric potential throughout a region. Examples of vector fields are the force on a unit mass due to gravity (gravitational field) or the force on a unit charge due to a configuration of charges (electric field). The vector field with which we will be concerned here is the magnetic field associated with the flow of electric current. *In fact, the magnetic field is a result of the flow of electric current.*

A.2 Magnetic Field Intensity [1,2]

The magnetic field intensity vector, $\overline{H}$, describes the magnetic field due to a configuration of current in a region. Given a specification of the currents in a region, we can in principle determine the magnitude and direction of the $\overline{H}$ vector at every point in the region. Since the magnitude of $\overline{H}$ drops off rapidly with distance, we need be concerned only with nearby currents.

It is an experimental fact that the magnetic field from a long line of current is as indicated by Figure A.1-1. The magnetic field is represented by circular lines centered on the current-carrying conductor.

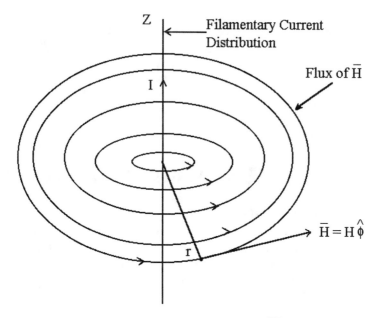

Figure A.1-1 The Magnetic Field Intensity Vector $\overline{H}$ Shown in a Plane Through Which a Long Filamentary Current Passes

The magnetic field intensity is a vector quantity $\overline{H}$ with the units of amperes/meter. Its direction is around the conductor. Its magnitude is given by:

$$H = \frac{I}{2\pi r} \qquad \text{amperes/meter,}$$

where I is the current magnitude (amperes) and r is the radial distance from the conductor in meters. In vector notation:

$$\overline{H} = \frac{I\hat{\phi}}{2\pi r} \qquad \text{amperes/meter,}$$

where $\hat{\phi}$ is a unit vector tangential to a circle of radius r with a positive direction as indicated in Figure A.1-1. (Right-Hand Rule)*

An important relationship in magnetostatics called Ampere's Circuital Law allows us to solve quite formidable problems in cases of symmetrical current

---

* By the Right-Hand Rule, if the thumb of the right hand is in the direction of the current, the fingers point in the same direction of the magnetic field. Similarly if the fingers are in the direction of current, the thumb points in the direction of the magnetic field intensity vector, $\overline{H}$.

distributions. Ampere's Circuital Law equates a closed line integral of $\overline{H}$ to the current enclosed by the line.

$$\oint_C H \bullet dl = I_{en} \qquad \text{in amperes,}$$

where the integral is taken along a closed path C encircling the current I_{en}. A few examples of the application of this law follow. This law can be called upon to answer many questions in magnetics. Every calculation and derivation in this text is based on Amperes' Circuital Law.

The magnetic field intensity in the vicinity of a long coaxial cable, as shown in

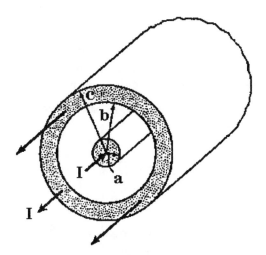

Figure A.1-2 Coaxial Cable

Figure A.1-2, is easily obtained using Ampere's Circuital Law. In the region between conductors:

$$H = \frac{I}{2\pi r} \qquad \text{amperes/meter,} \quad a < r < b$$

where r is the radial distance from the center of the inner conductor.

Beyond the outer conductor:

$H = 0$ for $r > c$, since the enclosed current is zero.

Similarly it can be shown that the magnetic field intensity at the center of a long solenoid coil, as shown in Figure A.1-3 is given by:

$$H = \frac{NI}{l} \text{ amperes/meter.}$$

Where I is the coil current, N is the number of turns of the coil and l is its length.

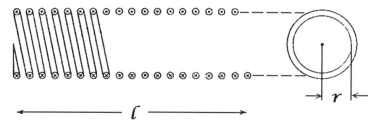

Figure A.1-3 Solenoid Coil

The direction of $\overline{H}$ is parallel to the axis of the coil as shown. Outside of the coil H = 0. $\overline{H}$ is essentially constant throughout the center of the solenoid, up to about two diameters from the end and two turn-to-turn distances from the perimeter.

The field of a toroidal coil, as shown in Figure A.1-4, can also be determined easily from Ampere's Law. In this case,

$$H = \frac{I}{2\pi r}, \qquad a < r < c, \qquad \text{otherwise } H = 0.$$

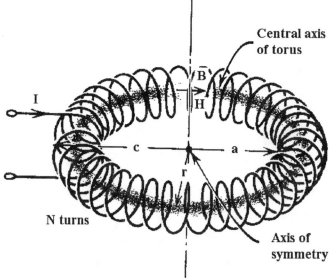

Figure A.1-4 Toroidal Coil

The direction of H is along the toroidal axis with polarity as determined by the Right-Hand Rule, counterclockwise in this example.

Ampere's Circuital Law applies to both static and time varying magnetic fields if I_{en} is defined to include both conductive and displacement current. (Displacement current flows in dielectrics as a result of a time-varying electric field.)

A.3 Magnetic Flux Density [1,2]

The magnetic field intensity vector $\overline{H}$ gives rise to a magnetic flux density vector represented by the vector $\overline{B}$. The relationship is:

$$\overline{B} = \mu \overline{H}, \qquad \text{tesla}$$

where μ is the permeability of the material in which the field exists. For most materials μ is a scalar and $\overline{B}$ is in the same direction as $\overline{H}$.

μ is often expressed as:

$$\mu = \mu_0 \mu_R \qquad \frac{\text{tesla} - \text{meter}}{\text{ampere}} \text{ or } \frac{\text{Henry}}{\text{meter}}$$

where μ_0 is the permeability of free space and μ_R is the relative permeability of the material.

$$\mu_0 = 4\pi \times 10^{-7} \text{ Henry/meter}$$

Representative values for μ_R a number of materials are shown in Table A.3-1.

As indicated in Table A.3-1, the ferromagnetic materials are characterized by large values of μ_R. The actual value of μ_R, however, depends not only on the material but the value of B in the material. This is illustrated in Figure A.3-1, which shows typical magnetization curves of two ferromagnetic materials.

Material	Type	μ_R
Silver	Diamagnetic	0.99998
Copper	Diamagnetic	0.999991
Vacuum	Nonmagnetic	1.00
Aluminum	Paramagnetic	1.00002
Cobalt	Ferromagnetic	250
Nickel	Ferromagnetic	600
Mild steel	Ferromagnetic	2,000
Iron	Ferromagnetic	5,000
Mumetal	Ferromagnetic	100,000
Supermalloy	Ferromagnetic	800,000

Table A.3-1 Representative Values of Relative Permeability μ_R

Referring to Figure A.3-1, the value of μ is the slope of the magnetization curve. Clearly μ is a function of B as well as the type of material.

Ferromagnetic materials exhibit hysteresis, a phenomenon illustrated in Figure A.3-2. If an alternating H field is applied to the material, the corresponding B field follows a hysteresis loop, the size of the loop depends on the maximum value of H as well as the magnetic material. Note that the magnetization curve, as shown in Figure A.3-1, is a single valued function while the hysteresis loop is double valued.

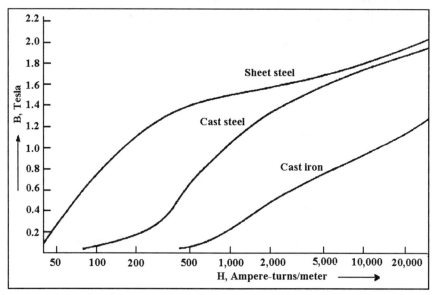

Figure A.3-1 Typical Magnetization Curves

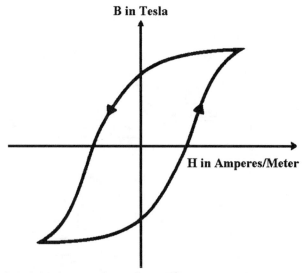

Figure A.3-2 Hysteresis in a Ferromagnetic Material

The μ of a ferromagnetic material, subject to an alternating H field, is time varying. An effective value for μ is often based on an average value over a complete cycle of H. It should be noted that many ferromagnetic materials have a relatively low effective value of μ at low values of B.

A.4 Magnetic Flux

The quantity of magnetic flux passing through an area A (m^2) is given by the integral:

$$\phi = \int_A \overline{B} \bullet d\overline{A} \quad \text{in webers,}$$

where $\overline{B} \bullet d\overline{A}$ indicates the scalar product of the vectors $\overline{B}$ and $\overline{dA}$ and $\overline{dA}$ is a differential element of area . If A is a plane area, entered normally by B, then $\phi = BA$.

A.5 Units [2]

The International System of Units (SI) is used in the above discussion. While the SI system is preferred, the Gaussian system is also commonly used. The conversion from SI units to Gaussian units is shown in Table A.5-1.

Quantity	Symbol	SI Unit	Gaussian Units
H	Magnetic Field Intensity	1 ampere/meter	$4\pi \times 10^{-3}$ Oersteds
ϕ	Magnetic Flux	1 weber	10^8 Maxwell
B	Magnetic Flux Density	1 weber/meter^2 = 1 tesla	10^4 gauss
μ_o	Permeability of Free Space	$4\pi \times 10^{-7}$ Henry/meter	1 gauss/Oersted

Table A.5-1 Comparison of SI and Gaussian Units

For example, the magnitude of the magnetic flux density vector in free space at a distance r (meters) from a long filamentary conductor carrying the current I (amperes) is given by:

$$B \qquad = 2 \times 10^{-7} \frac{I}{r}, \quad \text{tesla}$$

$$= \frac{2I}{r}, \quad \text{milligauss.}$$

A.6 Magnetic Circuits [3]

A magnetic circuit is a closed path for magnetic flux. An example is based on the toroidally wound coil of Figure A.1-4. The magnetomotive force is in terms of

188

ampere-turns, NI in this case. This MMF produces a magnetic field intensity H in the toroid as given by Ampere's Circuital Law. If the toroid is filled with a (core) material of permeability μ, the average value of resulting flux density B = μH or

$$B = \mu \frac{NI}{2pb},$$ where b is the mean radius of the core.

Let the cross section of the core be given by A. Then the total flux through the core is:

$$\phi = BA$$

$$\phi = \mu \frac{NIA}{2pb} = (NI)(\frac{\mu A}{2\pi b})$$

This is analogous to an electric circuit with:

Magnetomotive force (ampere-turns)	=	NI	~	Electromotive force (volts)
Magnetic flux (webers)	=	φ	~	Electric current (amperes)
Magnetic Reluctance (ampere-turns/weber)	=	$\mathcal{R}$	~	Resistance (ohms)

Ohm's Law of the Magnetic Circuit is: $NI = \phi \mathcal{R}$

The analogous circuit is shown in Figure A.6-1.

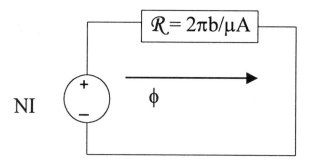

Figure A.6-1 Magnetic Circuit of a Toroid

The magnetic circuit approach is often very effective in analyzing magnetic field problems. It is particularly useful in visualizing the effect of shielding. As an example of this we consider the magnetic structure shown in Figure A.6-2

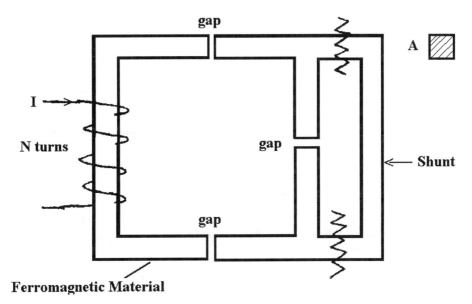

Ferromagnetic Material

Figure A.6-2 Magnetic Structure

There are three air gaps, each of length l_{gap}, in the magnetic structure. The structure has a uniform cross section of A $(m^2)^*$. The relative permeability of the ferromagnetic material is μ_R. The analogous circuit for Figure A.6-2 is shown in Figure A.6-3.

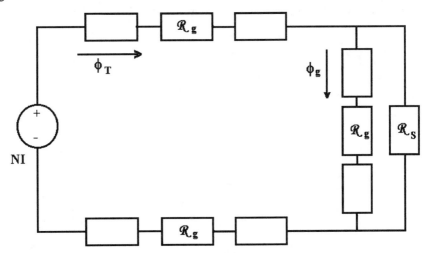

Figure A.6-3 Magnetic Circuit Corresponding to Figure A.6-2

The introduction of the shunt path, symbolized by the reluctance $\mathcal{R}_s$ can be very effective in reducing the flux density in the vertical gap. We will first simplify the

---

* Fringing flux is neglected

circuit, neglecting the reluctances in series with air gaps. This is justified on the basis that $\mu_R \gg 1$ in the ferromagnetic material.

Using circuit analysis techniques:

$$\phi_T = \frac{NI}{2\mathcal{R}_g + \dfrac{\mathcal{R}_s \mathcal{R}_g}{\mathcal{R}_s + \mathcal{R}_g}} = \frac{NI(\mathcal{R}_s + \mathcal{R}_g)}{2\mathcal{R}_g^2 + 3\mathcal{R}_s \mathcal{R}_g}$$

$$\phi_g = \phi_T \frac{\mathcal{R}_s}{\mathcal{R}_s + \mathcal{R}_g} = \frac{NI\mathcal{R}_s}{2\mathcal{R}_g^2 + 3\mathcal{R}_s \mathcal{R}_g}$$

The flux density in the vertical gap is therefore:

$$B_g = \frac{\phi_g}{A} = \frac{NI}{A}\{\frac{\mathcal{R}_s}{2\mathcal{R}_g^2 + 3\mathcal{R}_s \mathcal{R}_g}\}$$

We see from the last equation that B_g can be made as small as we like by making $\mathcal{R}_s$ small. This illustrates the principle of magnetic shielding. The flux density in a given path is reduced by providing a parallel low reluctance path as long as the total flux is not appreciably increased.

Example A.6-1   Magnetic Circuit Without and With A Shunt

Solution:

Refer to Figures A.6-2 and A.6-3 and assume: $l_{gap} = 0.001$ meters, $A = .001$ meter^2.

$$R_{gap} = \frac{.001}{\mu_o \times 10^{-3}} = \frac{1}{\mu_o} = \frac{10^7}{4\pi} = 3.18x10^6 \text{ amperes/weber}$$

$$\sum \mathcal{R}_{gap} = 9.54 \times 10^6 \text{ amperes/weber}$$

Let $NI = 500$ ampere turns

$$\phi = \frac{500}{9.54x10^6} = 52x10^{-6} \text{ wb}$$

$$B_g = \frac{52x10^{-6}}{10^{-3}} = 52x10^{-3} \text{ tesla} = 520 \text{ gauss}$$

This is the density in the vertical air gap without a shunt.

The reluctance of the shunt is:

$$\mathcal{R}_s = \frac{l_s}{\mu_o \mu_R A_s} = \frac{.6}{\mu_o \mu_R x 10^{-3}} = \frac{600}{4\pi x 10^{-7} \mu_R} \text{ , amperes/weber}$$

Where the length of the shunt is assumed to be 0.6 meters.

If $\mu_R = 100,000$, then

$$R_s = \frac{6 \times 10^4}{4\pi} = 4775 \text{ amps/weber}$$

$$B_g = \frac{500}{.01} \{ \frac{4775}{2(3.18 \times 10^6)^2 + (3x477.5 \times 3.18x10^6)} \}$$

$$= 1.18 \text{x} 10^{-5} \qquad \text{tesla}$$

$$= 1.18 \text{x} 10^{-3} \qquad \text{gauss}$$

$$= 118 \qquad \text{milligauss}$$

This is the flux density in the vertical air gap with the shunt.

$$\frac{B_{g(before)}}{B_{g(after)}} = \frac{520}{118 x 10^{-3}} = 4405 \text{ is the } \textit{Figure of Merit} \text{ for the reduction of the magn}$$

flux density in the vertical air gap by the addition of the shunt path for flux.

A.7 References

[1] *Electromagnetic Concepts and Applications*, by Marshall and Skitek, Prentice Hall, 3rd Edition.

[2] *Engineering Electro-magnetics*, by W. H. Hayt, McGraw Hill, 3rd Edition.

[3] *Electric Machinery Fundamentals*, by S. J. Chapman, McGraw Hill, 2nd Edition, 1991.

Appendix B   Modeling and Calculating the Transmission Line Magnetic Field

B.1   The Field Ellipse and the Resultant Field

The electric currents that produce power frequency magnetic fields have a fundamental frequency of 60 Hz. The resulting magnetic flux densities are time varying at the same frequency. The magnetic flux density vector from a single phase source oscillates along a discrete direction in space (linear polarization). On the other hand, the magnetic flux density vector due to multiple sources with different phase angles rotates in direction over each cycle (elliptical polarization). For example, the magnetic flux density ellipse at the point (x_1, y_1), due to the three phase currents of a transmission line may be as shown in Figure B.1-1.

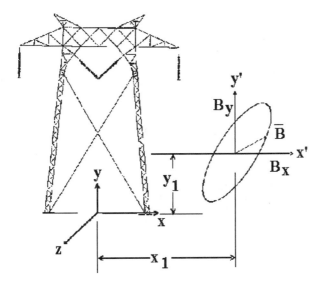

Figure B.1-1. Magnetic Flux Density Ellipse

Here (x', y') represents an axis system parallel to the (x,y) system but with its origin at $x = x_1$ and $y = y_1$. B_x and B_y are the root mean square (rms) X and Y components of $\overline{B}$ at the point (x_1, y_1).

The magnetic flux density ellipse is shown in more detail in Figure B.1-2. The tip of the magnetic flux density vector, $\overline{B}$, traverses the ellipse every 1/60 second. The ellipse itself is described by a semi-major axis vector $\overline{B}_{max}$ and a semi-minor axis vector $\overline{B}_{min}$. The resultant field, $B_{resultant}$, is a field magnitude computed from the equation:

$$B_{resultant} = \sqrt{B_{max}^2 + B_{min}^2} \quad \text{tesla , rms}$$

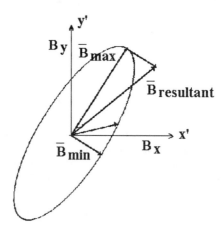

Figure B.1-2 Magnetic Flux Density Ellipse

AXIS	B	UNITS
Semi-Major	B_{max}	tesla, rms
Semi-Minor	B_{min}	tesla, rms

Notes:

B_{max} is the rms value read by orienting a single axis field meter until a maximum reading is indicated.

B_{min} is the rms value read by orienting a single axis field meter until a minimum reading is indicated.

In the illustration the ellipse lies in the x - y plane. More generally the magnetic field may have components along all three of the coordinate axes. Single axis field meters oriented along the x, y and z axes measure these components: B_x, B_y and B_z. The resultant magnetic field is then given by:

$$B_{resultant} = \sqrt{B_x^2 + B_y^2 + B_z^2} \quad \text{tesla rms}$$

The resultant value of the magnetic field cannot generally be measured directly by a single axis field meter. Instruments incorporating three single axis meters have been designed to register the resultant field. Unless otherwise specified, when values of magnetic fields are stated, the resultant value of the field is understood.

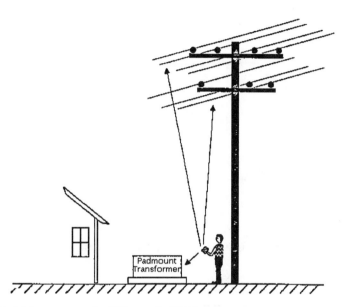

Figure B.1-3 Measurement of Magnetic Fields

Measurements of the magnetic fields associated with lines and equipment are usually carried out for a height of three feet above ground. As shown in Figure B.1-three, the field meter is located at this height. The field profile associated with a line is obtained by tabulating field readings for various lateral distances from the center of the line, at a height of three feet.

B.2   The Transmission Line Model [1]

The magnetic field of a transmission line is calculated using a two-dimensional analysis. This approach is deterministic (as opposed to a finite element method), assumes a flat uniform earth, with non-sagging uniform and parallel lines. The coordinate system is shown in Figure B.2-1. The calculation determines the magnitude and direction of $\overline{B}$ at any point (x_j, y_j).

Here the x-z plane is the plane of the earth's surface, the transmission line is parallel to the z-axis and intersects the x-y plane at the point x_i, y_i. The transmission line carries a sinusoidal current $\sqrt{2}(I_i)_{rms}\sin(\omega t + \theta_i)$ in the positive z direction. The effective current return is in the earth at a depth $(y_i + 2/\gamma)$ meters.

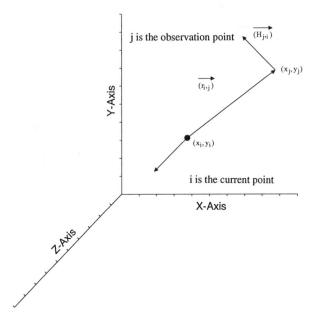

Figure B.2-1. Transmission Line Coordinate System

γ is given by the equation:

$$\gamma = \sqrt{J\,\omega\mu(\sigma + J\,\omega\varepsilon)} \quad (\text{meter}^{-1})$$

where:

σ is the earth conductivity ($\sigma \approx 0.01$ Siemens/meter)
ε is the earth permittivity ($\varepsilon \approx 9 \times 10^{-12}$ Farad/meter)
ω is the angular frequency of $I_i(t)$ (radian/second)

and $J = \sqrt{-1}$

For practical purposes, the complex number γ is given by:

$$\gamma = \sqrt{\omega\mu\sigma} \quad \underline{/-45^\circ}$$

At a frequency of 60 Hz, using the above values:

$$2/\gamma \approx 6000 \underline{/-45^\circ} \text{ (meter)}$$

Thus for a homogeneous earth with a conductivity of 0.01 Siemen/meter, the return current is at a depth of 6000 meters. As a result, the contribution of the ground return current to the magnetic field in the vicinity of the conductor for the assumed conditions is negligible.

With the assumption that $| x_j - x_i |$ and $| y_j - y_i |$ are of the order of 100m or less, we neglect the ground return current in calculating the magnetic field intensity at the observation point (x_j , y_j). Then, using Ampere's Circuital Law,

$$\overline{H}_{j,i}(t) = \frac{I_i(t)}{2\pi}\left[\frac{(y_j - y_i)\hat{x} - (x_j - x_i)\hat{y}}{(y_j - y_i)^2 + (x_j - x_i)^2}\right] \text{ amperes / meter} \qquad \text{(ampere/meter)}$$

where $\hat{x}$ is a unit vector in the x direction and $\hat{y}$ is a unit vector in the y direction.

$\overline{H}_{j,i}(t)$ is the magnetic field intensity at point j due to current in line i.

For n conductors,

$$\overline{H}_j(t) = \sum_{i=1}^{n}\overline{H}_{j,i}(t) \text{ ampere / meter}$$

The total magnetic field intensity is the vector sum of the fields due to the individual line currents.

The equation for $\overline{H}_{j,i}(t)$ can be viewed as a phasor equation if $I_i(t)$ is replaced by a current phasor $\overline{I}_i(\omega)$, where $\overline{I}_i(\omega) = (I_i)_{rms}\angle\theta_i$, then:

$$\overline{H}_{j,i}(\omega) = \frac{(I_i)_{rms}\angle\theta_i}{2\pi}\left[\frac{(y_j - y_i)\hat{x} - (x_j - x_i)\hat{y}}{(y_j - y_i)^2 + (x_j - x_i)^2}\right] \text{ amperes/meter rms}$$

For a three phase line,

$$\overline{H}_j(\omega) = \sum_{i=1}^{3}\overline{H}_{j,i}(\omega) \text{ amperes/meter rms ,and}$$

$$\overline{B}_j(\omega) = \mu_o\overline{H}_j(\omega) \text{ tesla rms, where } \mu_o \text{ is the permeability of free space (air),}$$

$$\mu_o = 4\pi \times 10^{-7} \text{ Henry/ meter.}$$

In terms of milligauss:

$$\overline{B}_j(\omega) = 4\pi\overline{H}_j(\omega), \text{ milligauss rms.}$$

The x component of $\overline{B}_j(\omega)$ is given by:

$$B_{jx}(\omega) = 2\sum_{i=1}^{3}\frac{((I_i)_{rms}\angle\theta_i)(y_j - y_i)}{(y_j - y_i)^2 + (x_j - x_i)^2} \text{ milligauss rms}$$

and the y component by:

$$B_{jy}(\omega) = -2\sum_{i=1}^{3}\frac{((I_i)_{rms}\angle\theta_i)(x_j - x_i)}{(y_j - y_i)^2 + (x_j - x_i)^2} \text{ milligauss rms}$$

Note that these components of $\overline{B}_j(\omega)$ can be written as:

$B_{jx}(\omega) = (B_{jx})_m \angle \theta_x$ and

$B_{jy}(\omega) = (B_{jy})_m \angle \theta_y$

where θ_x and θ_y are the phase angles of $B_{jx}(\omega)$ and $B_{jy}(\omega)$ and $(B_{jx})_m$ and $(B_{jy})_m$ are their rms values.

Since the phase angles θ_x and θ_y are not necessarily equal, the locus of values of $\overline{B}_j(t)$, in general, is an ellipse in the x - y plane. $\overline{B}_j(t)$ is the vector expression for the magnetic flux density given by:

$$\overline{B}_j(t) = \hat{x}(B_{jx})_m \cos(\omega\, t + \theta_x) + \hat{y}(B_{jy})_m \cos(\omega\, t + \theta_y)$$

$$B_j(t) = \left[\left((B_{jx})_m \cos(\omega\, t + \theta_x) \right)^2 + \left((B_{jy})_m \cos(\omega\, t + \theta_y) \right)^2 \right]^{1/2} \quad \text{is the rms}$$

magnitude of $\overline{B}_j(t)$

A representative locus of values of $\overline{B}_j(t)$ is shown in Figure B.2-2. This figure is traced out once each cycle of the electrical current frequency. Note that for a single phase transmission line $\theta_x = \theta_y$ and the ellipse collapses to a line.

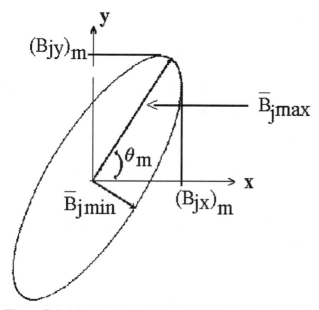

Figure B.2-2 Magnetic Flux Density Ellipse for a Three Phase Transmission Line

The following relationships hold for the ellipse of Figure B.2-2. θ_m is determined from the equations:

$$\tan 2(\omega t)_m = -\left[\frac{(B_{jx})^2_m \sin 2\theta_x + (B_{jy})^2_m \sin 2\theta_y}{(B_{jx})^2_m \cos 2\theta_x + (B_{jy})^2_m \cos 2\theta_y}\right]$$

and

$$\theta_m = (\omega t)_m + \theta_x + (m-1)\frac{\pi}{2} \text{ where m = 1, 2, 3, 4. The pair m = 1, 3 (and m = 2,}$$

4) correspond to two directions of the same axis. Further if m = 1, 3 corresponds to the semimajor axes, then m = 2, 4 corresponds to the semiminor axes and vice versa.

The magnitudes of the major and semi minor axes of the ellipse are given by:

$$B_{j\max} = \max_m\left[((B_{jx})_m \cos((\omega t)_m + \theta_x + (m-1)\frac{\pi}{2}))^2 + ((B_{jy})_m \cos((\omega t)_m + \theta_y + (m-1)\frac{\pi}{2}))^2\right]^{1/2}$$

and

$$B_{j\min} = \min_m\left[((B_{jx})_m \cos((\omega t)_m + \theta_x + (m-1)\frac{\pi}{2}))^2 + ((B_{jy})_m \cos((\omega t)_m + \theta_y + (m-1)\frac{\pi}{2}))^2\right]^{1/2}$$

while

$$B_{j\text{ resultant}} = \left[B^2_{j\max} + B^2_{j\min}\right]^{1/2}$$

B.3 Hand Calculation of the Magnetic Field of a Three Phase Transmission Line

A three phase transmission line is shown below. We will determine the field at a point of observation on the ground, directly under phase B as shown in Figure B.3-1, with the following phase currents:

Phase A current is $100 \angle 0°$ amperes rms
Phase B current is $100 \angle 240°$ amperes rms
Phase C current is $100 \angle 120°$ amperes rms.

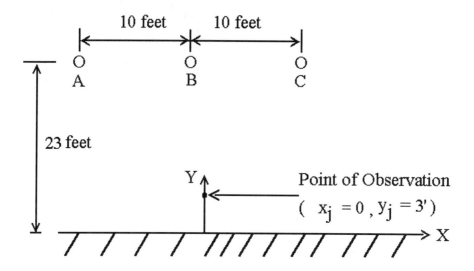

Figure B.3-1. Three Phase Transmission Line Geometry

The point of observation is at $(x = 0, y = 3)$ feet as shown. The coordinates of the phase currents are (-10, 23), (0, 23), and (10, 23) in feet. Using rms values for the current we will calculate rms values for the magnetic fields.

$$B_{jx}(\omega) = 2 \times 3.28 \times \left[\frac{100\angle 0°(-20)}{(20)^2 + (10)^2} + \frac{100\angle 240°(-20)}{(20)^2} + \frac{100\angle 120°(-20)}{(20^2 + 10^2)} \right]$$

$$= 6.56\angle 60° \quad \text{milligauss rms}$$

$$B_{jy}(\omega) = 2 \times 3.28 \times \left[\frac{100\angle 0°(-10)}{(20^2 + 10^2)} + \frac{100\angle 240°(0)}{(20)^2} + \frac{100\angle 120°(10)}{(20^2 + 10^2)} \right]$$

$$= 22.73\angle 150° \quad \text{milligauss rms}$$

(The factor 3.28 converts from the English to the SI units, i.e. 3.28 feet/meter)

$$\tan(2\omega t)_m = -\left[\frac{(6.56)^2 \sin(120°) + (22.73)^2 \sin(-60°)}{(6.56)^2 \cos(120°) + (22.73)^2 \cos(-60°)} \right] = 1.732$$

$$\theta_m = 30° + 60° = 90°$$

$$B_{j\max} = \sqrt{(6.56\cos(30° + 60°))^2 + (22.73\cos(30° + 150°))^2} = 22.73 \quad \text{milligauss rms}$$

$$B_{j\min} = \sqrt{(6.56\cos(30° + 90° + 60°))^2 + (22.73\cos(30° + 90° + 150°))^2} = 6.56 \quad \text{milligauss r}$$

These are the magnitudes of the semimajor and semiminor axes of the ellipse. The resultant has the value:

$$B_{resultant} = \sqrt{(6.57)^2 + (22.73)^2} = 23.66 \text{ milligauss rms}$$

B.4 Computer Calculation of the Magnetic Field of a Three Phase Transmission Line

There are several computer programs that have been developed to calculate the magnetic fields of transmission lines. [2-3] The programs of references [2] and [3] allow easy, interactive data entry and produce as outputs tabular data as well as field profiles. A profile is a plot of a magnetic field parameter versus lateral distance from the center of the line. The parameter may be the value of $B_{resultant}$, the values of B_{max}, B_{min}, $(B_{jx})_m$, $(B_{jy})_y$ or θ_m. The usual profile is that of $B_{resultant}$ versus lateral distance.

To illustrate the use of the FIELDS program of reference [2], we will use it to produce a field profile for the three phase transmission line of Figure B.3-1. The profile is for the parameter B_{max} versus lateral distance from the center of the line.

Several computer screens as produced by the FIELDS program follow. The first two screens show input data defining the solution region and the transmission line. The third screen shows a sample of the tabular input data. Note the following correspondences between the notation of the FIELDS report and this Appendix:

Fields Program	Appendix B
B Horz	$(B_{jx})_m$
B Vert	$(B_{jy})_m$
B Product	$B_{resultant}$
B Max	B_{max}

The fourth screen defines the parameters of the profile plot and the fifth screen is the plot of B_{max} versus lateral distance.

```
┌──────────────────────────────────────────────────────────────────┐
│  TITLE, PLOTTING, AND MISCELLANEOUS DATA                           │
├──────────────────────────────────────────────────────────────────┤
│                                                                    │
│   Main Title: Section B.4                                          │
│   Subtitle:   Mag Field 3 Ph OH Line                              │
│   Frequency (Hertz):                                     60.00     │
│   Soil Resistivity (Ohm-meter):                         100.00     │
│   Maximum Horizontal Distance From Reference (ft):      100.00     │
│   Step Size (ft):                                         2.00     │
│   Height For Field Calculation (ft):                      3.00     │
│   Left Coordinate of Right of Way (ft):                   0.00     │
│   Right Coordinate of Right of Way (ft):                  0.00     │
│                                                                    │
│                        < Continue >                                │
│                                                                    │
└──────────────────────────────────────────────────────────────────┘
```

FIELDS Screen 1

Phase Conductor Description

Phase Name	Phase Coordinates (ft) Horz	Phase Coordinates (ft) Height	SubConds. Per Bundle	Cond. Diam. (in.)	Bund. Diam. (in.)	Phase- Phase (kV)	Phase Curr. (Amp)	Phase Angle (deg)
A	-10.00	23.00	1	0.50	0.50	22.00	100.00	0.00
B	0.00	23.00	1	0.50	0.50	22.00	100.00	240.00
C	10.00	23.00	1	0.50	0.50	22.00	100.00	120.00

<Esc> - Done   <Insert>-Change number or conductors   Page 1 of 1

FIELDS Screen 2

MAGNETIC FIELD VALUES

DISTANCE (Feet)	B Horz (mG)	B Vert (mG)	B Product (mG)	B Max (mG)
-22.00	13.039	1.736	13.154	13.048
-20.00	14.099	2.866	14.387	14.246
-18.00	14.976	4.654	15.682	15.495
-16.00	15.547	6.901	17.009	16.762
-14.00	15.683	9.486	18.328	18.006
-12.00	15.282	12.260	19.592	19.178
-10.00	14.301	15.035	20.750	20.232
-8.00	12.786	17.603	21.756	21.125
-6.00	10.890	19.771	22.572	21.830
-4.00	8.893	21.397	23.171	22.332
-2.00	7.240	22.395	23.536	22.631
0.00	6.562	22.730	23.658	22.730
2.00	7.240	22.395	23.536	22.631

Page 4 of 8

< Previous >   < Next >   < Cancel >   < Print Table >

FIELDS Screen 3

```
─────────────────── Manual Graphing ───────────────────
                                                  Linear
  Maximum Magnetic Field Strength (mG)              25
  Minimum Magnetic Field Strength (mG)              0

  Maximum Horizontal Distance (ft)                  100
  Minimum Horizontal Distance (ft)                 -100

  Number of marks on an Axis                        50
  Label every nth mark on an Axis... n is:          10

                                               Major Minor
  Show Vertical Gridlines                        [X]  [ ]
  Show Horizontal Gridlines                      [X]  [ ]

  Show Wire Configurations                          [ ]

       < Graph >      < Save Defaults >     < AutoScale >
```

FIELDS Screen 4

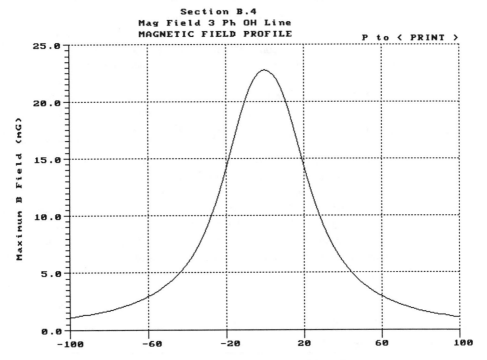

Section B.4
Mag Field 3 Ph OH Line
MAGNETIC FIELD PROFILE   P to < PRINT >

File: APNDXB-4.FLD Distance From Reference (Feet)

FIELDS Screen 5

The FIELDS program is available without charge from the Southern California Edison Company. An application to receive the program follows, for the convenience of the reader.

The *FIELDS 2.0* program is designed to calculate and plot the magnetic and electric fields produced by transmission and distribution lines. The program will compute fields for any combination of up to 60 conductors and 10 ground wires with voltage, current, and phase angle independently specified.

FIELDS 2.0 operates on IBM or compatible personal computers with DOS 3.0 or above, and with a minimum of 640 K of memory. Hewlett Packard laser printers and Epson dot matrix printers are supported with CGA, EGA or VGA graphics.

FIELDS 2.0 uses a file named FIELDS.INI to save some defaults. It will be in the same directory as the FIELDS.EXE automatically.

The program is available at no cost with the understanding that Southern California Edison (SCE) provides no warranty or guarantee of accuracy, suitability, or ongoing support. SCE does believe the FIELDS program to be accurate for the intended purpose. The undersigned User agrees to the above stated conditions, and to provide SCE's Research Division with comments and suggestions for improvements to the program.

Organization name _____

Street Address _____

City/State/Zip code _____

Phone number _____

Anticipated application(s) for FIELDS program: _____

_____

_____

User (Please print) _____

Signature

_____

Send comments to :

FIELDS 2.0 Program

Southern California Edison Co.

6090 North Irwindale Avenue

Irwindale, Ca 91702

B.5 References

[1]   *Transmission Line Reference Book 345 kV and Above,* Second Edition, Electric Power Research Institute, 1982.

[2]   *The FIELDS 2.0 Program*, Southern California Edison Company, 6090 North Irwindale Avenue, Irwindale, California, 1992.

[3]   *TL Workstation Code, ENVIRO Manual*, Version 2.2, Electric Power Research Institute Project 2472, July 1989.

Appendix C   The Magnetic Field of a Transmission Line at an Intermediate Distance from the Line

C.1 Theory of Reduced Magnetic Field Power Lines

As developed in Appendix B, the magnetic field phasor at a field point due to a multiconductor transmission line has the components:

$$B_{jx}(\omega) = 2\sum_{i=1}^{N} \frac{I_i(\omega)(y_j - y_i)}{\left(y_j - y_i\right)^2 + \left(x_j - x_i\right)^2} \quad \text{milligauss rms}$$

$$B_{jy}(\omega) = -2\sum_{i=1}^{N} \frac{I_i(\omega)(x_j - x_i)}{\left(y_j - y_i\right)^2 + \left(x_j - x_i\right)^2} \quad \text{milligauss rms}$$

Where $I_i(\omega)$ is the phasor current in the ith conductor, amperes rms and N is the number of current carrying conductors. This transmission line model is valid providing that the distance $\sqrt{\left(y_j - y_i\right)^2 + \left(x_j - x_i\right)^2}$ from the ith conductor to the field point is much less than the distance from its image current to the field point.

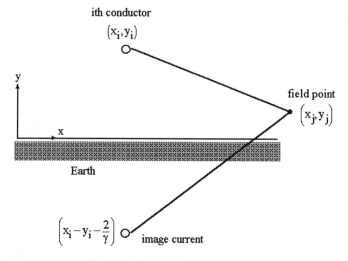

Figure C.1-1 Geometry for Magnetic Field Model

The geometry for this model is shown in Figure C.1-1. The coordinate system is placed in the earth's surface such that the geometric center of the conductors is at $x = 0$, that is, $\sum_{i=1}^{N} x_i = 0$. On the other hand, $\sum_{i=1}^{N} y_i = y_a$, the average height of the conductor.

We are interested in approximations to $B_{jx}(\omega)$ and $B_{jy}(\omega)$ for values of $|x_j|$ that are large as compared with distances between conductors yet small as compared to the depth of the image current. We will call these values of $|x_j|$ intermediate distances. For typical transmission lines intermediate distances are given (roughly) by:

$$100 \text{ meter} \leq |x_j| \leq 500 \text{ meter}.$$

The desired approximations were developed in [1] and are shown below:

$$B_{jx}(\omega) = \frac{2\sin\phi}{R_a} \sum_{i=1}^{N} I_i + \frac{2(1 + 2\sin^2\phi)}{R_a^2} \sum_{i=1}^{N} (y_i - y_a) I_i(\omega)$$
$$- \frac{\sin 2\phi}{R_a^2} \sum_{i=1}^{N} x_i I_i(\omega) + \dots \text{terms proportional to} \frac{1}{R_a^3} + \dots \text{ in milligauss.}$$

$$B_{jy}(\omega) = \frac{2\cos\phi}{R_a} \sum_{i=1}^{N} I_i(\omega) - \frac{2(1 + 2\cos^2\phi)}{R_a^2} \sum_{i=1}^{N} x_i I_i(\omega)$$
$$+ \frac{\sin 2\phi}{R_a^2} \sum_{i=1}^{N} (y_i - y_a) I_i(\omega) + \dots \text{terms proportional to} \frac{1}{R_a^3} \dots \text{ in milligauss.}$$

Where $R_a = \sqrt{y_a^2 + x_j^2}$

$$\phi = \tan^{-1} \frac{y_a}{x_j}, \ y_j = 0.$$

We see that these approximations are in the form of series expansions with decreasing terms,

$$\frac{1}{R_a}, \frac{1}{R_a^2}, \frac{1}{R_a^3}, \dots .$$

The distance relationships of fields that decrease with distance according to the inverse, inverse squared and inverse cubed rules are shown in Figure C.1-2.

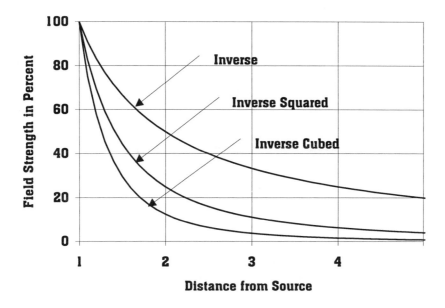

Figure C.1-2 Inverse Distance Relationships of Fields

The first order terms (inverse) become zero if: $\sum_{i=1}^{N} I_i(\omega) = 0.$ This condition occurs if there is no net transmission line current (no zero sequence current). In this case the magnetic field at intermediate distances decays at least as rapidly as $\dfrac{1}{R_a^2}$.

The second order terms (inverse squared) become zero if:

$$\sum_{i=1}^{N} x_i I_i(\omega) = 0 \text{ and } \sum_{i=1}^{N} (y_i - y_a) I_i(\omega) = 0.$$

These conditions occur if the transmission line currents are carried by six conductors in the double circuit low reactance arrangement as described in Section 5.2. Then the magnetic field at intermediate distances decays at least as rapidly as $\dfrac{1}{R_a^3}$ (inverse cubed). Of course this is only possible in a double circuit arrangement with ideal current splitting between circuits.

C.2 References

[1]   R. Olsen, V. L. Chartier and D. C. James, *The Performance of Reduced Magnetic Field Power Lines Theory and Measurements on an Operating Line*, IEEE Transactions on Power Delivery, Vol. 8, No. 3, July 1993.

Appendix D   The Science of Epidemiology[*]

D.1 Epidemiologic Methods

Epidemiology is generally defined as the study of the distribution of disease in human populations and the determinants of that distribution. Characteristics of people and their environment may be examined for possible causal associations with the occurrence of human disease.

Because epidemiology draws its conclusions from observations of the natural distribution of disease, it possesses both unique strengths and limitations. Since humans are the subjects of study, epidemiology avoids the problem of extrapolating from animal experiments in which both the exposure conditions and the appropriateness of the animal model are often questioned.

On the other hand, epidemiologic research generally provides less conclusive findings than laboratory research does. The inability of epidemiologic research to offer direct proof of a cause-and-effect relationship results from its observational methodology. In a laboratory investigation of a suspected harmful agent, it is assumed that the animals under study differ only on the basis of their exposure regimen. Any ensuing differences that are found between exposed and non-exposed animals can then reasonably be attributed to the exposure itself. Since obvious ethical and practical prohibitions on experimentation with humans exist, data must be collected on the "natural" occurrence of the disease and agent under study in human populations. However, human exposure to an agent is not a random phenomenon occurring among members of a homogeneous population. Exposed and non-exposed groups will differ in terms of age, residence, occupation, gender, and many other factors. Some of these

---

[*]   This material was published as Appendix B in *Health Effects of Exposure to Power-Line Frequency Electric and Magnetic Fields*, Report of Electro-magnetic Health Effects Committee, Public Utility Commission of Texas, Austin, Texas, March 1992. The material was initially prepared by Robert S. Banks, R. S. Banks Associates, Incorporated, for Electric Power Research Institute Seminar on New EMF Epidemiologic Results and Their Implications, October 16-19, 1990. Substantial revisions and additions were made by Boji Huang, M.D., and P.A. Buffler, Ph.D., University of Texas Health Science Center at Houston, School of Public Health, and R. A. Beauchamp, Texas Department of Health.

Only a portion of Appendix B is reproduced here.

variables are known to influence disease occurrence and can be accounted for in the design or analysis of a study.

Other factors associated with both the disease and the exposure may not be known to the investigator and, therefore, cannot be accounted for. Such "confounding" factors may lead to an incorrect interpretation of the relationship between the agent and the disease under study.

Other examples of the methodology problems that can alter or bias observational studies include uncertainties in determining the actual exposure status of individuals, variations in disease definitions and diagnoses in different geographic areas or in different hospitals, loss of study subjects who leave the area, unwillingness of subjects to participate, and inaccuracies in frequently used data sources such as death certificates and clinical records. Practical solutions to many of these problems have been developed by epidemiologists, although frequently these sources of bias are not adequately addressed.

Although each epidemiologic investigation poses its own unique problems and solutions, the overall approach of a study generally follows one of several basic study designs. The choice of a study design will depend on many factors such as time and cost limitations, frequency of the diseases to be studied, frequency of the exposure, intended use of the information, and the availability of required data. Several commonly used designs are described below, along with a brief consideration of their particular advantages and limitations.

The terms incidence and prevalence will be used in the following descriptions of epidemiologic study designs. These two commonly used measures of disease occurrence have distinctly different meanings in epidemiology. In a population-based study, the prevalence of disease is the proportion of individuals in a population with the disease at a given point in time. For example, the number of persons with lung cancer in a population of 100,000 on December 31, 1989 might be 30. The prevalence of lung cancer in this population at this point in time is 30/100,000 or, .0003. This figure would include all persons with lung cancer on December 31, regardless of whether the person has had the disease for one day or three years. Prevalence is dimensionless; that is, it has no units.

In a case-control study, on the other hand, disease defines the two types of individuals to be studied, those with disease (cases) and those without (controls). Here, the concern is to compare these two types of individuals with respect to the proportion having a history of an exposure or characteristic of interest. For example, in a case-control study of persons with lung cancer, it would be of interest to compare the proportion having a history of cigarette smoking in case and control subjects.

In contrast to prevalence, incidence is a measure of the *new* cases of disease occurring in a population in given time interval. If 15 of the 30 persons with lung cancer in the previous example were first diagnosed in 1989, the incidence of lung cancer in their population would be 15 per 100,000 per year (also written

15/100,000/year or 15 per 100,000 person-years). The dimension of time is the defining characteristic of incidence, making this measure of disease occurrence a rate. The term rate always implies that the measurement of disease or death in a population is related to a specified period of time. Thus, although some scientists refer to prevalence as prevalence rate, the expression is a misnomer.

Cross-Sectional or Prevalence Studies. Cross-sectional studies examine factors of interest in a defined population at a particular point in time. The study group may represent a random sample of a community, working at a particular occupation, or a sample chosen on the basis of some sociological or environmental variable. Through a questionnaire, physical examination and/or other means, the presence or absence or the disease(s) in question is determined for each individual, along with other characteristics or exposures of interest (for example, age, whether the person smokes, exercise level, blood pressure, diet).

Some of the advantages of cross-sectional studies include:

- They can generally be performed relatively quickly and inexpensively.
- They provide valuable descriptive information on the existing patterns of disease occurrence.
- They can examine a variety of factors and diseases simultaneously.

The major limitations include:

- The "snapshot" approach may not allow one to determine whether exposure actually preceded development of the disease.
- Diseases that generally have a longer duration are more likely to be detected than diseases with the same incidence rate but with shorter duration. Thus an association between an exposure and a disease of short duration may be missed.

Individuals who survive longer with a disease are more likely to be found than those with shorter survival times. Therefore, the cases with short survival will not be available for study; thus, remaining cases may not be typical of all cases and a potential association between exposure and disease may be masked or exaggerated.

Cohort Studies. Cohort studies start with the selection of groups of disease-free individuals on the basis of some exposure variable. Exposed and non-exposed individual are then followed-up to determine subsequent development of disease. There are two types of cohort or follow-up studies: prospective (or concurrent) and retrospective (or nonconcurrent or historical). These two types differ in terms of when exposure and disease occur in relation to the onset of the study.

Prospective (Concurrent) Cohort Studies. Prospective cohort studies are most similar to the classic laboratory study. These studies first identify a group of persons (cohort) who are currently free of disease, but who differ in terms of exposure to the agent under study. For example, the cohort may be a specified group of reproductive age women, and the exposure variable may be the use of oral contraceptives. The cohort is then "followed-up" at some future time to determine the occurrence of disease(s) in the cohort. How soon follow-up begins, or the length of time it must be conducted, depends on the disease outcome(s) of interest and their characteristics (for example, induction or latency periods). Incidence can then be compared in exposed and non-exposed groups. These rates, which must be adjusted for differences in age and other characteristics of the study subject, are typically expressed as a ratio or "relative risk" for the exposed group.

The advantages of prospective cohort studies are significant:

- They allow the direct determination of incidence among exposed and non-exposed groups. This permits calculation of the increased disease risk (relative risk) associated with the exposure. They also permit calculation of the "attributable risk" which is that portion of the incidence of a particular disease that is due to a specific cause.
- They may yield more extensive and more reliable data on exposure levels, as well as on confounding factors (for example, cigarette smoking).
- Many different disease outcomes can be investigated in a single study.
- Relatively rare exposures (or occupations) can be investigated.

Prospective cohort studies also have a number of limitations:

- They often require many years of follow-up since many diseases have long latencies.
- A very large cohort and/or a long follow-up period would be required to investigate relatively rare diseases, including most cancers.
- Substantial effort and expense are necessary to follow a large number of people over a long period of time.

Retrospective (Historical or Nonconcurrent) Cohort Studies. Retrospective cohort studies differ in that both the exposure and follow-up period have occurred prior to the onset of the study. These studies utilize data from existing records such as occupational records, professional registries, and death certificates to identify the cohort and conduct the follow-up.

Studies of the occupational groups are most often conducted using this approach. As with prospective studies, the cohort is first defined on the basis of exposure status. For example, the cohort may be defined as all members of a particular occupation of all employees at Company X as of some specified time in the past. The subsequent occurrence of disease in the cohort (up to the time of the study) is then ascertained, generally by using death certificates. To reduce costs, an unexposed cohort is usually compared to that experienced by the general population in the state, region or country from which the study population was derived.

The advantages of retrospective cohort studies include:

- Much less time and less cost is required to complete the study compared to a prospective study, since the disease outcome has already occurred.

- These studies are widely used in occupational settings, where personnel records, industrial hygiene data and other records can be used both to construct the cohort and establish some rough measures of exposure.

The limitations of retrospective cohort studies include:

- Past exposure cannot be defined as precisely as current exposures. For example, it may be difficult or impossible to estimate exposures to workers that occurred 40 years ago if no industrial hygiene data are available and work practices have changed over time.
- Little information may be available on confounding factors, such as smoking history.
- It may be difficult to select a suitable population to which the cohort can be compared. Frequently, the study results will differ depending on whether national, regional, or local disease rates are used as the comparison. The population of a suitable comparison population is avoided in large cohort studies in which internal comparisons can be made; that is, a particular category or workers within a cohort can be compared to a complete cohort.

Some cohort studies involve both prospective and retrospective components. For example, a cohort may be defined through personnel and other records as everyone who worked at Company X at least one month between 1945 and 1985. The mortality experience of this population as of 1986 (the time the study is undertaken) can then be determined using state and national mortality records. Additional follow-up of this cohort might then be conducted in 1990 or 1995, for example.

Case-Control Studies. This is the most common study design used in epidemiology. As outlined above, cohort studies first identify the exposure status of non-diseased individuals, then determine the subsequent incidence of disease in the cohort. In contrast, case-control studies begin by first identifying individuals who have developed the disease under study (cases) and individuals without the disease (controls). Cases can be selected through hospitals, disease registries, health maintenance organizations, physicians' practices, or even through death certificates. Controls may be selected from individuals who are "patients" at the same neighborhood as the cases or who live in the same hospital as the cases, or from other sources. An attempt is then made to compare the previous exposure experience of the cases with that of the control subjects. Certain factors that can influence disease rates (for example, age) must be taken into account in the design or analysis of these studies.

Case-control studies offer several advantages:

- They are generally much faster and less expensive to undertake than prospective cohort studies.
- Sample sizes can be much smaller than cohort studies, particularly in the case of relatively uncommon diseases. Whereas a cohort size of tens or hundred of thousands might be required to demonstrate some cancer risk, several hundred or even fewer subjects might be sufficient to reveal the

risk. For very rare diseases, the only practicable study design is a case-control study.

- A variety of previous exposure variables can be (and usually are) examined in a single study.

Case control studies have a number of limitations as well:

- The cases for inclusion in the study may not actually be representative of all those who develop the disease. For example, the selected cases may represent only those individuals who entered particular hospitals, and these cases may differ from non-hospitalized cases.
- It is often difficult to select an appropriate control group that is sufficiently comparable to the cases as well as representative of the general population from which the cases arise. Comparability and representativeness are both important yet sometimes mutually exclusive goals in selecting controls. In some studies, two different control groups have been utilized (for example, hospital controls and neighborhood controls). Many of the controversies in epidemiology arise from case-control studies in which the appropriateness of the controls is questioned.
- It is difficult or impossible to ascertain accurate exposures that have occurred in the past.
- These studies are inefficient for studying rate exposures.

Many EMF studies are of case-control design, including Wertheimer and Leeper, 1979; Fulton et al., 1980; Gilman et al., 1985; Savitz et al., 1988; and Nasca et al., 1988.

Proportional Mortality Studies. This study design is frequently used in exploratory or "hypothesis-generating" investigations, usually in occupational settings. The entire study is most often based only on death certificate information: age, sex, race, cause of death, residence, and in most states, usual industry and occupation. Proportional mortality study is conducted, when only the numbers and cases of deaths among the exposed group can be ascertained, but the structure of the population from which the numbers and cause of deaths is unknown.

In a proportional mortality study, the proportion of deaths from a specified cause relative to all deaths among the exposed group is compared with the corresponding proportion in the non-exposed group or a general population. This comparison is done independently of any relationship to the incidence rate of the disease in the exposed and unexposed groups, that is, only *numerator* data are used to make the comparison. For example, a researcher might wish to investigate hypothesis that EMF exposure is related to leukemia occurrence. Assuming that the job title "power lineman" is an adequate surrogate for exposure to electric and magnetic fields on the job, the researcher could plan a study to find out whether leukemia mortality in these workers was elevated in 1970-79. A relatively quick way to do this would be to identify all power linemen (from death certificates) who died in that time interval, determine what proportion died of leukemia, and compare that proportion with the proportion of person of similar age and sex in the general population of deaths due to leukemia. The quotient of the proportion in all power linemen divided by the proportion in the general population is called the proportional mortality ratio (PMR).

A similar type of study might also be done using incidence data from a disease registry such as a regional cancer registry. Such a study is referred to as a proportional incidence study, and a proportional incidence ratio (PIR) shows the observed-to-expected ratio of cases. The population of this type of study consists of those with cancers newly identified in a specified time period, and the study data is derived from the registry medical records. As with the mortality data, the proportions of cancers of different types can be compared among different occupational groups (if occupational data are available in the record).

PMR studies have the following advantages:

- They can be conducted very quickly and inexpensively, especially if the relevant data are already computerized (as they often are).
- Many different occupations and causes of death (or types of cancer) can be examined simultaneously.
- They can provide many useful leads for possible relationships between disease and occupation that can be investigated by more powerful study methods.

PMR studies also have some severe limitations:

- Unlike a cohort study, no information is collected on the population at "risk"; only on those already deceased. Therefore, no disease rates can be calculated. An elevation in the PMR may be due either to an increase in mortality from the cause of concern *or* to a reduction in mortality from another cause. For example, a PMR for leukemia may be elevated in power linemen because they are actually at increased risk of leukemia, *or* because they have lower risk of some other common disease such as heart disease.

Information on confounding factors (such as smoking) is usually not available.

- Accuracy and completeness of information obtained from death certificates or medical records is variable (see below)
- Study results can sometimes differ quite significantly when compared to results of more definitive studies such as cohort studies, especially if the overall mortality rates differ substantially among the groups (occupations) being compared.

Cluster Studies. A cluster of disease cases (for example, leukemia) is generally considered to be an unusually high number of cases appearing in the same setting (for example, neighborhood, town, work place) over a limited period of time. Considerable attention has been given to the study of disease clusters, particularly cancer clusters, and a number of statistical methods have been developed to analyze them. Nevertheless, no non-infectious agent has ever been consistently implicated in causation of any type of cancer by cluster analyses.

Cancer clustering is therefore generally thought to be the result of random (that is, nonuniform) distribution of disease cases in a population. Thus, even when a significant elevation of a disease is noted for a specified time and place, it is very likely a statistical artifact and not the result of the cases' exposure to an agent in their environment. For example, three new cases of leukemia might occur in a community of 4,000 persons in a five-year period and this incidence might be significantly ($p = 0.04$) elevated relative to the incidence in a comparison population for the same time period. But such a cluster might not be unusual. The calculations used to determine the statistical significance of the increased incidence of leukemia imply that if there are 1,000 communities of 4,000 persons in the country, 20 or 40 communities throughout the country (depending on the type of statistical test used) might have an elevated incidence of leukemia due to random statistical variation alone. Thus, such a cluster, while "statistically significant," would not be all that rare. It is also unlikely that similar environmental agents, leukemia subtypes, or age or gender distributions of the cases would be found in these 20 or 40 communities.

It is possible that a cluster of cancer cases may be related to the presence of an environmental carcinogen. The fact that no such relationship has been established through cluster analysis may in part be due to the fact that cancer clusters generally consist of so few cases that it is not possible to test a hypothesis relating an environmental carcinogen to cancer risk. Thus, the size of the population being studied is generally insufficient to yield statistically significant results.

D.2 Sources and Validity of Data

In any scientific study, careful attention must be given to the validity and reliability of the data. Unfortunately, all data collection methods involve some degree of inaccuracy and variability. These data quality problems are addressed in scientific studies by such means as full descriptions of data collections techniques, calculation of likely or potential measurement errors, validation procedures, and replication of measurements.

Because of the wide variety of epidemiologic data sources, professionals outside the field may find it difficult to judge the validity, utility, reliability, and limitations of these data. Data on exposure of study subjects, for example, may range from direct measurements of chemicals or their metabolites within the body to the relatively imprecise information on a decedent's exposure history taken from interviews with next-of-kin.

D.3 Comparability and Bias

A serious threat to the validity of any epidemiologic study is the possibility that its subject selection and data are biased. In epidemiology, "bias" does not imply any prejudice or prejudgment on the part of the study investigators. Rather, "bias" generally refers to systematic errors, that it, errors other than sampling variability that prevent the true value of a disease rate or other production variable from being obtained. The introduction of bias renders study groups noncomparable in some important way.

To understand how a study can become biased, it is useful to recall the degree to which the laboratory scientist strives to achieve comparability between exposed and unexposed organisms in his/her experiments. This is accomplished by such means as using a single strain of test organism, randomly assigning each organism to an exposure group, maintaining uniform environmental and dietary conditions during the course of the study, and using a consistent protocol for examination. The examination is performed with the investigator "blinded" to the subject's previous exposure history. Failure to achieve any of these major comparability elements can bias a study, and its conclusions must be considered suspect.

In epidemiologic research, utilizing an observational methodology, an equivalent degree of comparability cannot be achieved. The goal of the epidemiologist is to select from an existing population exposed and non-exposed (or diseased and non-diseased) groups that are fundamentally comparable and from which equivalent data can be obtained. Bias can be introduced in numerous ways, some of which cannot be known or controlled by the investigator.

Given the difficulty in recognizing and controlling all potential sources of bias, no study should be considered completely free of it. For example, rarely does a study attain 100% follow-up of its subjects. Critics can always attribute study findings to some form of bias since there are indeed so many potential sources of bias. However, it is difficult to actually demonstrate that some bias accounts for or even materially affects the study's findings. Strict adherence to established procedures and standards can reduce many, but not all, possibilities for bias. In many well-reported studies, the authors frequently discuss the possible sources of bias in their study and attempt to show, through logic and/or data, that their findings are not likely to be due to some bias. But there still remains the possibility that some unknown bias is operating.

D.4 Association or Causation

In laboratory research, a well designed experiment that results in a statistically significant effect (that is, one not due to chance variation) is usually interpreted as demonstrating a cause and effect relationship. The existence of a cause and effect relationship cannot be so readily inferred from observational epidemiologic studies. The epidemiologist, at best, can show that some association or relationship exists between an exposure (for example, chemical, radiation) and a physiological or health - related effect (for example, blood chemistry, disease, death). Typically, the epidemiologist further attempts to demonstrate that the association is unlikely to be due to chance and is not due to some third (confounding) variable.

Judging Positive Associations. Scientific "proof" of a cause and effect relationship cannot be obtained from a observational study. However, as a practical matter, explicit or implicit judgments of causality are frequently derived from such studies, and strongly influence public health policy. Therefore, it is important to consider epidemiologic findings from a variety of perspectives, such that a reasonable assessment can be made. Epidemiologists have not established hard and fast rules for determining when a positive association should be considered a cause and effect relationship. Different experts stress different factors in evaluating associations, and not all agree that certain items are particularly useful. However, certain guidelines arise frequently in discussions of causal relationships, and these are discussed below.

Strength of Association. The more strongly an exposure is associated with a disease, the more likely it is that the exposure causes the disease. Many different measures of association are utilized in epidemiologic studies, depending in part on the type of study conducted. Most are expressed as some ratio comparing or estimating disease risk in the exposed to disease risk in the unexposed. For example, the "relative risk" may be defined as:

- Risk of disease among those exposed
- Risk of disease among those not exposed.

The "strength" of the association refers to the magnitude of the risk ratio. Generally, the larger the magnitude of the relative risk, the stronger the association. As an example, smokers are ten times more likely to develop lung cancer than non-smokers (relative risk = 10). This greatly elevated relative risk makes it much less likely that some other variable (confounder) overlooked by the investigator(s) is actually responsible for the association. A very low relative risk, for example, less than 2.0, would have a greater probability of being due to some study bias or confounding factor.

Although a strong association is very suggestive of a causal relationship, it cannot be argued that a weak association is not causal. It might also be noted here that the strength of an association does not by itself reflect the overall public health impact or importance due to that risk factor.

Other crucial variables are the frequency of the disease associated with the risk factor and the number of people commonly exposed to the risk factor. A risk factor associated with a high relative risk for a relatively rare disease may have less public health impact than a risk factor associated with a lower relative risk for a more common disease. For example, occupational exposure to asbestos has caused many more deaths through lung cancer than through mesothelioma (a rare disease) although the relative risk for mesothelioma is much higher than the risk of lung cancer. And, of course, a risk factor commonly encountered by many people might result in more illness, even though it is associated with a low relative risk for disease, than a rarely experienced risk factor with a high relative risk for disease.

Consistency. An association consistently found by different investigators, in different populations and/or in different geographical areas, is more likely to be causal. Although the many studies that have examined lung cancer and smoking have involved many different investigators, study populations, locations, and study designs, have all shown a very strong positive association. It is extremely unlikely that such consistent findings can be the result of some overlooked bias or can occur by chance alone. Some investigators attach special significance to consistent associations found in both retrospective and prospective studies.

Temporal Relationship. Obviously, an exposure must precede a disease if it is even to be considered a possible causative agent. Prospective cohort studies most firmly establish this time sequence. In contrast, cross-sectional (prevalence) studies do not generally permit determination of whether exposures preceded disease development. It may also be difficult to establish the time sequence in some case-control studies in which, for example, a disease may have a very long pre-clinical (non-symptomatic) phase (for example, asbestos and lung cancer).

Dose-Response Relationship. The existence of a dose-response relationship between exposure level and disease incidence supports a causal interpretation. In other words, those who have the highest exposures should also have the highest disease risks. It has been clearly demonstrated that the number of cigarettes smoked correlates directly with the degree of lung cancer risk.

The absence of an apparent dose-response effect is not considered evidence against causality. Exposures may not have been ascertained accurately enough in studies, leading to a misclassification and bias that can obscure a risk gradient. It is also possible that some threshold of exposure is necessary for a given agent before an effect is observed. There is only weak evidence, for example, of a dose-response relationship between occupational asbestos exposure and mesothelioma risk, although there is no doubt about the causal nature of the relationship.

Coherence/Plausibility. A causal hypothesis is supported when an association is consistent with or supported by other known facts and observations. For example, a causative hypothesis is favored if there is some demonstrated or potential biological mechanism by which the effect can be explained. The cellular effects of ionizing radiation have long been recognized and offer a clear explanation for the health hazards of radiation. In the case of cigarette smoking, laboratory studies have identified a variety of organic compounds in inhaled smoke; a number of these compounds have been shown to cause cancer in animal studies. These findings are thus consistent with the human epidemiologic evidence.

The absence of a recognized biological mechanism does not necessarily contradict a causative interpretation. The lack of an apparent mechanism may only reflect an early stage of investigation, a situation well illustrated by the outbreak of toxic shock syndrome. Early epidemiologic findings clearly indicated that the highest risk group consisted of young, menstruating women who used a high-absorbency brand of tampon. However, it was found that some non-menstruating women and some men were also disease victims. Thus the use of high-absorbency tampons alone obviously could not account for all cases, and there was considerable doubt by some (including the manufacturers) that the tampons alone could be causally related to the disease. However, the absence of an explanation or recognized mechanism did not prevent withdrawal of the tampons from the market. Some time later an explanation was found: the disease was actually caused by a toxin from a relatively common bacterium. The toxin is only produced under certain physical and biological conditions, conditions that are more likely to occur in young women using high-absorbency tampons.

Specificity. When an association links exposure to a single disease rather than to a broad spectrum of diseases, a causal interpretation is favored. An example of high specificity is the association between occupational exposures to vinyl chloride and angiosarcoma (a rare form of liver cancer). The high specificity as well as the strength of this association leaves little doubt as to its causative nature.

A lack of specificity, however, does not necessarily argue against causality. For example, cigarette smoking has been associated with a wide range of diseases. In fact, the smoking history of study subjects is almost always considered in well-designed studies of other diseases. This lack of specificity, although sometimes still raised by the tobacco industry, is not particularly troublesome to epidemiologists since a great many components have been identified in tobacco smoke, and many of these components can be transported through the body to different sites. This rather broad exception to the concept of specificity has led some epidemiologists to consider this guideline useless in determining whether an association is likely to be causal or not.

Conclusions. It is important to emphasize that none of the above factors is sufficient either to prove or disprove that an association represents a true cause and effect relationship. They do, however, offer some reasonable guidelines with which both epidemiologists and non-epidemiologists may judge whether a positive association is likely to represent a true cause and effect relationship

D.5 Statistics: Risk Estimates

The following discussion addresses how rates and ratios are used to provide estimates of risk, with emphasis on the measures of risk derived from two major study designs: the cohort study and the case control study.

Risk Measures from a Cohort Study. A cohort study is similar to a laboratory study in terms of the time sequence of events. A study group (cohort) of healthy people is identified and each individual is then classified according to whether he/she is exposed or not exposed to the agent under study. At some later point in time, which may be many years later, the cohort is rechecked (the "follow-up"). Two risk measures from cohort studies will be described: the relative risk and the standardized mortality ratio.

Relative Risk. From the follow-up data an actual disease rate can be tabulated separately for both the exposed and non-exposed groups. Depending on the specific disease in question, the rate may be either a morbidity or a mortality rate. As an example, consider that the exposure under study is cigarette smoking and the disease in question is lung cancer. Data from an actual study showed that the lung cancer mortality rate for a particular age group of non-smokers was 19 deaths per 100,000 per year. In contrast, the rate for smokers in this same age group was approximately 190 per 100,000 per year.

These rates can be compared in several ways such that a quantified expression of risk can be obtained. The most common measure of risk in this type of study is the *relative risk* (RR). The relative risk indicates the increased (or decreased) degree of risk of disease among the exposed compared to the non-exposed. It provides a measure of the causative importance of the exposure under study. A relative risk with a value of one (1.0) indicates no association between the exposure and the disease.

The relative risk is as follows:

$$\text{Relative risk (RR)} = \frac{\text{rate in the exposed}}{\text{rate in the non-exposed}}$$

For smokers, the relative risk is:

$$\text{RR} = \frac{190/100,000/\text{year}}{19/100,000/\text{year}} = \frac{190}{19} = 10$$

This indicates that smokers have ten times the risk of dying from lung cancer compared with non-smokers.

Confidence limits can (and should) be computed for relative risks to determine if the risk is statistically significant, that is, not due to chance. Confidence limits are the range of the risk estimate that takes into account sample size and variability. If the range of the estimate does not include 1.0, it is recognized as being statistically significant.

Standardized Mortality/Morbidity Ratios. Standardized mortality ratios (SMR) are used for adjusting mortality rates in order to compare health outcomes between populations that may have different distributions of important variables such as age, sex, or race. Indirect standardizations (adjustment) involves applying mortality rates from some selected reference population, adjusted for age and possibly other factors, to the study population. This procedure generates the number of deaths that would be "expected" if the study population had experienced the same disease incidence as the reference population. Then, a common way to compute this expected number with the actual observed number is to compute the SMR. This is done by dividing the observed number of deaths by the expected number, and then multiplying the quotient by 100 to eliminate decimals:

$$\text{SMR} = \frac{\text{Observed deaths}}{\text{Expected deaths}} \times 100$$

An SMR of 100, then, means that the expected and observed deaths are essentially equal in number, and no excess risk is evident. An SMR of 120 means that there were 20 percent more deaths than expected, while an SMR of 80 would mean that the observed deaths were only 80 percent of the deaths expected based on the reference population. SMRs are used frequently in occupational studies.

In a typical SMR occupational study, the investigator has collected extensive information on who has worked in the industry, when, for how long, in what jobs, and if deceased, the cause of death. Thus both numerator data (deaths) and denominator data (person-years or persons at risk) are collected, and an actual mortality rate for each disease can be determined. This serves as the basis for the observed number of deaths.

A similar ratio can be calculated using morbidity data. For example, a standardized incidence ratio (SIR) can be determined using only incident cases.

Proportional Mortality Ratio. An entirely different ratio, which appears frequently and almost exclusively in occupational studies, is the *proportional mortality ratio* (PMR). As previously described, proportional mortality expresses the proportion of all deaths that are due to one cause. For example, of those who worked in a particular industry, 20 percent of the deaths may have been due to cancer, whereas heart disease may have accounted for 35 percent of the deaths.

In a PMR study, on the other hand, the data are often obtained completely from death certificates. Consequently, the investigator only has data on people who have already died. Recall that death certificates also list the usual occupation and other personal data such as age, race and sex. The investigator does not (and perhaps cannot) obtain denominator data, that is, the total person-years or persons at risk (most of whom may even still be living). Thus, a mortality rate cannot be determined; all that can be done is to compare, for example, the proportion of all deaths that were due to leukemia in one occupation with that proportion in another (reference) group.

Although the PMR and SMR appear superficially similar, they are quite different and are derived from different types of data. Because they are both widely used in occupational epidemiology, the distinctions need to be underscored here.

Risk Measures from a Case-Control Study. In a contrast to cohort studies that determine subsequent disease rates between exposed and non-exposed people, case-control studies first identify diseased and non-diseased persons, then ascertain their previous exposure history. This approach does not permit determination of actual disease rates. Thus, a relative risk cannot be determined. One can, however, compare "exposure ratios" between diseased and non-diseased groups. Under certain conditions, these exposure ratios can be used to estimate the relative risk by calculating an *odds-ratio*.

To illustrate the odds-ratio calculation, it is first useful to categorize case-control study data in a two-by-two table according to disease and exposure status.

	Exposed	Unexposed
Deceased persons (cases)	a	c
Controls	b	d

The letters represent the number of study subjects who fall into the four categories. Omitting its derivation, the odds-ratio (OR) is then calculated as:

$$OR = \frac{a \times d}{b \times c}$$

As an example, consider that the following data are obtained from a case-control study:

	Exposed	Unexposed
Deceased persons (cases)	85	15
Controls	40	60

The odds-ratio is then calculated as:

$$OR = \frac{85 \times 60}{15 \times 40} = 8.5$$

The odds-ration is interpreted exactly the same as the relative risk. If equal to one (1.0), it suggests no association between the exposure and disease. If greater than one, it indicates a positive association, and if less than one, a negative association or a protective effect. As with relative risks, confidence limits can be computed to determine if the odds-ratio is significantly different from a value of one.

In some case-control studies, the controls are individually "matched" to the cases during the selection process. For each case identified, a systematic approach is used to select a control who is in the same age bracket, of the same sex and race, etc. This matching process avoids having to account for these variables later in the analysis. When this type of matching is used in a study, the odds-ratio is calculated differently than above.

D.6 Statistics: p-Values, Confidence Intervals and Significance

The epidemiologist can use a variety of approaches and sources of information to obtain data on exposures and other factors that may be related to the development of disease. These data are then used to calculate rates that are compared between groups. In the case-control study, for example, the investigator identifies a group of people with the disease under study (cases) and another, presumably comparable group without the disease (controls). The investigator then seeks to assess and compare the exposure history of the cases and controls. The goal is to determine whether there is a higher proportion of those who are exposed among the cases than among the controls. If such a difference is found, its significance must then be evaluated.

A difference might be found for several reasons:

- There may actually be an association between the exposure and the disease;

- The difference (association) may be the result of bias in the study;

- The difference may be due to confounding; or

- The difference may be the result of random variation.

Unfortunately, it can never be proved that confounding and bias do not exist in a study. The best that can be done is to consider the potential sources of confounding and bias, and attempt to show that they are unlikely to exist of to have strongly influenced the study results.

An association between some exposure and a disease does not automatically demonstrate that the exposure causes the disease. It may be that the disease causes the exposure or that some other unknown factor causes both the exposure and the disease.

Before one attempts to judge the causal nature of an association, it is necessary first to determine whether the difference in rates (indicating an association) is likely to be real or just due to chance. Because of random variations that arise among population samples, the case and control groups are unlikely to have *exactly* the same proportion of exposed and non-exposed persons, even when there is no association between the exposure and the disease. The question then, is how large must a difference be to show convincingly that it is real and not due to chance. Two approaches are commonly used in epidemiology to determine whether a difference is likely to be the result of random chance: use of significance testing by the calculation of p-values, and use of confidence intervals. Both of these topics are covered in most introductory statistics texts.

D.7 References

The following works were used in preparing this appendix. The reader is referred to them for further information on the subject of epidemiologic methodology.

[1] Altman DG, Gore AM, Gardner MJ, Pocock SJ. Statistical guidelines for contributors to medical journals. *British Medical Journal* 1983; 286:1489-93.

[2]   Breslow NE, Day NE. *Statistical Methods in Cancer Research.* Vol. 1: *The Analysis of Case-Control Studies.* IARC Scientific Publication 32. New York: Oxford University Press, 1980.

[3]   Buechley R, Dunn JE, Lindon G, Breslow L. Death certificate statement of occupation: Its usefulness in comparing mortalities. *Public Health Reports* 1956; 71(11): 1105-11.

[4]   Cole P. The evolving case-control study. *Journal of Chronic Diseases* 1979; 32:15-27.

[5]   Feinstein AR, Horwitz RI. Double standards, scientific methods, and epidemiologic research. *New England Journal of Medicine* 1982; 307(26):1611-17.

[6]   Fleiss JL. *Statistical Methods for Rates and Proportions, 2nd Edition.* New York: John Wiley and Sons, 1981.

[7]   Friedman GD. *Primer of Epidemiology, 2nd Edition.* New York: McGraw Hill, 1980. ISBN 0-0-022434-X.

[8]   Gladen B, Rogan WJ. *Misclassification and the design of experimental studies.* American Journal of Epidemiology 1979; 109(5):607-16.

[9]   Glasser JH. The quality and utility of death certificate data (editorial). *American Journal of Public Health* 1981;71(3):231-33.

[10]   Greenberg RS, Kleinbaum DG. Mathematical modeling strategies for the analysis of epidemiologic research. *Annual Review of Public Health* 1985; 6:223-45.

[11]   Haines T, Shannon H. Sample size in occupational mortality. *Journal of Occupational Medicine* 1983; 25(8):603-08.

[12]   Hill AB. The environment and disease: Association or causation? *Proceedings of the Royal Society of Medicine* 1965; 58:295-300.

[13]   Horwitz RI, Feinstein AR. Methodologic standards and contradictory results in case-control research. *American Journal of Medicine* 1979; 66:556-64.

[14]   Kelsey JL, Thompson WD, Evans AS. *Methods in Observational Epidemiology.* New York: Oxford University Press, 1986.

[15]   MacMahon B, Pugh TF. Epidemiology: Principles and Methods. Boston: Little Brown and Company, 1970.

[16]   Mantel N, Haenszel W. Statistical aspects of analysis if data from retrospective studies of disease. *Journal of the National Cancer Institute* 1959; 22(4):719-48.

[17] McMichael AJ. Standardized mortality ratios and the healthy worker effect: Scratching beneath the surface. *Journal of Occupational Medicine* 1976; 18:165-68.

[18] Monson RR. *Occupational Epidemiology*. Boca Raton, FL: CRC Press, 1980.

[19] Percy C, Stanek E, Gloeckler L. Accuracy of cancer death certificates and its effect on cancer mortality statistics. *American Journal of Public Health* 1981; 71:242-50.

[20] Rim AA. *Basic Biostatistics in Medicine and Epidemiology*. New York: Appleton-Century-Crofts, 1979.

[21] Rothman KJ. *Modern Epidemiology*. Boston: Little Brown and Company, 1986.

[22] Sackett DL. Bias in analytic research. *Journal of Chronic Diseases* 1979; 32:51-63.

[23] Schlesselman JJ. Sample size requirements in cohort and case-control studies of disease. *American Journal of Epidemiology* 1974; 99:381-84.

[24] Steenland K, Beaumont J. The accuracy of occupation and industry data on death certificates. *Journal of Occupational Medicine* 1984; 26:288-96.

Appendix E Glossary

E.1 Glossary

$\overline{E}$	the electric field intensity vector usually expressed in volts/meter or kV/meter.
$\overline{H}$	the magnetic field intensity vector usually expressed in amp/meter.
$\overline{B}$	the magnetic flux density vector usually expressed in tesla (SI units) or milligauss (Gaussian units).
μ_0	permeability of free space $4\pi \cdot 10^{-7}$ henrys/meter.
μ_r	relative permeability of a material (dimensionless).
$\Re$	symbol used to represent magnetic reluctance expresses in amperes/weber.
σ	Greek symbol used to represent conductivity expressed in ampere-volt-meter.
ϕ	Greek symbol used to represent magnetic flux (webers) or phase (radians or degrees).
μ	Greek symbol used to represent permeability. $\mu = \mu_0\mu_r$
δ	Greek symbol used to represent skin depth expressed in meters.
ω	Greek symbol used to represent radian frequency expressed in radians/seconds.
$\angle$	The mathematical notation for angle. See **phase angle**.
D	lateral distance from a reference conductor expressed in feet. On a graph or diagram D is positive to the right and negative to the left.

h	height of a conductor above ground expressed in feet. h is positive for overhead conductors and negative for underground conductors.
I	electrical current in amperes
N	number of turns of conductor
$\hat{x}, \hat{y}, \hat{z}$	unit vectors in a rectangular coordinate system.
$\hat{r}, \hat{\phi}, \hat{z}$	unit vectors in a cylindrical coordinate system.
ABNORMALITY	The quality or state of being abnormal, i.e., deviating from the normal or average; markedly irregular; characterized by deficiency or disorder.
AC	the abbreviation for alternating current. See ALTERNATING CURRENT.
ACCLIMATIZATION	adaptation of an organism to a new environmental condition, e.g., temperature, altitude, climate, or situation.
ACCURACY	the degree to which a measurement, or an estimate based on measurements, represents the true value of the attribute that is measured.
ACTIVE SHIELDING	involves the use of current-carrying conductors whose magnetic field effectively cancels the applied field.
ACUTE LEUKEMIA	leukemia characterized by sudden onset and rapid progression of the disease. See LEUKEMIA.
ACUTE LYMPHOID LEUKEMIA (ALL)	See LYMPHATIC LEUKEMIA.
ACUTE MYELOGENOUS LEUKEMIA (AML)	See MYELOGENOUS LEUKEMIA.
ADRENAL	referring to the adrenal gland and its functions; complex endocrine organ(s) near the anterior border of the kidney consisting of a mesodermal cortex that produces steroids like sex hormones, and hormones concerned especially with metabolic functions and an ectodermal medulla that produces adrenaline.

231

| AGE-ADJUSTED | a statistical method used in rate calculations to minimize the effects of different age distribution among study subjects or populations. Only properly adjusted rates can be compared with each other. |

AGE-ADJUSTED — a statistical method used in rate calculations to minimize the effects of different age distribution among study subjects or populations. Only properly adjusted rates can be compared with each other.

ALTERNATING CURRENT (AC) — an electric current that reverses its direction at regularly recurring intervals, (e.g., 50-60 Hz). the abbreviation AC is commonly used to describe periodically varying electrical quantities.

ALTERNATIVE HYPOTHESIS — a numerical statement concerning the parameters of one or more distributions that is mutually exclusive to the null hypothesis. Sometimes called the research hypothesis because it is, in most cases, the hypothesis that the investigator would like to prove.

AMPERE — the unit used to measure electrical current. Sometimes abbreviated amp or A.

AMPERE'S CIRCUITAL LAW — physical law relating magnetic field intensity to a current distribution.

AMPLITUDE — the maximum departure of the value of an alternating current or wave from the reference value.

ANALYSIS OF VARIANCE (ANOVA) — widely used statistical methods that isolate and assess the contribution of categorical independent variables to variation in the mean of a continuous dependent variable.

APPARENT POWER — the product of volts times current in an electrical circuit. Units are volt-amperes.

ASSOCIATION — association refers to the statistical dependence between two variables, that is, the degree to which the rate of disease in persons with a specific exposure is either higher or lower than the rate of disease among those without that exposure. There are several factors to be considered in evaluating whether or not an association observed in an epidemiologic study is causal, these include:

1.   The strength of the association: A strong association is more likely to be causal than a weak one.

ASSOCIATION	2. Specificity of the association: The exposure to the supposed causative agent in every diseased patient tends to support causality.3. Dose-response relationship: In a causal relationship, increased exposure to the agent usually produces an increased risk in the exposed population.
	4. Consistency of the association: The repeated finding of the association in several studies of different populations tends to support causation.
	5. Time sequence: Clear antecedence of the exposure of interest to the outcome by a period of time is necessary to judge causality to be reasonable.
ASTROCYTOMA	a tumor intermingled with the essential elements of nervous system tissue especially the brain, spinal cord, and ganglia, but composed of the star-shaped cells (astrocytes) that are part of the supporting tissue of the nervous system. Astrocytomas in children and persons less than 20 years of age usually arise in a cerebellar hemisphere, and in adults they usually occur in the cerebrum, sometimes growing rapidly and invading extensively.
ATTRIBUTABLE RISK	the difference in disease rates between exposed and non-exposed groups. It can serve as a measure of the proportion of a disease in a population that can be explained by the exposure under study.
BEHAVIOR	anything that an organism does involving action, and response to stimulation; the response of an individual, group or species to its environment.
BENIGN	of a mild character; non-cancerous/non-malignant.
BIAS	Deviation of research results or inferences from the truth. Any trend in the collection, analysis, interpretation, publication, or review of data that tends to produce results that differ systematically from the "true values" of the population variables being

studied (e.g. disease rates). Many varieties of bias have been described. Unlike conventional usage, the term "bias" does not refer to a partisan point of view.

BIOPHYSICAL MECHANISMS
physical and/or chemical interactions of electric and magnetic fields with biologic systems.

BONE MARROW
a soft highly vascular modified connective tissue that occupies the cavities and cancellous part of most bones.

BUS
a conductor or group of conductors that serves as a common connection for two or more circuits in a switch gear assembly.

CABLE
either a single conductor or a combination of conductors insulated from one another.

CANCER
a disease characterized by malignant, uncontrolled growth of cells of body tissue; a malignant tumor of potentially unlimited growth that expands locally by invasion, and systematically by metastasis.

CANCER CLUSTER
a series of cancer cases that occur close together in time and/or location. The term is normally used to describe a grouping of relatively rare diseases, such as leukemia.

CANCER INITIATOR
a chemical substance or physical stimulus that causes or facilitates (makes easier) the beginning of cancer.

CANCER PROMOTER
a chemical substance or physical stimulus that furthers the growth or development of cancer.

CAPACITOR
a device made of conducting surfaces separated by insulation and capable of storing electric charge.

CARCINOGEN
a chemical, biological, or physical agent capable of producing tumor growth.

CARCINOGENESIS
a series of stages at the cellular level culminating in the development of cancer.

CARDIOVASCULAR
relating to or involving the heart and blood vessels.

CASE	in epidemiology, a person identified as having the particular health endpoint (e.g., disease) under investigation.
CASE-CONTROL STUDY	a type of epidemiologic investigation that begins with the identification of both a group of persons who have developed the disease under study, the cases, and a group of persons who have not developed the disease, the controls. An attempt is then made to compare the previous exposure experience of the cases with that of the controls to determine if the two groups differ significantly in the frequency or level of a particular exposure.
CAUSATION	a condition in which a situation, event, or agent produces an effect in an outcome variable of study.
CELL	a small, usually microscopic mass of protoplasm bounded externally by a semi-permeable membrane, usually including one or more nuclei and various nonliving products, capable alone or interacting with other cells of performing all the fundamental functions of life, and forming the least structural unit of living matter capable of functioning independently.
CELL MEMBRANE	the semi-permeable material forming the boundary of a cell that encloses and supports the cell, and controls efflux and influx of cell metabolites, nutrients, wastes, etc.
CHARGE	the electrical property of matter that is responsible for creating electric fields and electrical current, usually measured in Coulombs.
CHROMOSOME	a very long molecule of DNA, complexed with protein containing genetic information.
CHRONIC	a condition or situation marked by long duration or frequent recurrence.
CHRONIC LYMPHOID LEUKEMIA (CLL)	a type of leukemia that is not acute.

CIRCADIAN RHYTHM	the rhythmic biological cycle (of things like hormone concentrations in the body) that usually recurs at approximately 24-hour intervals.
CIRCUIT	a closed conducting path for the flow of electrical current.
CIRCULAR MIL	a unit of area equal to pi over 4 of a square mil. The cross-sectional area of a circle in circular mils is equal to the square of its diameter in mils (10^{-3} inches).
COAXIAL CABLE	a cable in which current flows in an inner cylindrical conductor and returns in an outer concentric cylinder.
COHORT	in epidemiology, an identified group of persons with some common point of reference, for example, birth year or place of employment, who are free of disease, but who have various degrees of exposure to the agent under study. The group is followed over time to determine the occurrence of disease among members of the cohort.
COHORT STUDY	a type of epidemiologic study in which the frequency of morbidity or mortality from a specific disease of interest in a group exposed to a suspected risk factor is compared to that in a group of unexposed people.
COMPONENT	a single piece of equipment, or a group of items viewed as an entity.
CONDUCTOR	a metallic wire that carries current. The wires on transmission lines are conductors.
CONFIDENCE INTERVAL	A range of values bracketing a risk estimate that is calculated in such a way that the range has a specified probability of including the true value of the risk.
CONFOUNDING	a situation in which an observed association between an exposure and a disease is influenced or distorted by other variable(s) that are associated with the exposure and affect disease occurrence.

CONTROL
in case-control studies, an individual in the group of people that has not developed the disease of interest. See CASE-CONTROL STUDY.

CONTROL GROUP
in experimentation, the group of subjects who are treated in a parallel experiment except for omission of the procedure or agent under test and that is used as a standard of comparison in judging experimental effects.

CORRELATION
a linear relationship between two or more sets of variables. A linear association. Correlation, like association, does not imply causality.

COULOMB
the unit of electric charge (C). One electron has a negative charge of about 1.6×10^{-19} Coulombs in magnitude.

CROSSARM CONSTRUCTION
See **horizontal configuration**.

CUMULATIVE EXPOSURE
the total exposure to an agent, such as magnetic fields, experienced by a person during a specified time period, e.g., one hour, one year, or a lifetime of work.

CURRENT
the organized flow of electric charge. Current in a power line is analogous to the rate of fluid flow in a pipeline. All currents produce magnetic fields. The unit for current is the **ampere**.

DC
the abbreviation for direct current. See DIRECT CURRENT.

DELTA CONNECTED
a method of connecting three phase circuits without a neutral. Also see WYE CONNECTED.

DEMOGRAPHIC INFORMATION
the characteristics of a population such as place of residence, age, sex , race, birth and death rates and socioeconomic conditions.

DEOXYRIBONUCLEIC ACID (DNA)
the nucleic acid molecule in chromosome that contains the genetic information.

DIAGNOSTIC CRITERIA
information, usually clinical information such as physical symptoms and laboratory test results, used to determine whether a person has a suspected disease.

DIAMAGNETIC MATERIALS	materials characterized by a relative permeability that is slightly less than 1.0 (e.g., silver).
DIELECTRIC	an insulator or non-conductor.
DIELECTRIC STRENGTH	the maximum electric field strength that a material can withstand without breaking down and conducting.
DIRECT CURRENT (DC)	an electric current flowing in one direction only and constant in value.
DISTRIBUTION LINE	A power line used to distribute power in a local region. Distribution lines typically operate at voltages of between 5 and 35 kV, much lower than the voltages of transmission lines. However, the currents on some distribution lines can be comparable to transmission line currents.
DISTRIBUTION SUBSTATION	a component of the distribution system. The distribution substation contains step-down transformers that transform power from a transmission voltage of, say, 230 kV to a primary distribution voltage of, say, 21 kV.
DISTRIBUTION TRANSFORMER	a transformer that converts primary distribution voltage to user level voltages.
DOSE	the amount of exposure of a kind that produces effects. Dose might be measured in milligrams of a specific chemical that enters the body. In the case of electromagnetic fields, it is often unclear what aspect of the field is involved in producing effects. Hence, it is not clear how to measure dose from electromagnetic fields.
DOSE-RESPONSE STUDY	an investigation that attempts to statistically define the functional relationship between a response (usually a disease incidence) and the dose of a specific agent.
DOUBLE-BLIND EXPERIMENT	an experimental procedure in which neither the subjects nor the experimenter know the makeup of the test and control groups during the actual course of the experiments.

DUCT	a pipe-like structure that contains electrical conductors or cable.
EARTH RETURN	the part of the return current that is not carried by the shield wire (or neutral) and is returned in the earth.
EDDY CURRENT	when a good electrical conductor is immersed in a time-varying magnetic field, electric currents are induced in the conductor by magnetic induction. These currents are commonly called eddy currents.
ELECTRIC FIELD	a vector field a force per unit charge. Electrical charges are a source of electric fields. The electric field from a power line is an alternating, 60-Hz field due to charges on the conductors. The intensity of the electric field is expressed in volts per meter (V/m) or kilovolts per meter (kV/m).
ELECTRICAL LOAD	the amount of power that is traveling through a line, transformer, substation, etc. The load is usually dependent on the number of end-users requiring power at a given time. Units are watts, kilowatts, megawatts.
ELECTROMAGNETIC FIELD (EMF)	a combination of both electric and magnetic fields.
ELECTROPHOBIA	the irrational fear of electromagnetic fields.
ELF	abbreviation for extremely low frequency. See EXTREMELY LOW FREQUENCY.
ELLIPTICAL POLARIZATION	with respect to fields, a field whose locus in space is an ellipse.
EMF	abbreviation for electromagnetic field. See ELECTROMAGNETIC FIELD.
ENDOCRINE SYSTEM	the glandular system that produces secretions that are distributed in the body through the bloodstream, and aid the nervous system in controlling and coordinating the body functions.
ENZYME	a protein molecule that acts as a catalyst in living organisms.

EPIDEMIOLOGY	the branch of medical science that studies the distribution and factors that cause health-related conditions and events in groups of people, often making use of statistical data on the incidence of disease or death.
EXPOSED GROUP	the experimental group of test organisms receiving a dose of a substance, to determine the effect(s) of the substance.
EXPOSURE	the joint occurrence in space and time of a person and the physical or chemical agent of concern, expressed in terms of the environmental level of the agent.
EXPOSURE-RESPONSE RELATION	a relationship between exposure and the effect produced by exposure. Response can be expressed either as the severity of injury or proportion of exposed subjects affected.
EXPOSURE ASSESSMENT	measurement or estimation of the magnitude, frequency, duration and route of exposure of an organism to environmental agents. The exposure assessment may also describe the nature of exposure and the size and nature of the exposed populations.
EXPOSURE METRIC	the means by which exposure to an agent of interest (e.g., magnetic fields) is measured or estimated. For example, in an occupational mortality study, the job title "electrician" listed on the death certificate might serve as the "exposure metric" to estimate the likelihood of exposure.
EXTRAPOLATION	an estimate of response or quantity at a point outside the range of the experimental data.
EXTREMELY LOW FREQUENCY (ELF)	an electrical frequency less than ten kilohertz. The standard 60 Hz frequency in the United States and the standard 50 Hz frequency found in Europe are both ELFs.
FARADAY'S LAW	a changing magnetic field induces a voltage in a conductor in that field.
FAULT	a component failure that may result in service interruption. Faults may be caused

by such things as a car striking a utility pole, lightening striking the electrical system or a transformer failure.

FEEDER TRUNK LINE — main feeder line as it leaves the substation.

FEEDERS — primary distribution lines emanate as feeders from the distribution substation. There may be several such feeders, each with a capacity on the order of 10 MVA, radiating from the substation. There is normally no other interconnection between feeders.

FERROMAGNETIC MATERIALS — materials exhibiting a strong magnetic moment that can be aligned with an applied magnetic field. Ferromagnetic materials exhibit large values of relative permeability (e.g., iron alloys).

FIELD — a set of values of a physical quantity occurring at different points in space.

FINITE ELEMENT ANALYSIS — an iterative method for solving differential equations such as Laplace's Equation. This method is useful for solving various electromagnetic field problems.

FOLLOW-UP — a process in epidemiology by which study subjects are tracked and observations of variables of interest are made over time. Follow-up has two critical features: completeness and duration. Completeness refers to the proportion of the study sample followed. Duration refers to the length of time the sample is followed.

FREQUENCY — the number of complete cycles of a periodic waveform per unit time. The units of frequency are hertz (Hz). One Hz equals one cycle per second.

GAUSS — a gaussian unit of measure for magnetic flux density. Abbreviated G. There are 10,000 gauss in one tesla.

GAUSSIAN UNITS — a system of electrical units. Examples are oersted, gauss, maxwell, etc.

GENE	the simplest complete functional unit in a DNA molecule. A linear sequence of nucleotides in DNA that is needed to synthesize a protein and/or regulate cell function.
GEOMAGNETIC FIELD	the earth's natural magnetic field.
GLIOMA	a type of cancer of the central nervous system, or a specific tumor composed of the supporting cells of the brain, spinal cord, or other nervous system tissue.
GROUND	an electrical connection to the earth.
GROUND CURRENTS	electrical currents that flow in the earth.
HARMONIC CURRENT	a current of a harmonic frequency. See HARMONICS.
HARMONIC DISTORTION	distortion of a sinusoidal varying wave due to the addition of harmonics.
HARMONICS	frequencies that are multiples of a fundamental frequency. For example, 120 Hz is the second harmonic and 180 Hz is the third harmonic of 60 Hz.
HEMATOPOIETIC CANCERS	cancers of the blood-making organs, specially the bone marrow and lymph nodes.
HERTZ	a unit used to measure frequency and abbreviated Hz. One hertz equals one cycle per second. In the United States, AC power has a frequency of 60 Hz. In most of Europe, AC power has frequency of 50 Hz. Radio waves have frequencies of many thousands or millions of hertz.
HORIZONTAL CONFIGURATION	the arrangement of conductors in a transmission line such that the conductors are spaced in a flat array parallel to the earth, also called cross-arm construction.
HORMONE	the secretions of endocrine glands that act as "chemical messengers", controlling and regulating the body's life functions.
HOT WIRE	a wire that is at a higher electrical potential than the neutral or grounded conductor.

HYPOTHESIS	a supposition. arrived at from observation or reflection, that leads to refutable predictions.
HYPOTHESIS TESTING	the systematic verification or rejection of a scientific proposition or argument.
HZ	the abbreviation for hertz. See HERTZ.
IMMUNE SYSTEM	the body's primary defense against abnormal growth of cells (i.e., tumors) and infectious agents such as bacteria, viruses and parasites.
IMPEDANCE	the electrical property of a conductor or circuit that resists the flow of an alternating electric current. Units are ohms.
IN PHASE	two sinusoidally varying currents are in phase if they have the same electrical frequency and reach maximum values at the same instant of time.
IN VITRO	describes studies that are done in the laboratory, literally "in glass", as distinct from those performed using living animals.
IN VIVO	experiments performed "in the living body" of a plant or animal.
INCIDENCE OF A DISEASE	the number of new cases (persons becoming ill) during a given time period in a specified population.
INCIDENCE RATE	the rate at which new events occur in a population. The numerator is the number of new events that occur in a defined period; the denominator is the population at risk of experiencing the event during this period.
INFERENCE	the act of making a decision or evaluation concerning one or more characteristics or properties of a population based on information obtained from a sample.
INITIATION	the first (initial) stage of carcinogenesis (onset of cancer) caused by carcinogenic agents (e.g., ionizing radiation, certain chemicals), wherein cellular genetic material (i.e., DNA) is irreversibly changed or mutated.

INITIATOR	any agent, such as ionizing radiation and some chemicals, that can start the process of turning normal cells into cancer cells.
INITIATOR EFFECT/INITIATION	the transformation of a normal cell of the body to a neoplastic (cancer) cell by means of a permanent change or mutation in the nuclear DNA. Initiation results from a limited exposure to a carcinogen, is accomplished rapidly, and is irreversible.
INSULATOR	a non-conductor of electrical current.
INTERNATIONAL CLASSIFICATION OF DISEASES (ICD)	the classification of specific health conditions by an international group of experts for the World Health Organization (WHO). Every health condition is assigned a specific numerical code. The complete list is periodically revised, in the *Manual of the International Statistical Classification of Diseases, Injuries and Causes of Death*. The Ninth Revision of the Manual (ICD-9) was published by WHO in 1977 after ratification of 1976.
INTERNATIONAL SYSTEM OF UNITS (SI)	a system of units adopted as the official system of measurement in 1960. Examples are ampere, meter, tesla, weber, etc.
INVERSE SQUARE LAW	a mathematical relationship in which the magnitude of a quantity decreases as the square of the distance from a reference point.
IONIZATION	the dissociation of compounds into ions through loss of electron(s).
IONIZING RADIATION	any electromagnetic or particulate radiation capable of producing ions directly or indirectly, in its passage through matter. Iodizing radiation possesses sufficient energy to remove electrons from the atoms or molecules it encounters, and is capable of causing injury to living cells. Examples of ionizing radiation are X-rays, gamma rays and alpha and beta particles .
IRRADIATED	exposed to the emission of radiant energy.

ISOLATED CONDUCTOR	an idealized conductor in space separated from all other current-carrying conductors. This concept is used as a model.
KILOVOLT (KV)	a kilovolt is a unit of measurement made equal to one thousand volts.
KILOVOLT AMPERES (KVA)	one thousand volt amperes. A measure of apparent power.
KNOB AND TUBE WIRING	an old-fashioned form of house wiring that produced higher fields than modern wiring because the wires are more widely spaced apart, so fields do not self-cancel as effectively.
KV	the abbreviation for kilovolt. See KILOVOLT.
KVA	the abbreviation for kilovolt amperes. See KILOVOLT AMPERES.
LAMINATIONS	thin flat layers of magnetic material used in various applications to minimize eddy current losses. Laminations are stacked to form transformer cores.
LATENCY, LATENT PERIOD	the delay between exposure to a disease-causing agent and the appearance of the disease. For example, after exposure to ionizing radiation there is an average latent period of five years before development of leukemia, and more than 20 years before development of certain other malignant conditions.
LATERALS	branch from a feeder trunk line in a primary distribution system. The feeder trunk line is three phase and carries hundreds of amperes of current, whereas laterals carry smaller currents. Principal laterals carry three phase currents and lesser laterals or sublaterals carry single phase currents.
LEUKEMIA	a general word used to refer to a number of different types of cancers of the blood forming tissues. Leukemia is an acute or chronic disease (cancer) in man and other warm-blooded animals characterized by an abnormal increase in the number of white blood cells.

LINEAR POLARIZATION	with respect to fields, a field that acts along a straight line in space.
LOW REACTANCE ARRANGEMENT	a double circuit horizontal arrangement of conductors (ABC/BCA) that results in a lower impedance per mile. With respect to magnetic fields, this arrangement produces a reduced field as compared to the ABC/ABC or super bundle arrangement.
LYMPHATIC CANCERS	acute and chronic cancers of all structures involved in the conveyance of lymph from the tissues to the blood stream. The lymph system includes the lymph capillaries, lacteals, lymph nodes, lymph vessels, and main lymph ducts.
LYMPHATIC LEUKEMIA/LYMPHOID LEUKEMIA	leukemia in which there is marked increase in the size of the spleen and lymph glands with great increase in white blood cells in the blood; acute forms occur in children and young adults.
LYMPHOMA	a general term for an abnormal (neoplasm) growth in the lymphatic system. Included in this general group are Hodgkin's disease, lymphosarcoma, and malignant lymphoma.
MAGNETIC FIELD	common usage for the magnetic flux density field, expressed in tesla (SI units) or milligauss (Gaussian units).
MAGNETIC FIELD METERS	either a single axis or three axis device. A single axis meter measures the magnitude of a magnetic field along a single direction. A three axis meter measures the fields along three orthogonal axes.
MAGNETIC FIELD PLOT	shows the lines of magnetic flux in a region.
MAGNETIC FLUX	is produced by the flow of electrical current and the units are webers or maxwells.
MAGNETIC FLUX DENSITY ELLIPSE	See ELLIPTICAL POLARIZATION.
MAGNETIC FLUX DENSITY FIELD	a vector field representing the magnitude and direction of the density of magnetic flux in a region. The units of measurement are the tesla or milligauss.
MAGNETIC FLUX LINES	a graphical representation of the relative density of magnetic flux in a given region.

MAGNETIC RELUCTANCE	a measure of the resistance in establishing a magnetic flux in a magnetic circuit. Units are amperes/weber.
MAGNETOMOTIVE FORCE (MMF)	the driving force in a magnetic circuit. Units are ampere-turns.
MALIGNANCY	a neoplasm or tumor that is invasive with a tendency to metastasize.
MALIGNANT	tending to produce death or deterioration through the process of infiltration, metastasis (spreading throughout the body) and destruction of tissue.
MANAGEMENT	in the sense of magnetic fields, management is mitigation of the magnitude of a field in a specified region.
MATCHED CONTROLS	in a case-control study, controls selected so that they are similar to the cases in specific characteristics such as age, sex, race, and socioeconomic status. (See CASE-CONTROL STUDY.)
MATCHING	the process of making a study group and a comparison group comparable with respect to extraneous or potentially confounding factors.
MATCHING VARIABLE	a characteristic such as age or sex used to select matched controls, such that those characteristics are similar in cases and controls. (See MATCHED CONTROLS.)
MEAN	one of several measures of central tendency of the distribution of a set of values. The arithmetic mean is the sum of all values in a set, divided by the number of values in the set. (See also MEDIAN.)
MEDIAN	one of several measures of central tendency of the distribution of a set of values. The median represents the middle figure when the measurements are arranged in ascending order. Half the measurements are below the median value, half are above.
MEGAVOLT AMPERES (MVA)	one million volt amperes. A measure of apparent power.

MELATONIN	a vertebrate hormone of the pineal gland that produces darkening of the skin by causing concentration of melanin in pigment-containing cells. Melatonin influences sleep, perception of pain, psychological depression, and social behavior.
METABOLISM	the biochemical reactions by which energy is made available for the use of an organism from the time a nutrient substance enters, until it has been utilized and the waste products eliminated.
METASTASIS	movement of bacteria or body cells (especially cancer cells) from one part of the body to another by means of the lymphatics or blood stream. In cancer cases the result of metastasis is a secondary growth arising from the primary growth in a new location.
MICROGAUSS (μG)	one millionth of a gauss, or 10^{-6} G.
MICROWAVES	Electromagnetic waves that have a frequency of between roughly 1 billion and 300 billion Hz (a wavelength of between roughly 30 centimeters and 1 millimeter). Microwaves have a frequency higher than normal radio waves but lower than heat (infrared) and light. In contrast to x-rays, microwaves are a form of non-ionizing radiation (See X-RAYS). Strong microwaves can produce biological damage by heating tissue; 60 Hz fields cannot do this.
MILLIGAUSS (MG)	one thousandth of a gauss, or 10^{-3} G.
MISCLASSIFICATION ERROR	the erroneous classification of an individual into a category other than that to which he or she could be assigned. In an epidemiologic study of EMF exposure, for example, including electricians who routinely work on dead circuits in the "exposed" group would result in misclassification error.

MMF	abbreviation for magnetomotive force. See MAGNETOMOTIVE FORCE.
MODEL	(1) Mathematical model. A mathematical representation of a natural system intended to mimic the behavior of the real system, allowing description of empirical data, and predictions about untested states of the system. (2) Biological model. A condition or disease in animals similar to the condition or disease in human being.
MORBIDITY	any departure from a state of physiological or psychological well-being, used in public health data to describe disease states.
MORTALITY	death; or the number of deaths in a given time or place, or the death rate.
MOTOR BEHAVIOR	movement of an organism in response to a stimulus.
MULTIPHASE CURRENTS	two or more currents that are not in phase. See THREE PHASE POWER.
MULTIPLE MYELOMA	a neoplastic disease characterized by the infiltration of bone and bone marrow by myeloma cells forming multiple tumor masses. It occurs commonly in the sixth decade of life and more frequently in males.
MULTIVARIATE ANALYSIS	a set of statistical techniques used when the variation in several variables is studied simultaneously.
MUTAGEN	any agent that causes genetic changes. Many medicines, chemicals, and physical agents, such as ionizing radiation and ultraviolet light, can be mutagens under the right conditions.
MVA	abbreviation for megavolt amperes. See MEGAVOLT AMPERES.
MYELOGENOUS LEUKEMIA/MYELOID LEUKEMIA	leukemia involving the bloodmaking bone marrow, especially that of the ribs, sternum and vertebrae. (See LEUKEMIA.)
NEGATIVE STUDY	a finding or study that confirms the null hypothesis, i.e., the association between exposure and disease is not different from

a measure of no association. Example: a study with a relative risk of one or a risk difference of zero.

NET CURRENT — the sum of the currents in a set of conductors.

NEUROLOGICAL — of or pertaining to the functioning of the nervous system.

NEUTRAL — a metallic conductor for return of current to the source.

NON-HODGKIN'S LYMPHOMA — See LYMPHOMA.

NONIONIZING RADIATION — radiation that does not transfer sufficient energy to remove electrons, or break chemical bonds, to form ions in the material it encounters. Examples of nonionizing electromagnetic radiation include ultraviolet and visible light, infrared and microwave radiation, radio and television waves and power-frequency fields (60 Hz).

NULL HYPOTHESIS — the statistical hypothesis that one variable has no association with another variable or set of variables, or that two or more population distributions do not differ from one another. In simplest terms, the null hypothesis states that the results observed in a study, experiment, or test are no different from what that might have occurred as a result of the operation of chance alone.

OBSERVED-TO-EXPECTED (O/E) RATIO — in an epidemiologic study, the ratio of the observed number of cases of a disease in the population under study to the number that would be expected on the basis of the disease experience of a reference group.

ODDS RATIO — a calculation used frequently in case-control studies to compare the exposure experience of diseased and non-diseased groups. The odds ratio can serve as an estimate of the relative risk of disease associated with the exposure.

OHM — the unit of electrical resistance and electrical impedance.

ONE-SIDED TEST	a statistical significance test based on the assumption that the data have only one possible direction of variability.
ORTHOGONAL	intersecting at 90-degree or right angles.
OVARIAN	Concerning the two glands in the female that produce the reproductive cell, the ovum, and two known hormones.
OVERHEAD (OH)	describes electrical transmission and distribution lines that are located on towers and/or poles.
P-VALUE	See STATISTICAL SIGNIFICANCE.
PAD MOUNTED TRANSFORMER	a distribution transformer utilized as part of an underground distribution system with enclosed compartments for high and low voltage cables entering from below, and mounted on a foundation pad.
PARAMAGNETIC MATERIALS	materials characterized by a relative permeability that is slightly larger than 1.0 (e.g., aluminum).
PARAMETER	in mathematics, a constant in a formula or model; in statistics and epidemiology, a measurable characteristic of a population. "True parameter values" are usually unknown and must be estimated from the data.
PASSIVE SHIELDING	involves the use of materials that interact with power frequency magnetic fields to reduce the field in a region.
PERIODICITY	the quality, state, or fact of being regularly recurrent (i.e., returning or happening time after time).
PERIPHERAL	located away from a central or central portion; involving the surface or external boundary of a body.
PERSON-YEAR	a unit of measurement obtained by summing the lengths of time (usually one year) for each person and used as a denominator in incidence and mortality rate calculations.

PERSON-YEARS AT RISK	the sum of the years that the persons in the study population have been exposed to the condition of interest. With this approach, each person contributes only as many years of observation to the study as he is actually observed at risk; if he leaves, contracts the disease under study, or dies after one year, he contributes one person-year; if after ten, ten person-years.
PHASE	the time of occurrence of the peak value of an AC wave form with respect to the time of occurrence of a reference waveform.
PHASE ANGLE	The angular displacement between two sinusoidal waves, measured in degrees (°) or radians.
PHASOR	a mathematical description of a steady-state sinusoidal quantity such as voltage or current. A phasor states the magnitude and phase angle of the sinusoid being represented.
PHOTON	a particle of electromagnetic energy.
PHYSIOLOGICAL	dealing with the functions and activities of life or living matter (as organs, tissues, or cells), and of the physical and chemical phenomena involved; the organic processes and phenomena of an organism or any of its parts or of a particular bodily process.
PINEAL MELATONIN	The endocrine hormone melatonin that is produced by the pineal gland in the brain. Melatonin is involved in the control of circadian rhythm in at least some animals.
PLASMA	the fluid part of blood, lymph, or milk as distinguished from suspended material.
PLASMA MEMBRANE	the membrane surrounding plant and animal cells.
POLE TOP TRANSFORMER	a distribution transformer that is located on a pole.
POPULATION	the entire set of subjects that are the object of a study or investigation.

POSITIVE FINDINGS/POSITIVE STUDY — a finding or study in which the association between exposure and disease is appreciably different from a measure of no association.

POTENTIAL — electrical potential energy, defined at a point by the work necessary to bring a unit positive charge to the point from an infinite distance. The difference in potential between two points is defined by the work necessary to carry a unit positive charge from one to the other. The unit is the volt (V).

POWER — the time rate at which work is done. Electrical power is proportional to the product of current, voltage, and power factor. The unit is the watt (W).

POWER FACTOR — the cosine of the angle of phase difference between the voltage and current.

POWER FREQUENCIES (50-60 HZ) — the frequencies of AC power systems in Europe and North America are 50 Hz and 60 Hz, respectively.

POWER FREQUENCY FIELDS — electromagnetic fields produced by power frequency currents.

PRECISION — the closeness of an observation to the mean derived from repeated sampling of the same population. Precision can be estimated by standard error or standard deviation.

PREDISPOSING FACTOR — a condition or characteristic that contributes to an individual's susceptibility to a disease.

PREVALENCE — the total number of existing cases of a disease or other condition in a specific population at a specific point in time.

PREVALENCE RATE — the total number of existing cases of a disease or other condition in a specific population at a specific point in time divided by the size of the population at that time.

PRIMARY DISTRIBUTION SYSTEM — distributes electrical power from the distribution substation to distribution transformers adjacent to customers. The primary distribution voltage is in the range from 4 to 35 kV; however, a range from 12 to 25 kV is the most common in the United States.

PRIMARY DISTRIBUTION VOLTAGE

See PRIMARY DISTRIBUTION SYSTEM.

PRIMARY VOLTAGE LEVELS

the voltage levels at which power is transmitted through the primary distribution system, usually in the range 4 to 35 kV.

PRINCIPAL LATERAL

See LATERALS.

PROBABILITY DISTRIBUTION

a mathematical function that assigns a probability to the occurrence of a specific value or range of specific values that can be assumed by the random variable.

PROFILE

a plot of the magnitude of the magnetic field as a function of horizontal distance, usually at a height of three feet from the ground plane.

PROGRESSION

a continuous and connected series (sequence) of events, actions, etc. In the three stage model of carcinogenesis, the state where a benign tumor becomes malignant.

PROMOTER

any agent, such as some chemicals, that can aid or accelerate the growth of cancer.

PROMOTER EFFECT

the facilitation of the growth of dormant cancer cells into tumors.

PROMOTION

The second hypothesized stage in a multistage process of cancer development. The conversion of initiated cells into tumorigenic cells.

PROPORTIONAL INCIDENCE RATIO (PIR)

the proportion of new cases of a specific disease among new cases of all disease in a given population, compared to the proportion of new cases of that disease among new cases of all diseases in a reference population. The PIR may be used, for example, to determine if compared to the occurrence of other diseases, leukemia occurs more frequently among electricians than among the general public. Calculated as follows: Percent new cases of disease due to leukemia in study population divided by percent of new cases of disease due to leukemia in reference population.

PROPORTIONAL MORTALITY RATIO (PMR)

the proportion of deaths due to a particular cause in one population, compared to the proportion of deaths due to that cause in another population. The PMR is used, for example, to determine if the proportion of deaths due to leukemia is greater among electricians than among the general public. The comparison of the PMRs from different populations can give rise to misleading conclusions if the populations have different distributions of causes of death. Calculated as follows: percent of deaths due to specific disease in study population divided by percent of deaths due to specific disease in reference population.

PROSPECTIVE STUDY

a type of epidemiologic study in which study subjects free of disease are followed into the future to determine their morbidity or mortality experience.

PROTEIN

any of numerous naturally occurring, extremely complex combinations of amino acids that contain the elements carbon, hydrogen, nitrogen, oxygen, usually sulfur, and occasionally other elements. Proteins are essential constituents of all living cells, and are synthesized from raw materials by plants but assimilated as separate amino acids by animals.

PRR (PROPORTIONAL REGISTRATION RATIO)

the proportion of a specific kind of cancer among all new cases of cancer registered for one group of people, compared to the proportion of that kind of cancer in another group. Similar to the PIR, the PRR is used to determine, for example, whether the proportion of leukemia compared to other cancer is higher among electricians than it is among the general public. (See also REGISTRY).

PRUDENT AVOIDANCE	a passive policy of limiting exposure to magnetic fields, due to the uncertainty of a link between magnetic fields and adverse health effects. This policy suggests that an easy and low-cost way of avoiding fields is the best. An example of prudent avoidance is to move the bed away from a wall that has a strong magnetic field associated with it .
QUADRAPLEX	four conductor secondary or service cable.
QUALITATIVE DATA	information collected in a study that can be classified into categories, such as sex, race, hair color and nationality.
QUANTITATIVE DATA	information collected in a study that can be classified on some continuous scale, such as age, height, weight and blood pressure.
RADIATION	any variety of forms of energy propagated through space. Radiation may involve either particles (for example alpha-rays or beta-rays) or waves (for example, x-rays, light, microwaves, or radio waves). Ionizing radiation such as x-rays carries enough energy to break chemical and electrical bonds; non-ionizing radiation like
	microwaves does not; most of the energy in the 60 Hz fields associated with power lines, wiring and appliances does not. Most of the energy in the 60 Hz fields associated with power lines, wiring and appliances does not propagate away from them through space. Hence, it is best not to refer to these fields as radiation.
RANDOM VARIABLE	a variable whose values follow a probability distribution.
RANDOMIZATION	a process by which study subjects are assigned to experimental test conditions such that each subject has an equal chance of being selected for each condition.
REFERENCE GROUP	a standard against which the disease experience of the population that is being studied is compared.

REGISTRY

in epidemiology, the term "registry" is applied to the file of all cases of a particular disease in a defined population such that the cases can be related to a population base, e.g., all cancer cases in the state of Iowa. With this information, incidence rates can be calculated.

REGRESSION

statistical procedures that allow the selection of the best numerical relationship among two or more variables. Given data on a dependent variable y (disease outcome) and one or more independent exposure-related variables, $x_1, x_2, x_3, ..., x_n$, regression involves finding the "best" mathematical model to describe y as a function of the $x_1, x_2, x_3, ..., x_n$. Types of regression include: simple linear regression that is a straightline fit to data points observed in a two-dimensional grid; multiple regression that is basically linear regression in n-space; polynomial regression that is a polynomial-curve fit to data points observed in a two-dimensional grid; nonlinear regression that fits functions other than polynomials to data; and logistic regression that is a categorical data method fitting functions to data adjusted by the logistic transformation.

RELATIVE RISK

the ratio of the risk of disease or death among an exposed group of study subjects to the risk among an unexposed group. A risk measure based on disease or death rates that is used frequently in cohort studies. The relative risk indicates the increased (or decreased) degree of risk among the exposed subjects compared to the non-exposed. A relative risk value of 1.0 indicates no association between the exposure and the disease. A relative risk of 2.0 would indicate that the exposed group is twice as likely as the non-exposed group to experience the health effect being studied (death or disease).

RELIABILITY

the degree to which the results obtained by a measurement procedure can be replicated. Lack of reliability may arise from divergences between observers or instruments of measurement, or from instability of the attribute being measured.

RESISTANCE

The electrical property of a conductor that resists the flow of an electric current without changing its phase. The unit of measurement is the ohm.

RESULTANT

the mathematical summation of two or more vectors that produces a single vector.

RETURN GROUND CURRENT

See GROUND CURRENT.

RIGHT-OF-WAY (ROW)

the legally defined corridor of land on which a transmission line is located.

RISK

the probability that an event will occur, e.g., that an individual will become ill or die within a stated period of time.

RISK ASSESSMENT

Dose-Response Assessment The determination of the relation between the magnitude of exposure and the probability of occurrence of the health effects in question.

Exposure Assessment The determination of the extent of human exposure.

Risk Characterization The description of the nature and often the magnitude of human risk, including attendant uncertainty.

RISK ESTIMATE

the quantitative estimate of the likelihood of adverse effects resulting from a specified exposure. The relative risk and odds ratio are examples of risk estimates.

RISK FACTOR

an aspect of lifestyle, environmental exposures, or an inherited characteristic, which on the basis of epidemiologic evidence is shown to be associated with adverse health effects.

RMS

abbreviation for root mean square. See ROOT MEAN SQUARE.

RNA

See RIBONUCLEIC ACID.

ROOT MEAN SQUARE	in space, the square root of the sum of the squares of the values of the spacial orthogonal components, such as the components of a magnetic field.
	in time, the square root of the average value of the square of an electrical quantity, such as current, voltage, or a magnetic field.
SAG	the distance measured vertically from a conductor to the straight line joining its two points of support.
SAMPLE	any selected subset from a population.
SAMPLE SIZE	the number selected (sampled) from a population to be the subjects of study.
SECONDARY AND SERVICES	that part of an electrical distribution system that distributes power from the distribution transformer to the customer service entrance.
SELECTION BIAS	error in the results of a study due to systematic differences in characteristics between those who are selected for study and those who are not. For example, selecting only HMO patients as study subjects might exclude persons who are not employed or employable. Thus, the study's results would not accurately reflect disease patterns in the general population, which includes unemployed people.
SEROTONIN	A phenolic amine ($C_{10}H_{12}N_{20}$) that is a powerful vasoconstrictor and is found especially in the blood serum and gastric (stomach) mucosa of mammals. Serotonin stimulates or inhibits many of the nerves and muscles, depending on the amount and phase of the organ in its function. It can stimulate or depress heartbeat, contract blood vessels, and change blood pressure. It prevents clotting, and provides reflexes such as coughing or hyperventilation. In humans serotonin also serves as a chemical transmitter in the brain. Serotonin and its product melatonin influence sleep, perception of pain, psychological depression and social behavior.

SERVICE AREA	territory in which a utility system is required or has the right to supply or make available electric service to ultimate consumers.
SERVICE DROP	the overhead conductors through which electric service is supplied, between the secondary and the point of their connection to the service facilities located at the building or other support used for the purpose.
SERVICE ENTRANCE	all components between the point of termination of the overhead or underground service and the building main disconnecting device, with the exception of the utility company's metering equipment.
SHIELDING	a management technique used in the mitigation or reduction of the magnitude of a field in a specified region.
SI	abbreviation for international system of units. See INTERNATIONAL SYSTEM OF UNITS.
SIBLINGS	children borne by the same mother.
SIEMENS	the SI unit of conductance.
SIGNIFICANCE LEVEL	The probability predetermined by the investigator) that the test statistic will assume a value that will lead to a rejection of the null hypothesis when the null hypothesis is true.
SINGLE PHASE	a single alternating current.
SINUSOIDAL	a time-varying sine wave.
SKIN DEPTH	a measure of the penetration of a material by electromagnetic waves. Units are meters.
SOLID STATE ELECTRONICS	electronic circuits that utilize semiconductor devices such as transistors.
SOLVENT	a substance, usually liquid, capable of dissolving or dispersing one or more other substances. Most industrial solvents are volatile organic compounds such as xylene, toluene, and perchloroethylene.

SPATIAL AND TEMPORAL VARIATIONS	with respect to a field, spatial variations refers to variations with respect to distance, temporal variations refers to variations with respect to time.
SPOT MEASUREMENTS	a one-time or snapshot measurement of a magnetic field.
STANDARD DEVIATION	a measure of the variation in a set of observations. The mean of the observed values indicates where the values for a group are centered. The standard deviation is a summary of how widely dispersed the values are around this central value.
STATIC FIELDS	electric and magnetic fields that do not vary in intensity or strength with time.
STATION BREAKER	a breaker located at the supply end of a primary feeder.
STATISTICAL SIGNIFICANCE	a finding of an epidemiologic study is considered to be statistically significant if, according to certain assumptions and based on a mathematical probability, the finding has a low likelihood of being due to chance or random variation. A test of statistical significance is a measure of whether a difference observed between the exposed and non-exposed groups in a study is statistically different from a chance occurrence. The probability of an observed difference being due to chance may be expressed as a "p"- value.
STATISTICS	the science of collecting, summarizing and analyzing data that are subject to random variation such that the uncertainty of inductive inferences may be evaluated. The term is also applied to the data themselves and to summarizations of the data, e.g., a statistic may be a value that is computed from sample data or any function of a random variable.
STIMULUS	an agent (as an environmental change) that directly influences the activity of living protoplasm (as by exciting a sensory organ or evoking muscular contraction or glandular secretion).

STRAY FLUX   magnetic flux that escapes from the
                                            intended magnetic circuit.

STRAY VOLTAGE   A condition that occurs when a distribution
                                            system has multiple grounds, which may
                                            result in ground currents, causing voltage
                                            differences between ground points. This
                                            situation may lead to small but perceptible
                                            electrical shocks. The problem can usually
                                            be fixed with proper grounding of the
                                            system.

SUBLATERALS   See LATERALS.

SUBSTATION   a station in which electric voltages are
                                            transformed or lines are switched. (Also see
                                            DISTRIBUTION SUBSTATION.)

SUBTRANSMISSION LINES   lines that are rated between 35 kV and 115
                                            kV.

SUPER BUNDLE ARRANGEMENT   a double circuit horizontal arrangement of
                                            conductors (ABC/ABC). With respect to
                                            magnetic fields, this arrangement produces
                                            a higher field as compared to the ABC/BCA
                                            arrangement.

SYMMETRICAL COMPONENTS   the components of three phase voltages or
                                            currents that describe their state of
                                            symmetry.

SYNERGISM   synergism exists between two
                                            environmental agents if the risk of disease
                                            that results from exposure to both agents is
                                            greater than the sum of the risks for each
                                            individual exposure.

SYSTEM   1. A regularly interacting or interdependent
                                            group of items forming a unified whole.

                                            2. A group of body organs that together
                                            perform one or more vital functions.

SYSTEMATIC ERROR   distortion of study results due to
                                            non-random events. (See BIAS.)

TESLA   the unit of magnetic flux density(T),
                                            equivalent to 10^4 Gauss (G) or 1
                                            Weber/meter^2.

TESTOSTERONE

a male hormone that is produced by the testes or made synthetically, and is responsible for inducing and maintaining male secondary sex characteristics, and is a crystalline hydroxy steroid ketone $\{C_{19}H_{28}O_2\}$.

THREE PHASE CURRENTS

normally sinusoidal currents flowing in a three phase circuit. The phase angles are separated by one third of a cycle or 120 electrical degrees. Units are amperes.

THREE PHASE POWER

power generated, transmitted or utilized in a three phase circuit. Units are watts, kilowatts or megawatts.

THRESHOLD

1. The point at which a physiological or psychological effect begins to be produced.

2. A level, point, or value above which something will take place, and below which it will not.

THYROID

a large endocrine gland of craniate vertebrates, lying at the base of the neck and producing especially the hormone thyroxine.

TIME-VARYING FIELDS

electric and magnetic fields that change in intensity or strength with time. Examples include 60 Hz and transient fields.

TIME-WEIGHTED EXPOSURE

a way of averaging an individual's exposure to an environmental agent over a specified period of time. Specifically, the integration of a monitoring curve of exposure to an environmental agent, divided by the total time the person was exposed. Time-weighted occupational exposures are generally standardized to an 8-hour workday.

TISSUE

an aggregate of cells usually of a particular kind together with their intercellular substance that form one of the structural materials of a plant or an animal.

TRANSFORMER

a device for changing from one set of voltage and current levels to another.

TRANSMISSION LINE	A power line used to carry large quantities of electric power at high voltage, usually over long distances, from the power station to load centers and substations. Transmission lines typically operate at voltages of between 115 and 765 kV. They are usually built on steel towers or very large wooden structures.
TRANSMISSION LINE VOLTAGE	range from 115 kV to 765 kV.
TRIANGULAR CONFIGURATION	the arrangement of conductors in a transmission line in which the conductors represent the three points of a triangle.
TRIPLEX	three conductor secondary or service cable.
UNDERGROUND (UG)	describes electrical distribution lines that are located below grade.
USER LEVELS	distribution line voltages are stepped down via distribution transformers that reduce voltages from primary levels to user levels. User levels typically range from 120 V to 460 V.
VARIABLE	any attribute, phenomenon, event, or measure that can assume different values.
VARIANCE	a measure of dispersion of a set of observations defined for a sample as the sum of squares of deviations from the mean divided by one less than the total number of observations. Standard deviation is the square root of the variance. (See STANDARD DEVIATION.)
VECTOR	a quantity that has a magnitude and a direction. A magnetic field is an example of a vector field.
VERTICAL CONFIGURATION	the arrangement of conductors in a transmission line in which the conductors are arranged in a linear orientation, perpendicular to the earth.
VERTICAL CONSTRUCTION	a type of construction using steel brackets to mount the insulators, the conductors being arranged vertically.

VOLTAGE — a measure of electric potential, the amount of work that must be done to move a charge from ground to a location in space such as a power line conductor. Voltage in a power line is analogous to pressure in a pipeline. Voltage is measured in volts, abbreviated V.

WINDOW EFFECT — for particular values of frequencies and intensities, some electric and magnetic field intensities produce an effect on test organisms/tissues, but others don't. Conversely, if an effect is observed at a particular value of field, it might be "tuned out" by changing the frequency of the field.

WINDOWED RESPONSE — effects found within bands or ranges of frequency or intensity separated by bands or ranges without effect; nonlinear exposure-response relation.

WIRE CODE — based on the thickness and configuration of the electricity distribution line. Exposure to power line frequency magnetic fields was found to be related in some way to the wire code.

WYE CONNECTED — a method of connecting three phase circuits, involving a common neutral. (Also see DELTA CONNECTED.)

X-RAYS — A form of electromagnetic waves similar to light but with a shorter wavelength (higher frequency). X-rays are a form of ionizing radiation. They can damage biological systems by breaking chemical or molecular bonds; sixty Hz fields cannot do this.

ZERO SEQUENCE COMPONENT — one of the three symmetrical components that are used to mathematically describe unbalanced three phase voltages or currents.

E.2 References

[1]   *Electric and Magnetic Fields from 60 Hertz Electric Power: What do we know about possible health risks?*, Department of Engineering and Public Policy, Carnegie Mellon University, Pittsburgh, PA 15213.

[2]   *Health Effects of Exposure to Powerline-Frequency Electric and Magnetic Fields*, Electro-Magnetic Health Effects Committee, Public Utility Commission of Texas, Austin, Texas, March 1992.

[3]   Gonan, Turan, *Electric Power Distribution Engineering*, McGraw-Hill Book Company, New York, NY, 1986.

Appendix F Personal Prudent Avoidance Actions

Though it's still uncertain that extremely low frequency (ELF) fields contribute to cancer, you may want to play extra safe and not take unnecessary risks. Granger Morgan, a professor of engineering and public policy at Carnegie-Mellon University, coined the term **prudent avoidance** to describe ways to limit ELF exposure without great cost or inconvenience.

As shown in the chart below, most people move in and out of fields of varying intensity all day. Many researchers believe that routine brief exposures, even if intense, probably do not put anyone at risk. However, it may be wise to avoid prolonged periods within a field. (Several studies have correlated cancer with long-term exposures above two to three milligauss.) The background level in most homes is around one milligauss. After Morgan used a gauss meter to measure fields in his own house, he moved his son's bed to a different part of the boy's room. In general, the young seem more susceptible than grownups.

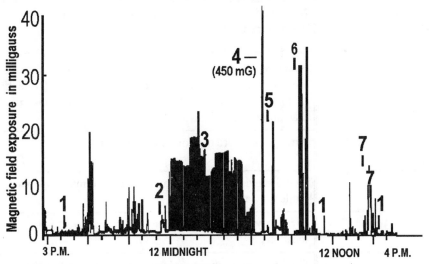

Typical Exposure to Magnetic Fields During a 24-Hour Period[*]

Chart Designation	Source
1	desktop computer
2	a TV set
3	an electric blanket
4	an electric razor
5	a microwave oven
6	a heating/cooling unit
7	a power line and substation he walked past

Typical Exposure Experienced by Typical Man During 24-Hour Period

Prudent avoidance applied to magnetic fields suggests adopting measures to avoid exposure when it is reasonable, practical, relatively inexpensive and simple to do. This course of action can be taken even if the risks are uncertain and even if health issues are unresolved. Prudent avoidance decisions about all sorts of risks are made by people every day. A variety of personal prudent avoidance actions with respect to magnetic fields are listed below. The list is not meant to be exhaustive but rather is given for illustrative purposes only.

Item	Possible Action
Bedroom	Move bed against wall with lowest magnetic field measurement.
	Move electric alarm clock or clock radio and any other motorized electrical devices that stay on continuously several feet away from your head while you are sleeping. (Note that motorized electrical devices emit fields of greater magnitude than nonmotorized devices.)
Living/Family Room	Arrange furniture to avoid the highest field areas.
Kitchen	Avoid standing very close to or in front of major appliances while they are in operation.
Play Areas	Arrange yard so that play areas are in the lowest field locations.
Toys	Use battery-powered toys instead of plug-ins whenever possible.

Item	Possible Action (continued)
Computers	If possible replace current monitor with low emission type. Keep monitor at least two to three feet away from face and don't sit within three feet of the sides or back of anyone else's. The fields tend to be more intense there. Keep main computer box at least two to three feet away from body.
Grounding	Extraordinarily high magnetic field readings around the home may be due to problems in power ground connections and may be correctable by a licensed electrician.
Pad Mounted Transformers	In a residential area with underground electric utilities, these may be located in a parking strip (parkway) or in a front yard. Children should not play on or around this type of equipment.
APPLIANCES **Electric blanket**	If possible use newer low-field electric blankets. Replace electric blankets with quilts. Turn on electric blanket to warm up bed and then turn it off when sleeping.
Water Bed	Turn on water heater to warm up bed and then turn it off when sleeping.
Electric Razor	Use battery-powered electric razor or replace with nonelectric razor.
Hair Dryers	Don't use hair dryers on children.
Can Openers	Replace electric units with manual ones.
Televisions	Stay back at least three feet while watching television.

Many utilities will measure the magnetic fields in your home or office and provide you with a written summary of these measurements. This service is provided free of charge. This information will be very useful in carrying out the following prudent avoidance actions. If your utility does not provide this service, you can buy or rent an easy-to-operate gauss meter and do it yourself. In addition, there are a growing number of independent experts who will do this for you for a fee; ask your local realtor or chamber of commerce for additional information.

Index